Praise for *Aid Effectiveness for Environmental Sustainability*

"Discussions of public policies to mitigate climate change and foster adaptation bring attention to the principle of 'common but differentiated responsibilities and respective capabilities,' enshrined in the 1992 United Nations Framework Convention on Climate Change. This principle of distributional equity highlights the importance of issues surrounding how the richest countries of the world might help finance activities in some of the poorest countries. This important new book brings together an excellent set of authors to address the wide range of issues that inevitably arise."

—Robert N. Stavins, *A. J. Meyer Professor of Energy and Economic Development, John F. Kennedy School of Government, Harvard University, USA*

"With the Paris Agreement the world has committed itself to climate aid. This book clarifies in detail how foreign aid can support sustainability and climate goals in the global south."

—Jeroen van den Bergh, *ICREA Research Professor, Universitat Autònoma de Barcelona, Spain and Professor of Environmental Economics, VU University Amsterdam, The Netherlands*

"This book is a superb collection of some exciting essays by leading experts on aid effectiveness for environmental sustainability. It presents theoretical issues supported by a wide range of applications and examples, which researchers and practitioners will find invaluable. It is bound to be an important contribution to the field. "

—Juzhong Zhuang, *Deputy Chief Economist and Deputy Director General, Economic Research and Regional Cooperation Department, Asian Development Bank, Philippines*

Aid Effectiveness for Environmental Sustainability

Yongfu Huang • Unai Pascual
Editors

Aid Effectiveness for Environmental Sustainability

palgrave
macmillan

Editors
Yongfu Huang
NDRC People's Republic of China
Beijing, China

Unai Pascual
Basque Centre for Climate Change
(BC3), Leioa, Spain

ISBN 978-981-10-5378-8 ISBN 978-981-10-5379-5 (eBook)
https://doi.org/10.1007/978-981-10-5379-5

Library of Congress Control Number: 2017959741

Printed on acid-free paper

This Palgrave Macmillan imprint is published by Springer Nature
The registered company is Springer Nature Singapore Pte Ltd.
The registered company address is: 152 Beach Road, #21-01/04 Gateway East, Singapore 189721, Singapore

Foreword

Faced with rising environmental degradation, inequality, and poverty, developing countries are far more vulnerable than developed nations. To achieve the UN Sustainable Development Goals, especially adaptation to and mitigation of the current and projected future impacts of climate change, developing countries are in need of financial assistance from developed nations. This highlights a vitally important role for foreign aid in the global arena.

Given the constrained supply of foreign aid, aid effectiveness has become a reoccurring theme in development discourse. This edited volume pays special attention to some key aspects of environmental sustainability in developing countries, especially the least developed countries, namely capacity building, biodiversity and forest conservation, sustainable agriculture, and the sustainability of the urban and energy sectors under climate change. It also identifies four common questions shared by all sectors: 'what works?', 'what could work?', 'what is scalable?', and 'what is transferrable?' By working through these four areas of enquiry across those different sectors, the book provides a consistent framework for analysing the role of actual and future foreign aid for environmental sustainability.

Although a rich literature exists on the relationship between aid and environmental sustainability, the authors extend it by using either cross-country quantitative approaches, or qualitative methods such as the case studies and fieldwork undertaken in several countries. This volume thus provides readers with fresh perspectives on how aid can be used to contribute to sustainable agriculture, green cities, sustainable energy, and so forth, and it will enhance the confidence of donors and other stakeholders in the effectiveness of aid for environmental sustainability.

I should like to express my sincere thanks to the editors, Yongfu Huang and Unai Pascual, for their professional skill in bringing this topical research work to fruition for our further reflection on the impact and design of foreign aid for environmental sustainability.

In completing the research that comprises this book, UNU-WIDER gratefully acknowledges specific programme contributions from the governments of Denmark (Ministry of Foreign Affairs, Danida) and Sweden (Swedish International Development Cooperation Agency—Sida) for the Research and Communication (ReCom) programme.

Director, UNU-WIDER — Finn Tarp
Helsinki, April 2017

Preface

The UNU World Institute for Development Economics Research (UNU-WIDER) was established by the United Nations University as its first research and training centre and started work in Helsinki, Finland, in 1985. The purpose of the Institute is to undertake applied research and policy analysis on structural changes affecting developing and transitional economies, to provide a forum for the advocacy of policies leading to robust, equitable, and environmentally sustainable growth, and to promote capacity strengthening and training in the field of economic and social policy-making. Its work is carried out by staff researchers and visiting scholars in Helsinki and via networks of collaborating scholars and institutions around the world.

United Nations University World Institute for Development Economics Research
(UNU-WIDER)
Katajanokanlaituri 6 B, 00160 Helsinki, Finland
www.wider.unu.edu

Contents

1 Introduction 1
Yongfu Huang and Unai Pascual
1.1 Setting the Landscape: Foreign Aid for Sustainability 1
1.2 Foreign Aid for Capacity Building, Energy Sustainability, Greening Urban Areas, Biodiversity Conservation and Climate Smart Agriculture 5
1.2.1 Foreign Aid for Capacity Building 5
1.2.2 Foreign Aid for Sustainable Energy Systems 6
1.2.3 Foreign Aid for Greening Urban Areas 7
1.2.4 Foreign Aid for Forest Conservation 8
1.2.5 Foreign Aid for Climate Smart Agriculture 9
1.3 Structure of the Book 10
References 15

2 Foreign Aid for Capacity Building to Address Climate Change 17
David Victor
2.1 Introduction 17
2.2 Setting the Scene: Key Definitions, Concepts and Fundamental Spending Patterns 19
2.2.1 Foreign Aid for Climate Change and Capacity Building 22
2.3 Question No. 1: What Works in Foreign Aid for Capacity Building with Regard to Climate Change? 23

2.4 *Question No. 2: What Kinds of Reforms Could Improve the Effectiveness of Climate Change Aid?* 26
2.5 *Question No. 3: What Scales?* 30
2.5.1 *Scaling from Perspective No. 1: Comparing Aid Sources—ODA, Climate Change-Related Aid, and Climate Finance* 33
2.5.2 *Scaling from Perspective No. 2: The CDM* 36
2.5.3 *Scaling from Perspective No. 3: Mainstreaming Climate* 37
2.6 *Question No. 4: What Lessons Are Transferable?* 39
2.7 *Analysis and Conclusions* 39
References 44

3 Lessons Learnt About Foreign Aid for Climate Change Related Capacity-Building 51
Zexian Chen and Jingjing He
3.1 *Introduction* 54
3.2 *What Works* 58
3.2.1 *Country-Driven and Demand-Driven Approaches* 60
3.2.2 *Education and Training* 61
3.2.3 *Results Monitoring* 62
3.2.4 *Data and Resources Sharing* 63
3.2.5 *Improved Decision-Making* 63
3.3 *What Could Work* 64
3.3.1 *Lack of Cooperation* 64
3.3.2 *Lack of Internal Capacity* 66
3.3.3 *Lack of Initiative* 66
3.3.4 *Lack of Flexible Aid Application Mechanism* 67
3.3.5 *Lack of an Integral Approach* 67
3.3.6 *Lack of a Long-Term Approach* 68
3.3.7 *Lack of Country Ownership* 68
3.4 *What Is Scalable* 68
3.4.1 *Enhance Awareness and Knowledge* 69
3.4.2 *Boosting International Negotiation Capacity* 71
3.4.3 *Green Technology Diffusion* 72
3.4.4 *Training Climate Change Professionals* 73
3.5 *What Is Transferrable* 74
3.5.1 *Transfer of Country-Owned and Demand-Driven Approach* 74

3.5.2 Transfer of Knowledge and Experience 75
3.5.3 Transfer of CDM Knowledge and Practice 76
3.6 Conclusion 77
References 79

4 The Effectiveness of Foreign Aid for Sustainable Energy and Climate Change Mitigation 81
H-Holger Rogner and Kei-Kit Leung
4.1 Introduction 81
4.2 Foreign Aid, Energy and Climate Change 83
4.3 Funding Mechanisms 87
4.3.1 Official Development Assistance 90
4.3.2 Multi-Lateral Agencies 91
4.3.3 Climate Finance and UNFCCC 92
4.3.4 Micro-Finance 93
4.3.5 Essential Instruments 94
4.4 Evaluation of Effectiveness 97
4.4.1 Measuring Effectiveness 98
4.4.2 Review of Effectiveness 101
4.4.3 Further Considerations 112
4.5 Conclusions 115
4.6 Appendix: Selected Studies 116
References 119

5 A Review of the Nature of Foreign Aid to the Energy Sector over the Last Two Decades 125
Luis Gomez-Echeverri
5.1 Introduction 125
5.2 Background 127
5.2.1 The Importance of Foreign Aid to the Energy Sector 127
5.2.2 The Challenges that Foreign Aid Needs to Tackle 127
5.3 Literature Review 135
5.3.1 History and Evolution of Foreign Aid 135
5.3.2 Who Gives Foreign Aid to Energy and in What Areas? 139
5.3.3 History of Efforts to Enhance Impact and Effectiveness 148
5.3.4 What Works 150

5.3.5 *What Could Work* 156
5.3.6 *What Is Scalable?* 160
5.3.7 *What Is Transferable* 163
5.4 *Discussion* 164
5.4.1 *Foreign Aid in Support of Transformative Rather Than Incremental Change* 164
5.4.2 *Foreign Aid That Promotes Integration Rather Than Fragmentation* 165
5.4.3 *Programme- Rather Than Project-Based Approaches to Foreign Aid* 166
5.4.4 *Foreign Aid in Support of Technology Development and Innovation Rather Than Equipment Supply* 166
5.4.5 *Foreign Aid in Support of Public Finance Mechanisms to Mobilize, Catalyse and Leverage Private Investment Rather Than Stand-Alone Project Funding* 167
5.5 *Conclusion* 168
References 179

6 An Analysis of the Links between Foreign Aid and Co_2 Emissions in Cities 185
Sandrine Kablan
6.1 *Introduction* 185
6.2 *Literature Review: Foreign Aid Effectiveness and Green City Procedures* 187
6.2.1 *Aid Effectiveness* 187
6.2.2 *Initiative of Sustainable Cities in Developing Countries* 189
6.2.3 *Scaling Up and Transferring Good Practices* 193
6.3 *Empirical Analysis* 194
6.3.1 *The Model* 195
6.3.2 *The Data* 198
6.3.3 *Results* 198
6.3.4 *Robustness Tests* 202
6.4 *Conclusion* 203
Appendices 205
References 212

7 **Foreign Aid, Urbanization and Green Cities** 215
Jun Li
7.1 Introduction 215
7.2 Literature Review 216
7.2.1 Introduction of Foreign Aid 216
7.2.2 Foreign Aid, Environment, and Development 219
7.2.3 The Pros and Cons of Foreign Aid, and the Lessons Learned 220
7.2.4 Foreign Aid, Urbanization and Green Cities 223
7.3 Analysis of Three Case Studies 231
7.3.1 Tianjin City, Northern China 232
7.3.2 Wenchuan and Beichuan Areas After the Earthquake 235
7.3.3 Curitiba in Brazil 238
7.4 Conclusion 243
References 250

8 **Opportunities and Conditions for Successful Foreign Aid to the Forestry Sector** 257
Unai Pascual, Eneko Garmendia, Jacob Phelps, and Elena Ojea
8.1 Introduction 257
8.1.1 Global Forest Loss and Degradation 257
8.1.2 Scope 259
8.2 What Works (and What Does Not)? 260
8.2.1 Forest Management for Multiple Objectives 260
8.2.2 Foreign Aid for SFM 262
8.2.3 Lessons from Forest Conservation and Management 263
8.3 What Could Work? 265
8.3.1 Emergence of REDD+ 265
8.3.2 Leveraging PES Instruments for SFM 267
8.3.3 Increasing Foreign Aid for the Forestry Sector 270
8.4 What Is Transferable? 274
8.4.1 Integrating Lessons Learned into REDD+ 274
8.4.2 Financing Mechanism: Building on Foreign Aid 276
8.4.3 Forest Governance and Institutions: Donor Engagement with Multiple Levels of Governance 279
8.4.4 Considering Multiple Performance Trade-offs 280

8.5 What Is Scalable? 285
8.5.1 Financial and Geographic Scaling Up 285
8.5.2 Scalability and Decision-Making 286
8.6 Conclusion 290
Annex 291
References 295

9 The Evolution, Paradigm Shift and Guidelines for Foreign Aid in Forestry 307
Pekka E. Kauppi
9.1 Introduction 307
9.2 What Has Worked in the Past? 310
9.3 How to Encourage Successful Forestry Aid? 314
9.3.1 Selecting Goals 314
9.3.2 Selecting Means 316
9.4 Opportunities for Scaling Up 317
9.4.1 Short Term 317
9.4.2 Mid-Term 319
9.5 Transferring Lessons: The Role of Universities 321
9.6 Discussion and Conclusions 323
References 326

10 Financing Sustainable Agriculture under Climate Change with a Specific Focus on Foreign Aid 329
Jikun Huang and Yangjie Wang
10.1 Introduction 329
10.2 Financing and Aid to Agriculture under Climate Change 332
10.2.1 Overall Financing and Aid to Agriculture 332
10.2.2 Overall Climate Change Funds 333
10.2.3 Climate Change Funds for Agriculture 334
10.2.4 Financing Measures to Battle Climate Change in Agriculture 337
10.3 What Will Work? 337
10.3.1 Major Areas of Involvement 337
10.3.2 Successful Experiences 341
10.4 What Measures Could Work 347
10.4.1 Major Areas 347

10.4.2 *Examples of Potential Areas for Foreign Aid Investment* 348
10.5 *What Can Be Scaled Up?* 353
10.5.1 *Expanding Pilot Projects for Reducing Carbon Dioxide Emissions Through Soil Carbon Sequestration* 353
10.5.2 *Scaling Up Existing Conservative Management Practices Aimed at Sequestrating and Reducing Carbon Dioxide Emissions* 356
10.5.3 *Scaling Up Climate Change Adaptation with Regard to Agricultural Irrigation* 356
10.6 *What Experiences and Concepts Are Transferrable?* 357
10.6.1 *Foreign Aid to Reduce Methane Emissions Through the Provision of Appropriate Technologies and Integration with Development Goals* 357
10.6.2 *Foreign Aid Assisting Agricultural Adaptation by Mainstreaming Climate Change Adaptation into National Development Programmes and Emphasizing Local Capacity Building* 358
10.7 *Conclusions* 361
References 362

11 Foreign Aid and Sustainable Agriculture in Africa 369
Siddig Umbadda and Ismail Elgizouli
11.1 *Introduction* 369
11.2 *Share of Foreign Aid to Agriculture* 371
11.3 *Aid to Agriculture in Africa* 372
11.4 *Agriculture in Development Thinking after Independence in African Countries and the Role of Aid* 374
11.5 *Recent Developments in Aid to Agriculture: Five Main Areas of Donor Consensus* 377
11.5.1 *Changing the Perception of Agriculture: Agricultural Sustainability* 377
11.5.2 *Market- and Private Sector-led Agricultural Growth* 378
11.5.3 *Improved Governance and Political Processes* 379

11.5.4 *Social Services and Empowerment* 381
11.5.5 *Aid Effectiveness* 382
11.5.6 *Scaling Up* 384
11.5.7 *Transferability* 386
11.6 *Recommendations* 389
11.7 *Conclusions* 391
References 392

12 The Global Partnership on Foreign Aid for Sustainable Development 397
Yongfu Huang and Muhammad G. Quibria
12.1 *Introduction* 397
12.2 *A Theoretical Model* 400
12.3 *An Empirical Framework* 404
12.4 *Data and Stylized Facts* 408
12.4.1 *Measures and Data on Sustainable Development* 408
12.4.2 *Measures and Data on Independent Variables* 411
12.4.3 *Stylized Facts* 413
12.5 *Evidence* 413
12.5.1 *Baseline Models* 414
12.5.2 *Full Models* 416
12.5.3 *Policy Discussion* 423
12.6 *Conclusion* 425
Appendix 427
References 436

13 Conclusion 441
Yongfu Huang and Unai Pascual
References 443

Index 445

Abbreviations

ACCCRN	Asian Cities Climate Change Resilience Network
CDM	clean development mechanism
CER	certified emissions reduction
CPA	country programmable aid
DAC	Development Assistance Committee
GHG	greenhouse gas
IEA	International Energy Agency
IPCC	Intergovernmental Panel on Climate Change
MACC	mainstreaming adaptation to climate change
MRV	measurable, reportable and verifiable
NAMAs	nationally appropriate mitigation actions
OA	official assistance
ODA	official development assistance
OECD	Organisation for Economic Co-operation and Development
UNFCCC	United Nations Framework Convention on Climate Change

List of Figures

Fig. 2.1	Bilateral climate change mitigation-related aid, 2006–10 (Source: OECD 2012)	22
Fig. 2.2	Sectoral breakdown of climate change-related aid (2010 commitments) (Source: OECD 2012)	24
Fig. 3.1	Three levels of capacity (Source: UNDP 2009)	55
Fig. 3.2	Capacity building process (Source: UNDP 2011b)	57
Fig. 5.1	Global Tracking Framework, 2017 World Bank	129
Fig. 5.2	Trends in energy-related aid (Source: OECD DAC CRS http://stats.oecd.org (Downloaded, 15 May 2017))	134
Fig. 5.3	Energy-related aid by sector (Source: OECD DAC CRS http://stats.oecd.org (Downloaded, 15 May 2017))	137
Fig. 5.4	Regional breakdown of aid to energy, 2003 to 2008 ('Regional Breakdown of Aid to Energy by All Donors; Development Cooperation Report 2012: Lesson in Linking Sustainability and Development', OECD 2012)	140
Fig. 5.5	China's Foreign Aid, 2014	144
Fig. 5.6	Africa's cooperation with the South (Source: UNCTAD 2010. 'South South Cooperation: Africa and the new forms of development partnerships', Economic Development of Africa report, New York)	145
Fig. 5.7	Sub-Saharan Africa: total exports and imports by partner (Source: IMF 2011 'Sub Saharan Africa Sustaining the Expansion', Regional Economic Outlook, World Economic and Financial Survey. Washington, DC)	146
Fig. 5.8	Results-based funding possibilities for the energy sector (Source: Reproduced with permission from ESMAP n.d.)	161

Fig. 7.1	Factors, conditions, interactions and the process of aid in ecocity development. Source: Author's illustration	241
Fig. 7.2	A recursive implementation process of an aid project. Source: Author's illustration	242
Fig. 7.3	Stakeholders, practices and goals in ecocity development. Source: Author's illustration	243
Fig. 8.1	Multiple dimensions shaping forest management in the context of multiple objectives. Source: Authors' elaboration	275
Fig. 9.1	A simple and robust cooking stove. Note: The stove quickly produces a hot flame using a small amount of wood. The idea is simple: cooking with an efficient stove spares fuelwood and saves forests. Source: Photo by the author	318
Fig. 12.1	Kernel density evolution of sustainable indicators. Note: This figure shows the kernel density plots of the distribution of two sustainability indicators in 1985 and 2010. Variables and data are described in the text. Source: Authors' own calculation	414

List of Tables

Table 2.1	Aid, climate change-related aid, and private-funded climate finance compared	35
Table 3.1	Examples of capacity building aid activities for 'what works'	59
Table 3.2	Examples of capacity building aid activities for 'what could work'	65
Table 3.3	Examples of capacity building aid activities for 'what is scalable'	69
Table 3.4	Examples of capacity building aid activities for 'what is transferrable'	75
Table 4.1	Total and energy-related gross ODA and OOF disbursements, 2002–10, in millions US$	85
Table 4.2	Energy and climate rated funding by the World Bank Group	88
Table 4.3	Funding for energy efficiency and renewables by the Asian Development Bank, in millions US$	89
Table 5.1	ODA in energy by sector: trends over 2006–10 (constant prices, 2011 US$ millions)	142
Table 5.2	ODA to the energy sector by donor	143
Table 6.1	Generalized method of moment regressions for the impact of foreign aid on carbon dioxide emissions in cities of developing countries	199
Table 6.2	Descriptive statistics	205
Table 6.3	Correlation matrix	206
Table 6.4	Some robustness checks on the impact of foreign aid on carbon dioxide emissions in cities of developing countries, whole sample	207
Table 6.5	Some robustness checks on the impact of foreign aid on carbon dioxide emissions in middle-income countries	209

Table 6.6	Some robustness checks on the impact of foreign aid on carbon dioxide emissions in least developed countries	210
Table 8.1	Major donor and recipient countries in REDD+ funds (percentage of total funds)	272
Table 8.2	Financial flows in REDD+ (in US$ million)	291
Table 10.1	Average annual bilateral and multilateral agricultural and total aid	331
Table 10.2	Multilateral adaptation funding channels	334
Table 10.3	Summary of existing climate change funding initiatives outside UNFCCC (non-convention funds)	335
Table 10.4	Estimated additional investment and financial flows needed for adaptation in 2030 (US$)	336
Table 10.5	Estimate of the reductions of emissions from non-carbon dioxide and soil carbon GHGs ($MtCO_2e$) and the investment needed to achieve these reductions (US$ billion) between 2000 and 2030 at a cost of US$30 tCO_2e (2000$)	338
Table 10.6	Potential areas for financing mitigation of climate change in agriculture in developing world through foreign aid	339
Table 10.7	Potential areas for financing adaptation to climate change in agriculture in developing world through foreign aid	340
Table 12.1	The effects of foreign aid (measured by NAT) on various sustainability indicators, 1985–2010	415
Table 12.2	Transmission channel: economic growth	417
Table 12.3	Transmission channel: natural resource exploitations	418
Table 12.4	Transmission channel: energy intensity	420
Table 12.5	Variables	427
Table 12.6	Names and country codes for the 70 aid recipient countries considered in this study	428
Table 12.7	The effects of foreign aid (measured by ODA) on various sustainability indicators, 1985–2010	430
Table 12.8	The transmission channel of economic growth (using ODA)	431
Table 12.9	The transmission channel of natural resources exploitation (using ODA)	432
Table 12.10	The transmission channel of energy intensity (using ODA)	433

CHAPTER 1

Introduction

Yongfu Huang and Unai Pascual

1.1 Setting the Landscape: Foreign Aid for Sustainability

Since the 1990s, climate change has become one of the most severe global policy challenges of our age. Rising temperatures and changes in precipitation disrupt food and water supplies, drive many plants and animal species to extinction and trigger massive sea-level rises, flooding the homes of hundreds of millions of people. Climate change affects agricultural production and the amount of arable land area available, and threatens the lives and livelihoods of the more than 6 billion people alive today (possibly 9 billion by 2050). Poor people are especially vulnerable to climate-induced rising sea levels, coastal erosion and natural disasters. In addition to the impact on environmental systems, climate change has become one of the most pressing international development issues and poses a permanent extremely serious threat to human development and prosperity.

Y. Huang (✉)
ICC, National Development and Reform Commission of PRC, Beijing, China

U. Pascual (✉)
Basque Centre for Climate Change (BC3), Leioa, Spain

Y. Huang, U. Pascual (eds.), *Aid Effectiveness for Environmental Sustainability*, https://doi.org/10.1007/978-981-10-5379-5_1

At a time when the world is faced with environmental degradation and rising inequity, developing countries are much more vulnerable to adverse situations than developed nations. This is for various reasons, such as low adaptation capacity, weak regulatory systems and disproportionate dependency on natural resources. Developing countries need financial assistance from developed countries to support their efforts towards a sustainable future. In this respect, foreign aid has played an important role in the global arena in attempts by developed countries to boost prosperity in developing countries.

Foreign aid (or foreign assistance) is the international transfer of capital, goods or services from a country or international organization for the benefit of a recipient country or its population. It can be humanitarian or development aid, official or private or non-governmental aid, and bilateral or multilateral. Development aid was defined by the Development Assistance Committee of the Organisation for Economic Co-operation and Development (OECD) in 1969 as the 'flows of official financing administered with the promotion of the economic development and welfare of developing countries as the main objective and which are concessional in character with a grant element of at least 25 per cent'. The history of foreign aid dates back to the days immediately after the Second World War when aid was used to address the impacts of war in Europe as well as other reconstruction efforts. As environmental degradation and inequality have reached alarming levels, the purpose of aid has expanded to include multiple goals such as the Sustainable Development Goals (SDGs) that focus on poverty, the environment, literacy, health, woman's right and so on. In other words, foreign aid targets the socioeconomic factors underlying poverty with the objective of promoting human development in a recipient country and generally includes development, humanitarian and food aid.

Given the constrained supply of foreign aid, aid effectiveness is a recurring theme in development discourses. The international community has taken serious steps to improve aid effectiveness, particularly after the adoption of the Millennium Development Goals (MDGs) in 2000 and of the successor Sustainable Development Goals (SDGs) in 2015. Major efforts include the 2005 Paris Declaration, the 2008 Accra Agenda for Action and the 2011 Busan Partnership for Effective Development Cooperation, important platforms for discussing mutual cooperation for achieving the development goals and increasing the effectiveness of aid.[1] The literature on aid effectiveness is voluminous and it focuses principally on the impact of foreign aid on advancing economic growth. However, systematic research on aid effectiveness for environmental sustainability is hugely lacking.

Foreign aid is key to achieve the SDGs as well as to facilitate intertwined development strategies that help with the adaptation to, and mitigation of, the current and future impacts of climate change. A number of foreign aid projects and programmes have been designed and established to integrate environmental sustainability and social inclusion into all aspects of development cooperation.[2] In fact, there is a widening recognition that the allocation of foreign aid in the form of official development assistance (ODA) needs to take into account and anticipate the effects of global environmental change, including climate change, on its effectiveness when measuring its impact on promoting human development. In this vein, foreign aid ought to be seen with a green economy lens, integrating a traditional development focus with global environmental protection.

Besides ODA, there are other financial mechanism including other official flows, as well as foreign direct investment, direct budgetary support, basket funding, and conditional and unconditional funding, which make up the bulk of foreign aid. Further, as pointed out throughout this book, the multilateral development banks or international financial institutions, such as the World Bank Group and regional development banks, as well as private-public partnerships through private sector funding and/or funding from foundations and non-governmental organizations (NGOs), offer other transaction channels for transferring foreign aid. There are also other channels specializing in environmental protection, and supporting programmes and projects to facilitate options to mitigate and adapt to the impact of climate change, such as the Global Environment Facility and various Climate Investment Funds.

Global (official) aid for development increased steadily from the 1960s, reaching a peak of US$68.7 billion (in 2010 US dollars) in 1992. After a decade of cuts in foreign aid, it started to increase again after the Group of eight most industrialized countries (G8) pledged to double aid to Africa by 2010 and triple it by 2015. In 2011 the global reimbursed aid had reached around US$149 billion (Chapter 2). About 25% of total ODA is now generally devoted to 'economic infrastructure' and 'production', which include transport systems, energy and agriculture. The reimbursed aid to the energy and agricultural sectors is now modestly rising, both having declined since the 1980s. In the agricultural sector there is a shift from enhancing food production towards broader human development needs.

The forestry sector is also seeing a revitalizing foreign aid interest through a mechanism aimed at reducing emissions from deforestation and forest degradation (REDD+), and through the conservation, enhancement and sustainable management of forest carbon stocks. In this context, a burgeoning literature exists on the effectiveness of foreign aid for economic development, including the investments in physical capital (often known as hard infrastructure for transport, etc.), human capital (mostly under health and education) and institutional capital (governance), but less so when related to investments in conserving natural capital. This volume introduces some of the key literature on this issue.

It is noteworthy that the share of climate-related foreign aid as part of total ODA is growing quickly, especially through bilateral agreements. The total amount of bilateral climate change-related aid for both mitigation and adaptation is about 15% or roughly US$22 billion (as per 2010) of total ODA in OECD countries, with two-thirds being directed at mitigation efforts and a third at adaptation (Chapter 2). ODA support for climate-change mitigation picked up momentum after CoP-13, held in 2007 in Bali, and at COP-15, held in Copenhagen, further agreements were achieved to scale up new and additional funding and improved access. In Copenhagen, developed countries pledged US$100 billion per year by 2020 to address the needs of developing countries (UNFCCC 2009). In this context, an important debate between multilateral agencies, donors, development and environmental NGOs, and country diplomats in general relates to whether the new environment (including climate change) aid will be new and additional or will simply be related to country programmable aid (which accounts for about half of total ODA), versus or together with the need to invest in institutional capacity for having performance-based monitoring and evaluation mechanisms.

In addition there is a growing consensus that the effectiveness of foreign aid projects related to environmental sustainability hinges on some common factors, including donor commitment, harmonization and donor-recipient cooperation, given the multiplicity of foreign aid initiatives and programmes in the various sectors, including energy, agriculture, biodiversity and urban areas to name some key ones and the focus of this volume. This book provides evidence and arguments regarding the effectiveness of foreign aid and climate finance regarding these sectors.

1.2 Foreign Aid for Capacity Building, Energy Sustainability, Greening Urban Areas, Biodiversity Conservation and Climate Smart Agriculture

1.2.1 Foreign Aid for Capacity Building

In this context, one area that needs urgent attention is the role of foreign aid in building capacity to address climate change by the least developed countries. Foreign aid for capacity building is related to building up social and administrative infrastructure, a developmental concept mostly coined during the 1990s, which amounts to about a third of global ODA (Chapter 2). While these countries tend to have relatively low levels of greenhouse gas (GHG) emissions, thereby having a lower level of responsibility for climate change, they are most vulnerable to climate change and are expected to bear the bulk of the associated cost, in addition to climate change intensifying their development problems such as poverty and rampant inequality (World Bank 2010). Given current climate projections, it is key that the least developed countries invest in their capacity to analyse and respond to the threats posed by climate change. This necessitates well-functioning international climate change aid programmes. It is well known that such programmes tend to work best in countries with robust institutional and governance systems, including well-functioning systems of public administration. According to the United Nations Framework Convention on Climate Change (UNFCCC), capacity building spans issues related to the root causes of climate change, to help create the incentive and governance structures for climate change mitigation and adaptation as well as to mainstream and prioritize them in development policies (UNFCCC 2012).

Given all forms (private, official, multilateral and bilateral) of climate finance, around US$97 billion per year is being provided to support 'low carbon, climate-resilient development activities' as per the latest data for 2009–10 (Buchner et al. 2011). This implies that reaching the goal of US$100 billion per year by 2020 in new and additional climate finance pledged in Copenhagen (UNFCCC 2009) will be challenging at best because this would require huge scaling in addition to leveraging of funds from the private sector (see Chapter 2 for what could be understood as additional in terms of climate finance).[3] In additional, still the massive share of foreign aid for mitigation (95%) is likely to start changing because it is likely that the demand for help from the least developed countries to build their strategies towards adaptation and resilience beyond what they are already achieving will be based on their national funding lines.

1.2.2 *Foreign Aid for Sustainable Energy Systems*

The energy sector is intrinsically linked to the goals of energy security, economic development and tackling global environmental challenges, among which climate change is the most notable expression. However, the current energy systems are not contributing effectively to the realization of most global development and climate mitigation goals. In other words, the mainstream energy systems which continue to be dependent mostly on fossil fuels are inherently unsustainable. For energy systems to help deliver on the promises of these goals (e.g. stabilizing GHG emissions at levels agreed on by the UNFCCC helping countries to achieve universal energy access by 2030 and better energy security (Chapter 5)) they would need to be transformed significantly. This transformation will need to be pursued on a globally cooperative basis, and foreign aid is a key component of this cooperative solution. In the energy sector, aid is being redirected from supporting the transition towards a more energy efficient green economy, most notably related to the production and distribution of electricity (Chapter 2). The energy-related funding in total ODA disbursements has increased over time, and the overall share of energy investments through ODA in renewables reached about 16% in 2010 (Chapter 4). Huang and Quibria (Chapter 12) provide evidence that foreign aid can be used to encourage technological innovations in energy systems or to reduce energy intensity, having a significant impact on sustainable development.

There are different avenues to transfer sustainable energy technologies and associated foreign aid flows that target the energy sector. Among them, the flexible mechanisms under the Kyoto Protocol joint implementation and Clean Development Mechanism (CDM), although the latter cannot be counted towards ODA when associated with governments directly using the CDMs to purchase certified emission credits (Chapter 4). The deployment of technologies that harvest renewable energy flows and offer efficiency improvements is the key to improving access to modern energy services and mitigating climate change. In fact, from a global perspective, current dominating energy systems, largely dependent on fossil fuels, are not socially, economically or environmentally sustainable. Further, they are a barrier to transforming economic structures that are responsible for climate change and other key environmental impacts across scales, such as biodiversity loss, in addition to societal problems associated with poverty and inequality in developing countries. While energy-related ODA is increasing, there is still a large gap in the funding required to redress the unsustainability of the dominant energy systems (Chapter 4), so it is crucial that the limited resources for sustainable energy are used as effectively as possible.

1.2.3 Foreign Aid for Greening Urban Areas

According to the United Nations (UN), by 2050 the number of people living in urban areas is expected to reach 6.4 billion out of a total population of 9.2 billion (Chapter 6), and most of this urban growth will occur in developing countries. This presents new challenges related to human development in cities, including the need for sustainable energy, security, water and food provision, education and health. In addition, cities are major contributors to climate change, currently producing more than 60% of all carbon dioxide and significant amounts of other GHG emissions, mainly through energy generation, vehicles, industry and biomass use. There is a need to move towards more sustainable and resilient urban development while confronting global climate change (in terms of both adaptation and mitigation). This involves new ways of planning, designing, building and retrofitting the cities of the future, such as mainstreaming environmental policies in the urban sector, restoring urban ecosystems and biodiversity, and promoting energy-efficient buildings and transport infrastructure. The challenges are not small, and foreign aid is seen as a key asset for developing countries, especially given their limited resources and growing urban populations (Cohen and Robbins 2011).

While the concept of the green city is still a novel idea in the context of foreign aid, and thus analyses of the link between foreign aid, urbanization and the green city are not well established in the literature, it is also clear that climate change adds a new dimension to this link in terms of investing in climate change mitigation strategies, most notably in energy efficiency in buildings and improving the energy intensity of transport systems and other infrastructure (water and waste treatments), as well as working towards the prevention of disasters related to climate variability, such as floods. But while the literature on sustainable cities has increased markedly over the last few years, there is little, both empirical and theoretical, that can shed light on how foreign aid could specifically help to green the cities of the future in developing countries, especially in the context of climate change. It is thus important to reflect on what works and could work in the future, as well as the scalability and transferability of current successful foreign aid initiatives.

1.2.4 Foreign Aid for Forest Conservation

Forests contribute directly to the livelihoods of more than 0.5 billion people, many of whom are resource poor, in tropical developing countries (Chhatre and Agrawal 2009). In addition, they provide a range of key indirect benefits, now usually termed ecosystem services in post-Millennium Ecosystem Assessment language, including services essential to climate change mitigation and adaptation (IPCC 2014a) and related to biodiversity conservation, provision of water resources and soil protection (MA 2005). Despite these well-known benefits, the forestry sector continues to be under intense pressure. Huge scientific effort is being devoted to identify land use changes in the forestry sector and what they represent in terms of GHG emissions. For instance, between 2000 and 2010, approximately 13 million hectares of forests were converted to other uses annually, or were lost through natural causes (FAO 2010). Currently this is the second largest source of GHG emissions, accounting for 10–20% of total anthropogenic carbon emissions globally (Harris et al. 2012).

While the results of sustainable forest management (SFM) initiatives have been mixed, and have generally lagged behind expectations, particularly in the tropics (Garcia-Fernandez et al. 2008), SFM in tropical developing countries receives considerable foreign assistance and attention from global multilateral institutions. Curbing deforestation and forest degradation poses enormous challenges both on the global and on other scales owing to the varying degrees of governance structures that exist across countries to remediate this problem.

Amidst the key role of SFM, REDD+ is emerging as the principal foreign aid mechanism for promoting climate mitigation in forest areas as well as conserving biodiversity and associated ecosystem services. REDD+ payments an potentially offer great foreste area coverage based on achieved results, in contrast to project-based mitigation activities (e.g. under the CDM of the Kyoto Protocol) (IPCC 2014a).

The REDD+ mechanism's function and financing are under negotiation through the UNFCCC, as well as through a number of non-governmental, national, and bilateral and multilateral donor initiatives. This volume devotes some space to analysing the current scale and potential role of foreign aid through REDD+ programmes to addresses the question of how to improve coordination and cooperation, between international donors/service buyers and local ecosystem service providers, by first acknowledging the trade-offs in what each of these actors consider priorities within REDD+ (e.g. carbon trade-offs with human development goals). There are expectations that REDD+ as a foreign aid instrument may contribute to curbing deforestation while conserving biodiversity and ecosystem services, as well as enhancing sustainable rural development under well-defined social and biodiversity conservation safeguards, including those forest areas where high carbon content and high biodiversity are not necessarily coincidental (Phelps et al. 2011). However, it is important to realize that win–win outcomes for carbon emissions, human development and other ecosystem services are not automatic, and they involve the interplay of the complex foreign aid architecture. Leveraging effective foreign aid through REDD+ critically relies on matching such potentially confronted goals.

1.2.5 Foreign Aid for Climate Smart Agriculture

Agricultural development continues to face huge challenges in meeting global food security and is expected to face even greater difficulties as a result of climate change. As the WGII 5th Assessment Report of the IPCC states, 'Without adaptation, local temperature increases in excess of about 1 degree centigrade above pre-industrial is projected to have negative effects on yields for the major crops (wheat, rice and maize) in both tropical and temperate regions' (IPCC 2014a). Global and regional weather conditions are also expected to become more varied, with increases in the frequency and severity of extreme events such as cyclones, floods, hailstorms and droughts (IPCC 2007, 2012, 2014a, b). Such weather conditions are expected to increase the variability of crop yields and food supplies, adding to the increased risk of food insecurity.

Agriculture is severely underfunded, especially in the developing world (Islam 2011), and foreign aid has not increased sufficiently to assist developing countries in transforming agricultural systems to achieve the joint objectives of development (including associated food security) and the mitigation of and adaptation to climate change (Huang and Wang, this volume). Huang and Wang (this volume) shows that the share of aid from OECD countries to the agricultural sector from total aid increased from 13% in 1973–5 to 23% in 1979–81, but that this has declined since the mid-1980s, just reaching about 6% within the 2006–8 period. The actual level of aid directed at transforming unsustainable agricultural practices into more sustainable ones is hard to quantify. However, the existing climate-related financial flows specifically targeted at agriculture in the developing nations cover a very small part of the total climate-change funds (Nelson et al. 2010) and are unlikely to match the cost of mitigation in the agricultural sector (Huang and Wang, this volume). So far, the modest efforts by foreign aid have been used to reduce methane emissions from paddy fields, increase carbon sequestration through land-use changes, and limit carbon dioxide emissions from agricultural soils. Yet there is scope to enhance the area of intervention, including control of nitrous oxide emissions from crop production or limiting carbon dioxide emissions through energy-saving technology (Huang and Wang, this volume). In addition, foreign aid can be disbursed to help the most vulnerable small farmers to adapt to climate change, ranging from investments in sustainable ecological intensification approaches, such as through the deployment of agrobiodiversity, hard infrastructure (e.g. irrigation) and subsidizing agricultural insurance products.

1.3 Structure of the Book

This volume is organized into twelve chapters, including the present introduction, that focus on the role of foreign aid towards achieving environmental sustainability, broadly defined, underpinned by the challenge of climate change. It pays special attention to areas that span key aspects related to climate change in the least developed countries, including investments in capacity building, biodiversity and forest conservation, food and agriculture, and the sustainability of the urban and energy sectors. Notably, while each of these four key focal areas have specific challenges and opportunities associated with foreign aid under a climate change context, towards achieving environmental sustainability goals, all share the four common questions of 'what works?', 'what could work?', 'what is scalable?' and 'what is transferable?'. All of the chapters deal with the analysis of how effective foreign aid is at present with regard to capacity building, biodiversity and forest conservation, promoting sustainable agriculture, and the sustainability of the urban and energy sectors as a result of climate change. These four questions that run through this volume across the four key sectors provide a consistent framework for analysing the role of actual and future foreign aid for environmental sustainability.

More specifically, the book addresses how foreign aid can be used to enhance *capacity building* to address climate change (Chapters 2 and 3), to improve access to sustainable energy and promote more efficient uses of energy resources (Chapters 4 and 5), to move towards greening urban areas (Chapters 6 and 7), to support the conservation of forests in developing countries (Chapters 8 and 9) and to move towards a climate smart agriculture (Chapters 10 and 11).

In chapter 2, David Victor focuses on the effectiveness of foreign aid regarding capacity building by the least developed countries, which are least able to respond and adapt to climate change. He addresses an important point—that, as countries try to expand climate aid quickly, bilateral aid, which is easier for donors and recipients to control, is likely to expand much more than multilateral aid, as suggested by the historical patterns of other aid programmes. He discusses a fundamental problem related to the idea of an 'aid paradox'. Such a paradox is that the conditions of national capacity under which aid is most likely to be effective are least likely to be present in the countries that are most in need of foreign aid. This is because they cannot raise the needed funds on their own to invest in institutional and governance assets that make foreign aid effective and adaptive. Victor concludes that in the case of foreign aid to tackle climate change, including capacity building, it will be slow and difficult to scale the lessons that have been learned from other foreign aid experiences, especially as a result of the need for tailoring aid to each country setting and owing to the difficulty for donor countries to make credible long-term aid commitments. Both aspects are essential to build the necessary capacities to invest in climate change mitigation and adaptation in the least developed countries. Zexian Chen and Jingjing He's contribution (Chapter 3) concurs with the idea that despite the need of developing countries for more financial and technical resources to tackle the social, economic and environmental challenges posed by climate change, emphasis should be placed on investing in the institutional scaffolding that underpins the effectiveness of the use of foreign aid. It looks into various case studies related to actual projects on capacity-building programmes and finds that foreign aid for capacity building is more successful when the projects are demand-driven. Further, to make aid work and to allow it to be scaled up, emphasis is placed on the need to strengthen cooperation among climate change aid projects and raise public awareness, among various other key conditions.

The extent to which foreign aid is being effective in promoting a transformation of the energy systems for sustainability is addressed by Luis Gomez-Echeverri (Chapter 5) and H-Holger Rogner and Kei-Kit Leung (Chapter 4). In chapter 5, Gomez-Echeverri portrays an increasingly complex foreign aid–energy landscape, which is becoming increasingly fragmented, where many new actors, both public and private, are gaining in prominence, in conjunction with foreign direct investment. He points out that this emergent landscape is making coordination more difficult and the management of aid by those receiving it more challenging, more costly and with heavier demands on scarce national financial and human resources. Gomez-Echeverri offers a detailed picture of the evolution of foreign aid related to the energy sector, emphasizing the emergent south–south financial flows, especially from emergent countries such as China, India and Brazil to Africa, on top of the traditional north–south ones, as well as the involvement of new NGOs and private foundations in reshaping this landscape. In this context he also adds to the discussion of what initiatives have worked best, and what conditions are necessary to make foreign aid for sustainable energy more effective, scalable and transferable in the future. He points out that the Sustainable Energy for All initiative of the UN secretary general is a good example of a coordinated effort from which to derive important lessons in this regard. Then in chapter 4, Rogner further reviews the key enabling conditions to transform the energy system and foreign aid policy prerequisites towards it which would determine to a large extent the impacts on global climate change as well as how such systems can adapt to a context of growing energy scarcity. He argues that support of sustainable energy is the stepchild of foreign aid and its efficacy has been questioned. His review suggests that there is still much room for improving the effective use of foreign aid for sustainable energy and climate protection. He also points out that foreign aid has been instrumental in promoting many renewable and efficiency projects that might not have been implemented in developing countries without ODA flows.

The next two chapters focus on the role of foreign aid in greening urban areas in developing countries. In chapter 6, Sandrine Kablan looks at what kinds of foreign aid practice have the potential to work towards the achievement of green cities. The chapter reflects on how innovative and promising are the programmes, initiatives and practices aimed at greening cities, in order to be able to scale them up and make them transferable across countries focusing on various urban sectors: urban design and public policy on buildings and infrastructure, transport, pollution and waste treatment; and energy supply and water supply and sewage. She employs an econometric panel data analysis covering 144 developing countries during the period 2002–11 to check the effectiveness of foreign aid in curbing urban the intensity of carbon dioxide emissions from buildings through investment in renewable energy sources. Kablan finds that foreign aid helps in curbing these emissions, especially those associated with residential buildings, as well as commercial and public services, thus contributing to greener cities. In chapter 7, Jun Li, by means of three case studies, focuses on the aspects of what works and could work, and the degree of scalability and transferability in terms of greening urban areas through foreign aid. The case studies are based on the Chinese urban areas of Tianjin, Wenchuan and Beichuan, as well as Curutiba in Brazil. These cities provide the opportunity to compare two already planned ecocities (Tianjin City and Curitiba) and two areas that are recovering from an earthquake that took place in 2008: Wenchuan and Beichuan. From these case studies, Li derives the conditions for successful planning, design and implementation of future ecocities to combat and respond to the challenges of climate change and how this vision can be catalyzed through foreign aid.

Chapters 8 and 9 turn their attention to the role of foreign aid in the conservation of biodiversity and forests, especially in tropical developing countries. In the first, Unai Pascual and colleagues address the link between sustainable forest management initiatives, the climate change policy arena and foreign aid. They look into the way mechanisms for REDD+ are supporting sustainable forest management in the context of complex, multiple stakeholder interests and negotiations, including those of international foreign aid donors and recipient countries. Pascual et al. discuss the role of foreign aid in helping to achieve sustainable forest management, framing this as the condition for delivering multiple ecosystem services, and considering the potential for donor support for the forestry sector associated with new climate finance. More specifically, the chapter examines recent policy developments through REDD+ as a potentially scalable initiative across the developing world, and considers how it can integrate lessons from previous forest conservation instruments. It explores the conditions for promoting forest conservation through foreign aid, taking into account the varying interests of multiple actors. The authors warn that while the way in which REDD+ financing, catalyzed by foreign aid, has the potential to move beyond traditional sustainable forest management efforts, the mechanism still faces uncertainty regarding the long-term sustainability of financing, thus affecting the scalability of the mechanism. Chapter 9 by Pekka Kauppi looks at how forestry aid has evolved since the 1960s, when the emphasis of foreign aid was to incentivize industrial forestry up to the present where one of the main focus is on REDD+. He argues that while investments in forestry are still largely associated with the value of timber as a global commodity, the environmental and social concerns are gaining greater attention. Kauppi discusses the history, scalability and transferability of tree planting and initiatives aimed at improving cooking stoves to save trees and to improve human health. He also points to the important role of local universities in catalyzing research on forest conservation in developing countries, especially in Africa where university education is limited.

The next two companion chapters focus on the role of foreign aid in supporting efforts towards sustainable agriculture, which in the context of climate change is generally known as 'climate smart agriculture'. In chapter 10, Jikun Huang and Yangjie Wang provide an overview and case studies of the connection between agriculture, foreign aid and climate change, and they argue that agriculture is heavily underinvested and that foreign aid needs to be scaled up significantly to tackle the great challenge of food security in a climate change context. They also point at how to make climate finance through foreign aid more effective in tackling mitigation and adaptation in agriculture across different types of mitigation and adaptation measures. Chapter 11 by Siddig Umbadda and Ismail Elgizouli, directs the focus to agriculture in the African continent, which is undoubtedly highly aid dependent and where for the majority of the population small-scale farming is the anchoring livelihood. The authors review the recent history of foreign aid directed to agriculture in Africa and identify the strategic areas to which aid ought to be directed to support sustainable agriculture and food security in Africa.

Chapter 12 by Yongfu Huang and Muhammad G. Quibria examines whether foreign aid, together with other economic, social and environmental factors, contributes to sustainable development. It starts with an illustrative theoretical growth model where foreign aid promotes sustainable development by protecting the environment. Using factor analysis and newly developed estimation methods for a dynamic panel data model with endogenous regressors, the empirical section of the chapter finds evidence that foreign aid has had a significantly positive influence on sustainable development in aid-receiving countries. This effect is very likely to go through channels related to growth and resources as well as a technology channel with respect to energy intensity. This research has important implications for the 2030 development agenda on international collective action with regard to a sustainable future.

Notes

1. The 2005 Paris Declaration on Aid Effectiveness adopted five principles to strengthen aid effectiveness and 13 targets to measure their implementation, which were to be achieved by 2010. The principles and targets were confirmed in Accra in 2008. The Fourth High-Level Forum on Aid Effectiveness in Busan in 2011 shifted the focus from purely aid effectiveness to a more holistic approach that looks at the contribution that effective development cooperation can make to overall development effectiveness, marking a turning point in the international consideration of development cooperation.
2. For example, in 1992 the World Conference on Development and Environment initiated the Global Environment Facility, a mechanism to facilitate aid for the environment. The OECD's Green Growth Strategy, announced in 2009, aims to help developing countries achieve economic growth, job creation, environmental protection and the development of more equitable societies.
3. In this book the term 'climate finance' is applied both to the financial resources devoted to addressing climate change globally and to financial flows to developing countries to assist them to address climate change (IPCC 2014b).

References

Buchner, B., A. Falconer, M. Hervé-Mignucci, C. Trabacchi, and M. Brinkman. 2011. *The Landscape of Climate Finance*. CPI Report. Venice: Climate Policy Initiative.

Chhatre, A., and A. Agrawal. 2009. Trade-Offs and Synergies between Carbon Storage and Livelihood Benefits from Forest Commons. *Proceedings of the National Academy of Sciences* 106 (42): 17667–17670.

Cohen, N., and P. Robbins. 2011. *Green Cities: An A to Z Guide*. London: Sage Publications.

FAO. 2010. *Global Forest Resources Assessment*. FAO Forestry Paper 140. Rome: FAO.

Garcia-Fernandez, C., M. Ruiz-Perez, and S. Wunder. 2008. Is Multiple-Use Forest Management Widely Implementable in the Tropics? *Forest Ecology and Management* 26: 1468–1486.

Harris, N., S. Brown, S.C. Hagen, S.S. Saatchi, S. Petrova, W. Salas, et al. 2012. Baseline Map of Carbon Emissions from Deforestation in Tropical Regions. *Science* 336: 1573–1576.

IPCC. 2007. *Climate Change 2007. Impacts, Adaptation and Vulnerability.* Contribution of Working Group II to the Fourth Assessment Report of IPCC. Cambridge: Cambridge University Press.

———. 2012. *Managing the Risks of Extreme Events and Disasters to Advance Climate Change Adaptation.* Special Report of the Intergovernmental Panel on Climate Change. Cambridge: Cambridge University Press.

———. 2014a. *Climate Change 2014: Impacts, Adaptation, and Vulnerability.* IPCC Working Group II Contribution to AR5. http://ipcc-wg2.gov/AR5/report/final-drafts

———. 2014b. *Climate Change 2014: Mitigation of Climate Change.* IPCC Working Group III Contribution to AR5. http://mitigation2014.org/

Islam, N. 2011. *Foreign Aid to Agriculture—Review of Facts and Analysis.* IFPRI Discussion Paper 01053. Washington, DC: IFPRI.

Millennium Ecosystem Assessment. 2005. *Ecosystems and Human Well-Being: Synthesis.* Washington, DC: Island Press.

Nelson, G.C., M.W. Rosegrant, J. Koo, R. Robertson, T. Sulser, T. Zhu, C. Ringler et al. 2010. *The Costs of Agricultural Adaptation to Climate Change.* WB Discussion Paper 4: Development and Climate Change. Washington, DC: World Bank.

Phelps, J., E.L. Webb, and L.P. Koh. 2011. Risky Business: An Uncertain Future for Biodiversity Conservation Finance through REDD+. *Conservation Letters* 4: 88–94.

UNFCCC (UN Framework Convention on Climate Change). 2009. *Report of the Conference of the Parties on Its Fifteenth Session, Held in Copenhagen from 7 to 19 December 2009.* Copenhagen: United Nations.

———. 2012. *Synthesis Report on the Implementation of the Framework for Capacity-Building in Developing Countries.* Bonn: UNFCCC.

World Bank. 2010. *World Development Report: Development and Climate Change.* Washington, DC: World Bank.

CHAPTER 2

Foreign Aid for Capacity Building to Address Climate Change

David Victor

2.1 Introduction

Since the mid-1990s, the diplomatic community has been engaged in the problem of global climate change (Victor 2011). In part, its efforts have involved the creation of special multilateral funds to pay for many activities, including capacity building in the least developed countries. These efforts build on a long history of including foreign assistance programmes in international environmental agreements (Keohane and Levy 1996; Gutner 2002). At the same time, many governments have proffered substantial bilateral assistance for the same purposes.

This chapter examines whether and how those multilateral and bilateral funds—what, together, I call 'foreign aid'—have actually worked. While significant sums of money have been allotted for these purposes over the last two decades, ever since the 2009 Copenhagen Accord, governments have promised much larger expenditures by public agencies as well as private firms on climate change in the future. While those new funds will be used for many functions, it is widely known that capacity building is an essential function. As this expansion occurs, which lessons should guide

D. Victor (✉)
School of Global Policy and Strategy, UC San Diego,
San Diego, CA, USA

Y. Huang, U. Pascual (eds.), *Aid Effectiveness for Environmental Sustainability*, https://doi.org/10.1007/978-981-10-5379-5_2

the effort? To explore that issue, I review and assess the existing literature, focusing on four questions:

- What has worked best in terms of delivering aid, and why?
- What kinds of reforms could improve effectiveness, and what are the best models for implementing them?
- What kinds of programmes could be scaled up—such as the many-fold increase in aid envisioned under the Copenhagen Accord? A large part of the answer here hinges on which programmes have been able to leverage other sources of funding, such as from private industry.
- What lessons are transferrable?

The aim here is to answer these four questions apropos one area of foreign assistance in particular: capacity building. Capacity building is important because it helps to create the right conditions for much larger public and private funding that will be crucial to long-term solutions to the climate change problem. The focus in this chapter is climate change, but any serious assessment of foreign aid in this area, along with guidance for the future, must recognize that these questions have been the subject of extensive analysis more generally in the field of foreign aid, and so this study includes some attention to this broader field of foreign aid research.

This chapter makes two broad arguments. One is that the planned rapid expansion of climate change aid requires careful attention to lessons about how best to make aid 'work'. The other main argument is that much of what has been learned, so far, has concentrated on building capacity for mitigation of emissions, but diplomacy around foreign aid for climate change—especially regarding the countries that are least developed and thus also most likely to need foreign aid—is shifting quickly to focus on adaptation. These least developed countries generally have low emissions (and thus there is little role for mitigation) but are highly vulnerable to climate change. For them, capacity building is largely concerned with the capacity to analyse and respond to the changing climate.

This chapter begins with an explanation of key definitions and concepts, as well as discussing fundamental trends in foreign aid generally and capacity building in particular. Then I turn to the key patterns in the subset of foreign aid that is directly related to climate change, including capacity building for climate change. With that foundation in place, the focus is then on the four analytical questions that organize this study, starting with the question of what actually works.

2.2 Setting the Scene: Key Definitions, Concepts and Fundamental Spending Patterns

We begin with some definitions that help to chart the landscape.

Foreign aid (or the equivalent term, 'foreign assistance') can be classified into different types according to the main objective, including development aid, humanitarian aid, military aid, and food aid. Development aid, the focus of this review, is defined by the Development Assistance Committee (DAC) of the Organisation for Economic Co-operation and Development (OECD) as 'flows of official financing administered with the promotion of the economic development and welfare of developing countries as the main objective and which are concessional in character with a grant element of at least 25 per cent' (OECD 2003).[1] Aid money is not always free, but it is always concessional and is often fully granted. Most environmental aid, including that related to climate change, falls into the broad category of 'development aid'. However, in the future, some portion of climate change assistance might also involve military or humanitarian assistance if, for example, climate change creates large numbers of refugees.

The DAC further separates development aid (which consists of project aid, programme aid, and technical assistance) into three categories. *Official development assistance* (ODA) is aid provided by donor governments to low- and middle-income countries, which accounts for 80–85% of the total development aid,[2] and is also the one that most people have in mind when they think of 'aid'. *Official assistance* (OA) is aid provided by governments to richer countries with per capita incomes higher than approximately US$9000 and to countries that were formerly part of the Soviet Union or its satellites. *Private voluntary assistance* includes grants from non-government organizations, religious groups, charities, foundations, and private companies. This latter category is growing both in terms of quantities and in importance as some foundations, such as the Bill and Melinda Gates Foundation, take on particular aid-related topics in strategic ways. The Gates Foundation, for example, disbursed US$1.8 billion in grants in 2009 alone to improve health in developing countries, marking it as the third-largest international donor of aid for health, after only the United States and the Global Fund to Fight AIDS, Tuberculosis and Malaria (OECD 2011a). However, most of the statistics about foreign aid, the best of which come from DAC, do a better job of tracking government-based expenditures.

Global ODA increased steadily from the 1960s, until it reached a peak of US$68.7 billion (converted to 2010 US dollars) in 1992, just after the end of the Cold War, and then declined sharply to just under US$55.4 billion in 1997. It began to rebound in the late 1990s and had a sharp increase in 2005 when the heads of state of the Group of Eight industrialized countries ('G8') pledged to double aid to Africa by 2010 and triple it by 2015 (VandeHei and Blustein 2005). The practical effect of these pledges has been modest; while global ODA measured as a share of donor income fell sharply during the 1990s, it has rebounded only slightly since 2005. Total global ODA reached around US$149 billion in 2011.

Aid is devoted to many different purposes. In 2010, about 25% of total ODA was directed to 'economic infrastructure' and to 'production', which include agriculture, energy, and transport systems that are likely to be vulnerable to changes in climate. Another 13% was apportioned to multi-sector programmes, many of which address activities that could be affected by climate.

There are two striking trends relevant to climate change that have emerged over time. First, since the mid-1980s, aid to agriculture has fallen by almost half, although since 2005 there have been some modest increases partly due to the renewed emphasis on aid to Africa led by the G8 countries.[3] The other trend is a similarly steep decline in aid for energy, which also fell by half since the mid-1980s and is now rising modestly.[4] Within the field of energy, the most striking trend has been a steep reduction in fossil fuel spending, as donors have shifted their resources to renewable energy and from the production side of the energy sector (e.g. large infrastructure projects) to capacity development and energy efficiency. Electricity dominates energy-related spending; electrical transmission/distribution and the energy policy subsectors both account for more than half of the resources allocated by donors in recent years (OECD 2010a). The importance of electricity is consistent with the significant role of electricity more generally in the world's energy system—the most recent projections from the International Energy Agency (IEA) see almost half of world investment in energy as anchored in the power sector (IEA 2011).

About half of ODA takes the form of country programmable aid (CPA), which is a measure of the aid that donors and multilateral development banks can reasonably influence.[5] CPA is an important consideration when thinking about aid for climate change because it is this portion that might be redirected towards climate change purposes in the coming years, and it is precisely the prospect of such a redirection of funds that leads many diplomats to demand that climate change assistance be 'new and additional'—that is, over and above current aid spending.

Our focus in this chapter is 'capacity building'. Also referred to as 'capacity development', this is the long-term continual process of development that involves all stakeholders, including ministries, local authorities, non-governmental organizations, professionals, community members, academics, and more (UN 2006). The term 'capacity building' has evolved from past terms such as 'institutional building' and 'organizational development'; it emerged as a leading developmental concept in the 1990s.

As a practical matter, capacity building can take place on three levels. First, on an individual level, capacity building requires the development of conditions that allow individuals to build up and enhance existing knowledge and skills and to engage in the process of learning and adapting to change. Programmes of this type include, for example, training of government officials. Second, on an institutional level, programmes can modernize existing institutions (or build new ones) and support the formation of sound policies, organizational structures, and effective methods of management and revenue control. Third, on a societal level, capacity building can help create a more informed and engaged society—one that better holds government institutions accountable and is more fully aware of and engaged in how those institutions operate.[6]

The standard systems for accounting do not include a category for 'capacity building' that covers the whole range of activities included in the concept. However, DAC statistics suggest that the broad category of social and administrative infrastructure accounts for more than one third of all aid expenditure. Those same data show that 11.9% of all aid is spent on capacity building in the government sector. These fractions are a benchmark for what to expect as the field of climate change finance matures; perhaps around one third of resources will be devoted to capacity building broadly, and one third of that amount will be focused on the government.

2.2.1 *Foreign Aid for Climate Change and Capacity Building*

Now that we have set the scene generally with regard to foreign aid we can focus specifically on the numbers related to climate change and explore, in particular, the interactions between capacity building and the overall effectiveness of foreign aid.

The developed countries that signed the three Rio Conventions in 1992 committed themselves to assist developing countries in the implementation of these Conventions. Over time, those commitments have come to include mitigation and adaptation, and, within those broad categories, capacity building is a component.[7] Total bilateral climate change-related aid given by members of the OECD's DAC was US$22.6 billion in 2010, representing about 15% of total ODA. Of this total, roughly two thirds were for mitigation, and one third for adaptation.

Climate change-related aid is increasing rapidly. The 'upper bound' estimate of mitigation-related aid, for example, exceeded US$17.6 billion in 2010, an increase of 76% from 2009 (see Fig. 2.1). In terms of rate of growth, climate change is now one of the fastest growing major areas of foreign aid, although precise comparisons to other issue-areas are not possible since climate change (and other aid-related topics) typically cover

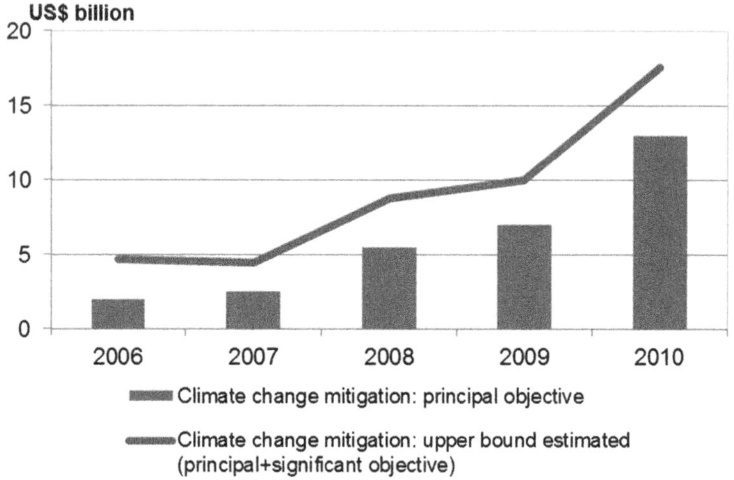

Fig. 2.1 Bilateral climate change mitigation-related aid, 2006–10 (Source: OECD 2012)

several categories and there is a lot of double counting in the statistics. In addition to bilateral assistance, total multilateral climate change-related aid, including DAC members' contributions to specific climate funds (except Climate Investment Funds) plus the climate-related share of DAC members' core contributions to multilateral organizations, was US$727 million in 2010. *That is, bilateral climate change assistance is more than 20 times larger than multilateral funds.* Other studies that have compared multilateral and bilateral sources witness much less of a disparity—with bilateral funding accounting for a smaller multiple of the multilateral source (Buchner et al. 2011b).[8] No study has carefully assessed the portions of multilateral and bilateral assistance programmes that are strictly devoted to capacity building, but it is likely that the proportions of total spending between multilateral and bilateral assistance also apply to capacity building.

It is important to note that many large donors have started to move away from a project-based approach and towards programmatic financing, with funds being allocated more for budget support and supporting national development plans, thus making it more difficult to track the sector of destination (Corfee-Morlot et al. 2009). Figure 2.2 shows the sectoral breakdown of aid activities targeting climate change mitigation and adaptation, respectively. In value terms, more than three quarters of aid targeting climate change mitigation concerns energy supplies, transport, and general environmental protection. For adaptation, three quarters of aid was reported in the sectors of general environmental protection, water, and agriculture and rural development (OECD 2012). Geographically, most climate change ODA is allocated to Asia (51%). The shift to programmatic support probably has increased the share of aid that is devoted to capacity building, including long-term capacity programmes that are likely to be more effective than one-off projects.

2.3 Question No. 1: What Works in Foreign Aid for Capacity Building with Regard to Climate Change?

Surprisingly, there is almost no research that addresses this question systematically. That problem arises for two reasons. First, the field of climate change assistance is relatively young. Second, one of the main challenges that has confronted scholars who have assessed foreign aid in other areas

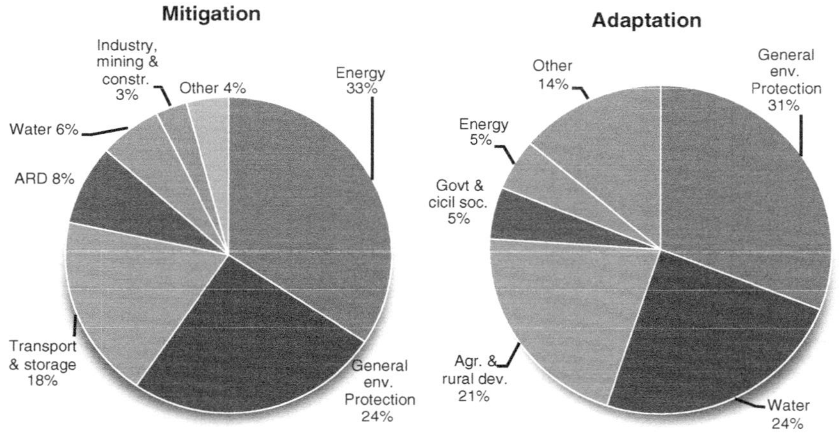

Fig. 2.2 Sectoral breakdown of climate change-related aid (2010 commitments) (Source: OECD 2012)

is that agreeing on which goals can most usefully be employed as a yardstick (Radelet 2006) is not settled.

Some efforts, such as the Bali Action Plan in 2007, try to standardize efforts. The Action Plan introduced the phrase 'measurable, reportable and verifiable' (MRV) in the context of both 'nationally appropriate mitigation actions' (NAMAs) and finance, technology, and capacity building to support mitigation actions. The concept of NAMAs recognizes the need to tailor financial support to implement actions that align with constraints in scale and the institutional capacity for each locale. In practice, though, the language of the Bali Action Plan does not address important questions, such as what is the relationship between mitigation action and mitigation support or how to measure, report, and verify mitigation support and action (Corfee-Morlot et al. 2009).[9]

However, it is possible to triangulate some answers to this question by looking, initially, at what has been learned through studying aid in other settings—in particular, aid related to capacity building. Then we evaluate which of those lessons might apply to climate change programmes, while focusing on the few studies that have looked directly at this area.

There is a voluminous empirical literature on the effectiveness of foreign aid in other fields, notably its role in promoting economic development.[10] This field of research blossomed first in the 1990s and has evolved into two camps. Some scholars (I call them 'aid pessimists') have disparaged most aid as unproductive and perhaps even counter-productive; it squanders resources on projects that are inappropriate for the context and rewards bad behaviour by governments (Bauer 1972; Moyo 2009). While much of this work is polemical in nature, some is rooted in careful econometric analysis that finds only very narrow conditions under which aid truly works (Easterly 2003, 2007, 2009; Rajan and Subramanian 2008). Others discover a closer relationship between aid and growth, such as through allowing for diminishing returns or by testing for conditional relationships.[11] Put differently, this more pragmatic scholarship sees a role for making aid effective when it is applied in the right circumstances with the right managers.[12] One lesson from this strand of the literature is that aid should be made conditional upon the presence of certain circumstances that lead to more effective use of aid. Exactly how conditionality should be designed and applied is a hotly debated topic (Stokke 1995; Crawford 1997; Scholl 2009). One of the few areas where pessimists and pragmatists often agree is that aid generally has been successful in some countries, where that aid has helped improve health by supplying essential medicines; better health, especially for people reaching or in their economic prime, leads to the accumulation of human capital in ways that can sustain economic growth (Levine and What Works Working Group with Molly Kinder 2004). Healthiness helps increase the overall capacity of a society to address a range of challenges, although health assistance is normally not called 'capacity building'—a term that is applied more narrowly to creating administrative systems and infrastructures that improve the ability of societies to govern themselves.

Developing countries that are party to the United Nations Framework Convention on Climate Change (UNFCCC) have, themselves, identified several priority areas where they see the greatest need of funding for capacity building. On mitigation, many report a need to develop human and institutional capabilities to prepare mitigation project proposals for funding, to facilitate data collection and analysis, to prepare national communications, and to manage climate change programmes. For adaptation, in addition to the need for greater human and institutional capabilities, many parties also saw a need to improve scientific research, particularly in modelling and for training in planning and implementing adaptation activities (UNFCCC 2007).

2.4 Question No. 2: What Kinds of Reforms Could Improve the Effectiveness of Climate Change Aid?

A clear answer to 'what works' helps set the agenda for reforms that might improve the effectiveness of climate change aid. But since we have comparatively little research looking at what works in this area, here too we must look to the broader literature on foreign aid. Fortunately, such questions have been given extensive analytical attention and there has also been sustained political thought on this question. Of particular note—because they involved most major donors and are recent—are the Paris Principles adopted in 2005.[13] In February 2005, the international community came together at the Paris High Level Forum on Aid Effectiveness, hosted by the French government and organized by the OECD. The event led to endorsement of the Paris Declaration on Aid Effectiveness. Looking at the Paris Principles, the larger analytical studies on aid effectiveness, and applying some logic to how these topics might be relevant to climate change, leads to five major suggestions on how to improve aid effectiveness. All five apply to climate change, yet few of them have so far been actively considered in that context—although there are some exceptions, such as Thornton and colleagues' analysis of 11 case studies of climate change finance at the national level, assessed with an eye to the Paris Principles (Thornton 2010; Norrington-Davies 2011; Cameron 2011; Norrington-Davies and Thornton 2011a; Grant 2011; Norrington-Davies and Thornton 2011b) and a few others (Brown and Peskett 2011; Bird 2011; Hedger 2011).

First is the idea that aid should primarily be provided to countries with good policies and robust institutions. The relevant analytical work, starting with the World Bank's 'Assessing Aid' study in the mid-1990s—probably the most influential study of aid impacts ever done—shows that aid has a big effect on growth and poverty reduction in recipient countries that have a good policy environment (World Bank 1998). For economic growth, that good policy environment includes having a balanced public budget (so that aid money is truly additional), low levels of corruption, and institutions such as a free press that can hold the government accountable. As applied to issues of climate change, the basic logic of a 'good policy environment' probably also includes accountable and transparent government, prices for fossil energy and other things linked to climate change that are representative of real scarcity in the economy, and a clear relationship between climate policy and the other functions of government. The idea of 'good institutions' is centrally about a society's own abilities to govern itself in a fair and accountable manner.

Second, ownership by national governments—also known as country ownership—is emphasized by many analysts in achieving aid effectiveness (Radelet 2006). (This idea is a prominent part of the Paris Principles but has not been measured systematically in the analytical literature.) Donor countries used to dominate in setting aid priorities, designing programmes, and implementing projects. The call for recipient countries rather to take a lead or joint-lead position in agenda setting is intended to help eliminate some of the problems in the long chain of principal–agent relationships that earlier studies had shown led to failure (Martens 2002). Applied to climate change, this insight affects both mitigation and adaptation. For example, research on the Clean Development Mechanism (CDM) has tried to explain the relatively low level of technology transfer in some countries (e.g. India) and the high level in others (notably China, which accounts for more CDM projects than any other country), and finds that one of the major explanatory factors is involvement by the central government (Dechezleprêtre et al. 2009; Popp

2011). Looking beyond the statistical associations, such research finds that 'ownership' by the central government has steered investment into particular kinds of CDM projects, helped to reduce regulatory barriers, allocated the benefits from projects to favoured technologies and regions, and created a coherent policy that channels benefits broadly into sustainable development—all functions that result in various forms of government capacity. Examples include efforts to reduce emissions from deforestation and forest degradation, which have spawned programmes to build capacity to design and analyse forestry and land use projects as well as to build capacity in monitoring methodologies and identification of best practices.[14] Through UN-based initiatives on deforestation and forest degradation, for example, Germany provided US$105.1 million to Brazil to build local capacity in monitoring climate-relevant biodiversity and recording bush fires, improve management of nature reserves, and strengthen forest monitoring systems, among other things.[15]

Third, beyond 'ownership' is the need to engage local participation in beneficiary countries, particularly non-government stakeholders such as development organizations, charities, religious institutions, and the private sector (World Bank 1998; Radelet 2006). Multilateral donors have recognized this need. For example, in 2011, the World Bank implemented a study to test the application of the Adaptation Coalition Framework in Latin America and the Caribbean. The aim of the Coalition was to provide local communities with the knowledge, organizational tools, and strategies to mobilize the essential resources necessary to adapt. By the end of the study, the majority of the communities showed greater awareness of climate change and its risks, increased ability to form coalitions with other groups, and, it appears, a greater ability to gain access to resources to adapt to climate change (Ashwill et al. 2011). Evidence shows that beneficiary participation could achieve better aid effectiveness. An evaluation of 121 rural water supply projects financed by donors and non-governmental organizations in 49 countries shows that among projects with a high level of beneficiary participation, 68% were highly successful, while only 12% of those projects in which there was little beneficiary involvement were highly effective (Narayan 1995). The insight that engagement with local stakeholders is important resonates with the central finding in aid effectiveness: accountable and responsive government is essential. For example, one study found that investment projects have been more effective in countries where citizens enjoy civil liberties, where people have greater freedom to express their views including through a free press, freedom of association and assembly, and the freedom to petition governments.[16] Another study has examined this issue as applied to climate change in some detail and found, looking at 11 countries across Asia and Africa, that 'none of them had a dedicated forum for dialogue where funding partners, recipient government, and other stakeholders such as civil society could meet around climate change assistance and financing' (Thornton 2010). This third point suggests that capacity building activities are best pursued when they augment not just governmental capacity—the traditional focus—but also the ability of government to interact with other stakeholders.

Fourth, the Paris Principles focus on the need for harmonization and coordination among donors. Managing aid flows from many different donors is a huge challenge for recipient countries, since different donors may insist on using their own unique processes for initiating, implementing, and monitoring projects. Recipients can be overwhelmed by requirements for multiple project audits, environmental assessments, procurement reports, financial statements, and project updates. In Tanzania, for instance, health workers in some districts spent almost 25% of their working days writing reports for different donors (Deutscher and Fyson 2008). Aid fragmentation and uncoordinated aid donations have led to numerous suggestions for donors to more closely coordinate their activities—to build the capacity to harmonize their systems and 'pool' their funds (Kanbur and Sandler 1999; Deutscher and Fyson 2008; OECD 2011b). My assessment of the literature is that the need for coordination has been a frequent conclusion among aid workers and some bureaucrats in aid agencies (who often seek bureaucratic solutions to more fundamental problems). In one study that looked at donors and recipients of climate change finance side by side, the author surveyed the experiences in 11 recipient countries and found there are 'specific institutional requirements of the external funds from donors, which may be out of step with the roles and responsibilities of institutions in recipient countries'. To some degree, this conclusion may reflect the fact that most climate change assistance has historically been driven by interest in mitigation, and the goal of mitigation has been foregrounded by donors rather than recipients. However, serious scholarly research on coordination is notably scarce, and in the area of climate change, there has been no systematic research on this issue.

Indeed, scholarship on the international institutions related to climate change suggests that competition among donors and institutions might actually lead to more effectiveness.[17]

The fifth major suggestion is results-based management with stronger monitoring and evaluation (Deutscher and Fyson 2008). The Paris Declaration on Aid Effectiveness outlines an indicator to track to what extent partner countries have established transparent and monitorable performance assessment frameworks. The indicator looks at three dimensions, including the quality of information generated, stakeholder access to that information, and coordinated country-level monitoring and evaluation systems. The results of the 2011 survey indicate that partner countries are making important progress in developing results-oriented frameworks, but still only one fifth of the countries surveyed are considered to have relatively strong results-oriented frameworks (OECD 2011b). Exactly the same logic applies to climate change, although in this area, relatively little progress has been made, for two possible reasons. One is that it has been difficult to agree on goals for climate change aid generally and thus results-based monitoring is essentially impossible. The other is that the major decisions about levels of aid and strategy are made through an intergovernmental process that is steeped in suspicions about the unwillingness of donors to fulfil aid commitments, which leads most diplomacy to focus on demands for new and additional resources, rather than building the institutions needed for performance-based monitoring and evaluation. This point applies to all climate-related aid projects but is perhaps especially relevant for climate change capacity building programmes, since those programmes are particularly likely to be tied to the intergovernmental process.

2.5 Question No. 3: What Scales?

Earlier in this chapter, we reviewed the total size of ODA and the proportion of it that is related to climate and that which is more purely assigned to climate change purposes. If one assumes that total ODA represents an expression of the will by donors to transfer resources internationally for all purposes, then it is clear that reaching the goal of US$100 billion per year in new climate finance will require massive upscaling and also the leveraging of private-sector funds. To date, the most systematic analysis of all forms of climate finance is reported in two studies by Buchner et al. and the Climate Policy Initiative (Buchner et al. 2011a, b). Their research suggests that in 2009/10 at least US$97 billion per year was provided to support 'low carbon, climate-resilient development activities'. (These studies do not specifically distinguish the smaller portion of that total which is devoted to capacity building.) By their estimate, private finance already equals US$55 billion per year, much larger than the US$21 billion that comes from public sources. Of the total US$97 billion per year of climate change finance, so far only US$4 billion flows to climate change adaptation measures (Buchner et al. 2011b). This very small fraction of the total probably reflects that adaptation is a relatively new topic and that serious adaptation efforts at present are largely effected within countries. There have been particular projects that are funded internationally—for example, an advanced weather and flood forecasting system that has helped communities in Pakistan and Bangladesh get early warnings about floods, which in turn has helped radically reduce the cost of flood-related damage—but most adaptation is an internal matter and not well captured in international statistics on climate funding (Webster 2013). The research is also a reminder that measuring climate finance is extremely sensitive to method and their published estimates for climate finance just one year later put the total at US$350 billion per year (with all but US$14 billion going to mitigation)—a huge increase, due partly to the scaling up of climate finance and mainly to a broader scope to the analysis (Buchner et al. 2012).

A smaller subset of the US$100 billion per year pledged in Copenhagen would be dedicated to capacity building, but the same logic applies to these activities, perhaps with even greater difficulties attached. Capacity building activities must be tailored to individual countries and thus are difficult to scale quickly. And the private sector, for the most part, devotes funds to incentive-compatible investments rather than broad-based capacity building, whose benefits are hard for any particular firm to appropriate.

The importance of increasing expenditure by involving other actors—also known as scaling—is particuarly important with the private sector, which plays a central role in most sectors of the economy that might be affected by climate change mitigation, impacts, or adaptation. For example, consider the insurance industry. It has seen weather- and climate-related losses that have more than doubled each decade since the 1980s, today averaging US$50 billion a year and has long realized the incentive of making greater efforts to manage climate change-related risks. There is much literature offering suggestions on how the industry can integrate mitigation and adaption measures while bolstering its own profitability (Mills 2005; Hecht 2008; Bals et al. 2005; The Geneva Association 2009). Capacity building in the private sector includes creating a robust and adaptive insurance industry.

Three global initiatives—the United Nations Environment Programme Finance Initiative (1995), ClimateWise (2007), and the Kyoto Statement (2009)—have pulled together 129 insurance firms from 29 countries to commit to activities such as supporting climate research, raising awareness on climate change, reducing in-house emissions, quantifying and disclosing climate risks, incorporating climate change into investment decisions, and engaging in public policy. The industry has become a significant voice in world policy forums through its collaboration with scientists on the latest three Intergovernmental Panel on Climate Change (IPCC) assessments and participation in the international climate negotiation process. Additionally, it has invested at least US$23 billion in emissions-reduction technologies, securities, and financing for specific projects. In the past decade, the industry has engaged in 1148 initiatives in 51 countries, representing US$2 trillion of industry revenue, focused on climate change adaptation and mitigation. These activities offer massive leverage on the actions of virtually every sector of the economy as firms and individuals make decisions related to investment and behaviour. Yet even in this industry that is already on the front lines of climate change policy discussions, there has been no systematic documentation of how leveraged funds and activities affect capacity building versus the impact on actual mitigation and adaptation projects and the wide array of private-sector activities implicated by climate change.

Here I investigate the logic of scaling from three perspectives. First, I look at the status of climate finance and the relationship between public and private funds. Most scaling will probably need to occur with private financing, and building the private capacity to raise and manage climate-related financing effectively is essential. Second, I look at the one area where already there has been some substantial scaling through private funds: the CDM. While the CDM is centrally about investment in mitigation—not capacity building—it helps to reveal what might be possible if a set of incentives were created to leverage private finance. Third, I very briefly look at the potentials for climate change 'mainstreaming', which, in theory, is another area of potential scaling. If climate change issues could be mainstreamed to a greater degree in capacity building programmes linked to private and public finance then lots of scaling could occur. So far, however, most evidence of mainstreaming seems to be in the sphere of public financial transfers (ODA) and there has been little attention to mainstreaming in private finance.

2.5.1 Scaling from Perspective No. 1: Comparing Aid Sources—ODA, Climate Change-Related Aid, and Climate Finance

ODA and climate change-related aid have both been defined in earlier sections of the chapter. It is worthwhile, also, to define 'climate finance' in order to understand questions such as how to evaluate and improve climate change-related aid and how to leverage other sources of funding, such as from private industry. While there is no widely accepted international definition at present, the term 'climate finance' broadly refers to the whole range of financial resources that catalyse low-carbon and climate-resilient development (World Bank 2011). One of the many areas of debate concerns whether those sources must be 'new and additional' or whether climate finance covers anything that relates to climate. The Copenhagen Accord calls for a collective commitment by developed countries to provide 'new and additional resources' to fight climate change, with a goal of mobilizing jointly US$100 billion dollars per year by 2020 to address the needs of developing countries (UNFCCC 2009).

Since it remains unclear how additionality is defined in the Copenhagen Accord when it promises 'new and additional resources' and how such large sums of money are going to be raised, at least four major definitions of climate finance have emerged (Brown et al. 2010).

- Definition 1: Climate finance is classified as aid, but it is additional to (over and above) the 0.7% ODA target;
- Definition 2: Climate finance is classified as aid. The 2009 ODA disbursements on climate change should be set as the reference level. Any new ODA finance going to climate change measures above the reference level can be considered as additional;
- Definition 3: Climate finance is classified as part of traditional aid but limited to a certain portion (obviously, in addition to the target set,

other non-ODA sources of finance will be needed to meet climate change needs);
- Definition 4: Climate finance should come from other sources of finance not categorized as ODA.

Three of these (all but #4) define climate finance as a form of aid. In my assessment, this debate over the correct definition seems unlikely to deliver a climate finance that allows for scalability, since total aid is limited in size (see Fig. 2.1) and thus hard to scale much beyond the already growing provision of climate aid. The most fruitful approach, therefore, is to see climate finance as the sum of traditional climate-linked foreign aid and other non-aid sources of finance. None of these definitions is specifically tailored to aid for capacity building.

At present, all international climate funding instruments rely on ODA, with three exceptions:

- finance linked to certified emissions reduction (CER) credits issued by the clean development mechanism (CDM). The CDM is by far the largest non-ODA source of climate finance. Despite a recent sharp decline, the volume of primary CERs still reached approximately 91 million tonnes of carbon dioxide equivalent in 2011, with a total value of US$990 million (World Bank 2012);
- the Kyoto Protocol's Adaptation Fund, which is financed through a 2% levy on CDM proceeds; and
- part of the German International Climate Initiative, which is financed through a national auction of emissions allowance units.

Table 2.1 compares similarities and differences among aid in general, climate change-related aid, and private-funded climate finance. The information gathered in the table suggests that on one hand, the 2005 Paris Principles of Aid Effectiveness might offer a useful framework to help steer climate change-related aid (and private-funded climate finance) to outcomes that are effective, efficient, and equitable; on the other hand, any assessment of climate change-related aid (and private-funded climate finance) through the lens of aid effectiveness will deliver only a partial result and therefore should take account of the consensus within the UNFCCC negotiations on the principles appropriate for climate finance. At the time of writing, the literature has not settled on practical solutions for how climate aid and finance should be governed, nor the right propor-

Table 2.1 Aid, climate change-related aid, and private-funded climate finance compared

	Aid	*Climate change-related aid*	*Private-funded climate finance*
Paradigm	A voluntary paradigm	Yet to determined?	Yet to be determined?
Sources	Focus on budgetary contributions from donor governments	Focus on budgetary contributions from donor governments	Rely on private flows and innovative sources
Objective	Present imperative of poverty reduction	Dealing with an uncertain future	Dealing with an uncertain future
Leadership	OECD-DAC leadership	OECD-DAC leadership?	UNFCCC leadership
Partnership	Aid conditionality set by donor countries	Commitments expected from both contributor and recipient countries	Commitments expected from both contributor and recipient countries
Effectiveness	Aid effectiveness has been a retrospective exercise after many years of delivery	Delivery at scale has just begun?	Delivery at scale is yet to begin
Principle of effectiveness	The 2005 Paris principles • National ownership • Alignment • Harmonization • Managing for results • Mutual accountability	Both principles: Paris and UNFCCC?	The UNFCCC Convention principles • Polluter pays • Additionally • Transparency • Accountability • Equitable representation • National ownership • Timeliness • Appropriate • Fair distribution • Complementarity

Source: Adapted by author from Bird and Glennie (2011)

tions of aid that should be devoted to capacity building. Absent that kind of governance, it seems unlikely that private finance will scale except in areas where there is a clear signal to deploy resources and also a direct reward flowing back to private investors—so far, there is only one area where that has happened, the CDM.

2.5.2 *Scaling from Perspective No. 2: The CDM*

A substantial literature has emerged to assess one area of climate change finance: the CDM. This, fortuitously, is also the area where the most scaling has been observed. Less fortuitously, the CDM is not centred on capacity building, although it has had some effects on capacity. Governments have designated and invested in national authorities to manage the flow of CDM projects within their borders, and the flow of revenues from the CDM has encouraged private investors to build capacity while also generating streams of income for a variety of government purposes. These insights apply not just to national administrative capacity but also to variations in subnational administration, which are particularly striking in India (Benecke 2009).

So far, the research on the CDM has not looked much at capacity building, although the current research has led to one major conclusion about national administrative capacity and strategy: they are important. As indicated earlier in this chapter, countries that foster strategic and competent national CDM policies and administrators attract more investment (and more technology transfer) than those that do not. The experiences of China and India are notable contrasts.[18]

There have been at least two major studies of CDM investment patterns (Dechezleprêtre et al. 2008; Seres et al. 2009). Those studies lead to many conclusions; among them is the fact that investments are more efficient and lead to more technology transfer when projects are larger in size. Furthermore, in some sectors, transfers correlate with the degree of a country's technological advancement—outside of agriculture, technology transfer is more likely (and projects more numerous) when the host country's own technological skills are relatively significant. These insights suggest that scaling is most likely to occur in the countries that are probably least likely to need foreign assistance. They also suggest—although here I am speculating by extending the logic of this research to capacity building—that scaling depends critically on the pre-existing presence of 'capacity' in a country.[19] Perhaps because the CDM is so young, we have not yet observed in the empirical studies of large numbers of CDM projects much actual transfer or investment in 'capacity'. Studies looking across the whole range of policy implications for the CDM find that national capacity is important and that the concept of 'capacity' is quite broad and must include not only administration but also clear local goals and guidance for CDM investors (Ockwell et al. 2007).

It is important to keep perspective on the size of the CDM relative to other financial flows, notably from the private sector. In the most recent Climate Policy Initiative tracking of carbon finance, which covers annual flows around 2010–11, offsets (dominated by the CDM) accounted for just 1% of total climate finance worldwide (Buchner et al. 2012).

2.5.3 *Scaling from Perspective No. 3: Mainstreaming Climate*

At present, it is difficult to see how climate change capacity building programmes scale easily. Capacity building is an activity tailored to individual governments and unlikely to be undertaken by the private sector on its own. Thus, perhaps there is an opportunity to scale capacity building by mainstreaming climate change into other aspects of foreign assistance. Particularly notable has been the World Bank's efforts, since the mid-1990s, to mainstream climate change (and other international environmental missions) into its main lending and grant portfolios. So, the World Bank has shifted its lending policies on fossil fuel investments and forests, for example, with the goal in part of reducing the emissions impacts of its activities. (Most major bilateral ODA programmes have done something similar.) This effort has focused mainly on mitigation, although mainstreaming as concerns adaptation will be discussed below. Mainstreaming around mitigation has largely looked at the opportunities for low-cost (or even negative-cost 'win-win') opportunities for cutting emissions. For example, the Global Environment Facility has assessed investments related to urban infrastructure—focusing in particular on the World Bank's urban lending and grant programmes since 1995. That review has shown that efforts to mainstream climate change measures have been particularly successful in solid waste management and with some progress in building climate change concerns into urban infrastructure planning—for example Bus Rapid Transit. Less success has been achieved in water supply, buildings, and other infrastructure where the potential for climate change mitigation is more diffuse and difficult to administer. This effort by the World Bank, which is typical of climate change mainstreaming, has mainly focused on projects rather than administrative capacity. However, one insight from this work is that the key means of mainstreaming climate change mitigation into capacity building arise through building awareness of climate change issues among urban planners and other officials (GEF 2011).

A much larger literature has focused, almost since the late 1980s when climate change first appeared on the international agenda, on the potential benefits of mainstreaming climate change concerns into other important goals such as protection of public health (Epstein 2005). Independent assessments have concluded that while the logic of mainstreaming is important, the actual extent of its practice has been quite limited.[20] Essentially, all these efforts at systematic assessments have focused on the multilateral institutions; assessments of bilateral programmes are more scattered.

Perhaps even more important than mitigation will be the mainstreaming of climate change concerns related to adaptation. For the purposes of the present study, adaptation is particularly interesting because all countries will have a self-interest in adaptation investments and the scale of investments affected could run to many trillions of dollars per year. All activities in agriculture and in mountainous areas and along coastal zones—among other places that are vulnerable to climate change—could be impacted. No ODA programme will affect more than a tiny percentage of all these activities, and thus the greatest leverage may be found through building national capacity. While this logic has been articulated in several places, so far actual implementation is quite limited.[21] One example is the Africa Adaptation Programme, launched by the United Nations Development Programme in partnership with other UN agencies with US$92.1 million in funding support from Japan. The central goal of this programme is to mainstream adaptation efforts while focusing on poverty reduction. It is mostly concerned with capacity building, notably in data and information management, building institutions and leadership, and creating the capacity for improved analysis, implementation, and management of adaptation projects.[22] Several detailed handbooks have been prepared to help local officials examine the implications of mainstreaming adaptation issues into development planning (UNDP-UNEP Poverty-Environment Initiative 2011; Lebel et al. 2012). Some regional development organizations have developed projects to mainstream climate adaptation but these are mainly small and relatively recent in vintage—for instance, the Mainstreaming Adaptation to Climate Change (MACC) project in the Caribbean region is based on a US$5 million grant from the Global Environment Facility (GEF). Whereas mitigation studies emphasize the potential of direct action to control emissions, these handbooks that focus on adaptation are principally concerned with capacity building in the planning process. To my knowledge, there has been no systematic examination of whether these mainstreaming for adaptation efforts have actually been successful, let alone the capacity building elements of those programmes.

2.6 Question No. 4: What Lessons Are Transferable?

Most of the lessons learnt about foreign aid concern country programmes and individual projects. Indeed, most aid money is devoted to such activities, and a smaller portion (perhaps about one third of total ODA) is used for the kinds of administrative infrastructure building that might be called 'capacity building'. The lessons from aid programmes and projects pertaining to the particular task of capacity building are difficult to transfer directly because capacity building, by design, is intended to create the very conditions that allow for effective use of foreign assistance. Apropos climate change, in particular, this role for capacity building is particularly apparent, since much (maybe most) of such funds are devoted to tasks such as building the national administrative authorities needed to participate in international talks, the IPCC process, manage funding linked to climate change from multilateral and bilateral donors, participate in the CDM, and other such activities. Capacity building should be an inward-looking activity, as even in the poorest countries most of the funds for development usually come from internal sources; as a practical matter, especially in areas like climate change where aid is targeted for a particular purpose, capacity building tends to be an outward-oriented (often aid-oriented) endeavour. In theory, it should be possible to transfer lessons from other projects to capacity building by making funds for capacity building contingent upon host countries credibly putting into place the conditions needed to make effective use of foreign aid. Where that is done, capacity building programmes can help create the conditions that, in turn, would allow foreign funds to be used effectively for projects. As far as I know, no major donor programme related to climate change has followed this strategy.

2.7 Analysis and Conclusions

It is hard to answer the question of aid effectiveness by looking to climate change development assistance alone. Its history is too short, and careful analytical studies are too few. But answers to the question of 'what works?'

can be found by considering the experience with foreign assistance more generally. The central lessons from that experience are embodied in the Paris Principles. At present, very little of the discussion about foreign assistance and climate change is rooted in application of these Principles, perhaps because they imply making aid conditional not just on the need for assistance (a key aspiration of recipient countries, especially the least developed ones) but also the ability to spend funds wisely. One of the central lessons from the CDM experience is that the countries that are most able to utilize foreign funds are generally those that least need it.

Making foreign assistance more effective requires identifying the conditions that favour effectiveness and then targeting funds to countries and markets that meet those conditions. On this front, the Paris Principles are not that helpful because they are broad and do not automatically lead to an actionable programme. However, the string of studies starting with the World Bank's 'Assessing Aid' programme are helpful guides. When aid is delivered to countries that have 'good governance', the money tends to be assigned to activities that promote public welfare, notably sustainable economic growth. For example, in countries with low levels of corruption, accountable systems of government, and good management of public budgets, the injection of new aid money tends to be reinvested in the country for productive activities. While no study has looked in detail at whether exactly those conditions must also hold for climate change programmes, it is likely to be true. Climate change spending is likely to be most effective when it is devoted to activities that yield broad-based public benefits rather than narrower supply of rents to particular elites. Delivering on all the elements of the Paris Principles and on the basic insights of 'Assessing Aid' requires a central focus on capacity building.

Donors and recipients alike are thus faced with what might be called an 'aid paradox'. The countries that have the conditions in place needed to make aid effective are probably able to address on their own many of the same challenges that aid is supposed to help. Thus, for example, when the US government created the Millennium Challenge Corporation under the George W. Bush administration, with the aim of putting 'effective aid' into practice, it faced the problem that very few countries met the rigorous standards for making good use of aid and also that they needed large amounts of aid. The same is probably the case for most climate change aid.

Overcoming the aid paradox could prove to be a particularly great problem as more countries focus on capacity building for climate change adaptation. Funds actually dedicated to adaptation—such as hardening of transport infrastructures or investing in more resilient forms of agriculture—are probably many thousands of times smaller than total spending on those same activities.[23] Thus, adaptation, in particular, requires that foreign assistance leverage other sources of investment. Transferring the lessons from foreign-funded mitigation projects to climate change adaptation is therefore likely to lead to many errors, since the role of self-funded investments is greater in adaptation. Put differently, most effective adaptation is likely to come from mainstreaming climate change concerns into the normal process of investing in climate-sensitive infrastructures, rather than in particular, discrete adaptation projects. By contrast, most mitigation funding for projects has been devoted to discrete projects, with the extreme example of that mode being the CDM, where projects are funded only if they are discrete.

Looking to the future, what kinds of programmes could be scaled? The Copenhagen Accord envisioned a manyfold increase in aid; overall, diplomacy has also shifted from emphasizing mitigation to a larger role for adaptation. My assessment of the literature is that it will be slow and difficult to scale the lessons that have been learned, especially where this concerns capacity building. If the central goal of capacity building is to rework (and build) local institutions so that countries are better able to manage their own affairs and also better able to use foreign aid, then the relevant guidelines must be tailored to each local circumstance and must be credibly sustained over a long period of time. The history of aid shows that many countries (including the largest donors, such as the USA) have a hard time making credible long-term commitments and that the local tailoring process is time consuming. This basic insight is now evident, as a number of localities explore, for example, ways to adapt to rising sea levels—one of the most likely (and possibly most dangerous) effects of climate change. In Asia, a consortium of cities known as the Asian Cities Climate Change Resilience Network (ACCCRN) is trying to use foreign funds (in this case, mainly from foundations—notably the Rockefeller Foundation) to help vulnerable cities in four Asian countries adapt to sea level rise and other climate impacts. Their work, so far, has underscored that every locality is different and tailoring is perhaps even more important than directly transferring lessons from one setting to another (ACCCRN 2009). These challenges are hardly unique to the least developed countries. California, for example, faces a wide array of likely impacts of climate change and is still in the early stages of planning comprehensive adaptation responses because each must be tailored to the particular effect and to local institutions.[24]

Acknowledgements I am grateful for assistance from Linda Wong, Fang Rong, and especially Yongfu Huang at UNU-WIDER.

Notes

1. The OECD's DAC is a forum for selected OECD member states to discuss issues surrounding aid, development, and poverty reduction in developing countries. There are 24 members of DAC, including the European Union, which acts as a full member of the committee.
2. OECD/ODA database.
3. On one hand, the rise and fall in the aid to agriculture during this period followed broadly the same pattern as that in total aid to all sectors; on the other hand, it reflects that fact that the emphasis of worldwide development strategy shifted from the narrow concepts of food security, in terms of adequate and stable food supplies, to broader human and social development. In 2007–8, total annual average aid commitments to agriculture amounted to US$7.2 billion; the largest donors (among DAC members) were the United States, Japan, and France. The largest recipients are primarily sub-Saharan Africa and South and Central Asia. For statistics, see OECD (2010b).
4. The decline is considered a consequence of the 'Helsinki package', an agreement that came into force in 1992 and prohibits (with some exceptions) the provision of tied-aid loans to high-income countries (based on World Bank per capita income) and for commercially viable projects.
5. CPA excludes non-programmable items such as humanitarian aid, debt relief, and in-donor costs like administration costs and refugees in donor countries. Over the past five years, CPA has corresponded to roughly half of DAC donors' gross bilateral ODA. For more details on CPA, see www.oecd.org/dac/cpa
6. This three-part definition is based on United Nations (2006).
7. Climate change mitigation-related aid is defined as activities that contribute 'to the objective of stabilization of greenhouse gas (GHG) concentra-

tions in the atmosphere at a level that would prevent dangerous anthropogenic interference with the climate system by promoting efforts to reduce or limit GHG emissions or to enhance GHG sequestration'. Climate change adaptation-related aid is defined as activities that aim 'to reduce the vulnerability of human or natural systems to the impacts of climate change and climate-related risks, by maintaining or increasing adaptive capacity and resilience'. See OECD (2010c). In detail, mitigation activities include those that contribute to '(i) the mitigation of climate change by limiting anthropogenic emissions of GHGs, including gases regulated by the Montreal Protocol; or (ii) the protection and/or enhancement of GHG sinks and reservoirs; or (iii) the integration of climate change concerns with the recipient countries' development objectives through institution building, capacity development, strengthening the regulatory and policy framework, or research; or (iv) developing countries' efforts to meet their obligations under the Convention'. The third category is directly related to capacity building, while the others may also implicate capacity-building. See OECD (2012).

8. Buchner et al. (2011b) find that bilateral institutions are distributing US$24 billion per year, while multilateral agencies distribute US$15 billion.
9. In addition, see a study by Neuhoff et al. (2009) on financing options for NAMAs.
10. In addition to Radelet (2006) see also Tsikata (1998); Clemens et al. (2004); Riddell (2008); Rajan and Subramanian (2008); Doucouliagos and Paldam (2009); and Krasner (2011).
11. Some examples include Hadjimichael et al. (1995); Hansen and Tarp (2000); Lensink and White (2001); Dalgaard et al. (2004); Clemens et al. (2004); Doucouliagos and Paldam (2008); Clemens et al. (2012); and Clausen and Schürenberg-Frosch (2012).
12. In the non-technical literature, see Collier (2008) and Sachs (2006).
13. In the 2005 Paris Declaration on Aid Effectiveness, developed and developing country governments pledged joint supports to five key commitments to improve aid effectiveness.
14. On the UN-REDD Programme, see www.un-redd.org/
15. On Voluntary REDD+Database, see www.reddplusdatabase.org/
16. On civil liberties, see Isham et al. (1997).
17. For example, see Keohane and Victor (2011) on how competition and fragmentation can lead to more effective climate change coordination and policies.
18. See above, and also Ganapati and Liu (2009).

19. For a formal model that points to similar conclusions see Bayer and Urpelainen (2013).
20. For example, see Nakhooda et al. (2005).
21. For an articulation, see Victor (2011).
22. African Adaptation Programme available at www.undp-aap.org/about-us
23. While such data are not collected, a rough order of magnitude calculation is possible. In a typical year, a typical country will spend at least about 2–3% of GDP on infrastructure investments that are plausibly sensitive to changes in climate—for example, roads, river diversions, agriculture, etc. Many countries spend more. World GDP is about US$70 trillion, suggesting that climate-sensitive investments total about US$200 billion. While there is no comprehensive source of information on adaptation funding, total cash transfers per year under the 'Adaptation Fund' have been around US$30 million per year for the last two years (see Financial Status of the Adaptation Fund Trust Fund as of 31 December 2011). The Adaptation Fund is managed by the Global Environment Facility and oversees spending of the 2 % tax that is levied on CDM transactions (see Adaptation Fund official website at www.adaptation-fund.org). This funding source—so far the only credible multilateral adaptation programme created under the UNFCCC—is thus 0.02 % of total world spending on infrastructure. Of course, the Adaptation Fund focuses its disbursements on the least developed countries (LCDs), and looking just at that subset of countries (which has more than 12 % of the world population but accounts for less than 1 % of world GDP) the fractions are still miniscule: 0.4 %. By contrast, the net ODA disbursement to these countries, together with the net debt relief, has continued to increase and reached a record level of US$40 billion in 2009, the equivalent of about 8 % of their GDP (see 'The Least Developed Countries Report' 2012 at www.unctad.org/en/pages/PublicationArchive.aspx?publicationid=188)
24. On the likely effects, see for example Cayan et al. (2012).

References

ACCCRN. 2009. *Responding to the Urban Climate Challenge*. Boulder, CO: Asian Cities Climate Change Resilience Network.

Adaptation Fund Trust. 2011. *Financial Status of the Adaptation Fund Trust Fund*. www.adaptation-fund.org/sites/default/files/AFB.EFC_.8.7%20Financial%20Status%20of%20the%20AF%20Trust%20Frund.pdf

Ashwill, M., C. Flora, and J. Flora. 2011. *Building Community Resilience to Climate Change: Testing the Adaptation Coalition Framework in Latin America*. Washington, DC: World Bank.

Bals, C., I. Burton, S. Butzengeiger, A. Dlugolecki, E. Gurenko, E. Hoekstra, P. Höppe, et al. 2005. *Insurance-Related Options for Adaptation to Climate Change*. Munich: The Munich Climate Insurance Initiative.

Bauer, P.T. 1972. *Dissent on Development: Studies and Debates in Development Economics*. Cambridge, MA: Harvard University Press.

Bayer, P., and J. Urpelainen. 2013. External Sources of Clean Technology: Evidence from the Clean Development Mechanism. *The Review of International Organizations* 8 (1): 81–109.

Benecke, G. 2009. Varieties of Carbon Governance: Taking Stock of the Local Carbon Market in India. *Journal of Environment and Development* 18 (4): 346–370.

Bird, N. 2011. *The Future of Climate Finance in Nepal*. ODI Report. London: ODI.

Bird, N., and J. Glennie. 2011. *Going Beyond Aid Effectiveness to Guide the Delivery of Climate Finance*. ODI Background Note. London: ODI.

Brown, J., and L. Peskett. 2011. *Climate Finance in Indonesia: Lessons for the Future of Public Finance for Climate Change Mitigation*. EDC 2020 Working Paper 11. Bonn: EADI.

Brown, J., N. Bird, and L. Schalatek. 2010. *Climate Finance Additionality: Emerging Definitions and Their Implications*. Climate Finance Policy Brief 2. London: ODI.

Buchner, B., J. Brown, and J. Corfee-Morlot. 2011a. *Monitoring and Tracking Long-Term Finance to Support Climate Action*. Paris: OECD/IEA.

Buchner, B., A. Falconer, M. Hervé-Mignucci, C. Trabacchi, and M. Brinkman. 2011b. *The Landscape of Climate Finance*. CPI Report. Venice: Climate Policy Initiative.

Buchner, B., A. Falconer, M. Hervé-Mignucci, and C. Trabacchi. 2012. *The Landscape of Climate Finance 2012*. CPI Report. Venice: Climate Policy Initiative.

Cameron, C. 2011. *Climate Change Financing and Aid Effectiveness: Ghana Case Study*. Paris: OECD.

Cayan, D., M. Tyree, D. Pierce, and T. Das. 2012. *Climate Change and Sea Level Rise Scenarios for California: Vulnerability and Adaptation Assessment*. A White Paper prepared by Scripps Institution of Oceanography for the California Energy Commission. California Energy Commission.

Clausen, V., and H. Schürenberg-Frosch. 2012. Aid, Spending Strategies and Productivity Effects: A Multi-sectoral CGE Analysis for Zambia. *Economic Modelling* 29 (6): 2254–2268.

Clemens, M.A., S. Radelet, and R.R. Bhavnani. 2004. *Counting Chickens When They Hatch: The Short Term Effect of Aid on Growth*. CGD Working Paper 44. Washington, DC: Center for Global Development.

Clemens, M.A., S. Radelet, R.R. Bhavnani, and S. Bazzi. 2012. Counting Chickens When They Hatch: Timing and the Effects of Aid on Growth. *The Economic Journal* 122 (561): 590–617.

Collier, P. 2008. *The Bottom Billion: Why the Poorest Countries Are Failing and What Can Be Done About It?* New York: Oxford University Press.

Corfee-Morlot, J., B. Guay, and K.M. Larsen. 2009. *Financing Climate Change Mitigation: Towards a Framework for Measurement, Reporting, and Verification.* Paris: OECD/IEA.

Crawford, G. 1997. Foreign Aid and Political Conditionality: Issues of Effectiveness and Consistency. *Democratization* 4 (3): 69–108.

Dalgaard, C.-J., H. Hansen, and F. Tarp. 2004. On the Empirics of Foreign Aid and Growth. *The Economic Journal* 114 (496): F191–F216.

Dechezleprêtre, A., M. Glachant, and Y. Ménière. 2008. The Clean Development Mechanism and International Diffusion of Technologies: An Empirical Study. *Energy Policy* 36 (4): 1273–1283.

———. 2009. Technology Transfer by CDM Projects: A Comparison of Brazil, China, India and Mexico. *Energy Policy* 37 (2): 703–711.

Deutscher, E., and S. Fyson. 2008. Improving the Effectiveness of Aid. *Finance and Development* 45 (3): 15–19.

Doucouliagos, H., and M. Paldam. 2008. Aid Effectiveness on Growth: A Meta Study. *European Journal of Political Economy* 24 (1): 1–24.

———. 2009. The Aid Effectiveness Literature: The Sad Results of 40 Years of Research. *Journal of Economic Surveys* 23 (3): 433–461.

Easterly, W. 2003. Can Foreign Aid Buy Growth? *Journal of Economic Perspectives* 17 (3): 23–48.

———. 2007. *The White Man's Burden: Why the West's Efforts to Aid the Rest Have Done So Much Ill and So Little Good.* New York: Penguin Press.

———. 2009. Can the West Save Africa? *Journal of Economic Literature* 47 (2): 373–447.

Epstein, P.R. 2005. Climate Change and Human Health. *New England Journal of Medicine* 353: 1433–1436.

Ganapati, S., and L. Liu. 2009. Sustainable Development in the Clean Development Mechanism: The Role of the Designated National Authority in China and India. *Journal of Environmental Planning and Management* 52 (1): 43–60.

GEF. 2011. *Mainstreaming Climate Change Mitigation in Cities.* Washington, DC: The World Bank.

Grant, P. 2011. *Climate Change Financing and Aid Effectiveness: Morocco Case Study.* Paris: OECD.

Gutner, T.L. 2002. *Banking on the Environment: Multilateral Development Banks and Their Environmental Performance in Central and Eastern Europe.* Cambridge, MA: MIT Press.

Hadjimichael, M.T., D. Ghura, M. Mühleisen, R. Nord, and E. Murat Ucer. 1995. *Sub-Saharan Africa: Growth, Savings, and Investment 1986–93.* IMF Occasional Paper 118. Washington, DC: IMF.

Hansen, H., and F. Tarp. 2000. Aid Effectiveness Disputed. *Journal of International Development* 12: 375–398.

Hecht, S.B. 2008. Climate Change and the Transformation of Risk: Insurance Matters. *UCLA Law Review* 55 (6): 15–59.

Hedger, M. 2011. *Climate Finance in Bangladesh: Lessons for Development Cooperation and Climate Finance at the National Level.* EDC2020 Working Paper 12. Bonn: EADI.

IEA (International Energy Agency). 2011. *World Energy Outlook 2011.* Paris: IEA/OECD.

Isham, J., D. Kaufmann, and L.H. Pritchett. 1997. Civil Liberties, Democracy, and the Performance of Government Projects. *World Bank Economic Review* 11 (2): 219–242.

Kanbur, R., and T. Sandler. 1999. *The Future of Development Assistance: Common Pools and International Public Goods.* Washington, DC: Overseas Development Council.

Keohane, R.O., and M.A. Levy, eds. 1996. *Institutions for Environmental Aid: Pitfalls and Promise.* Cambridge, MA: MIT Press.

Keohane, R.O., and D. Victor. 2011. The Regime Complex for Climate Change. *Perspectives on Politics* 9 (1): 7–23.

Krasner, S.D. 2011. Review Article: Foreign Aid: Competing Paradigms. *Journal of Intervention and State Building* 5 (2): 123–149.

Lebel, L., L. Li, C. Krittasudthacheewa, M. Juntopas, T. Vijitpan, T. Uchiyama, and D. Krawanchid. 2012. *Mainstreaming Climate Change Adaptation into Development Planning.* Bangkok: Adaptation Knowledge Platform and Stockholm Environment Institute.

Lensink, R., and H. White. 2001. Are There Negative Returns to Aid? *Journal of Development Studies* 37 (6): 42–65.

Levine, R., and The What Works Working Group with Molly Kinder. 2004. *Millions Saved: Proven Successes in Global Health.* Washington, DC: Center for Global Development.

Martens, B., ed. 2002. *The Institutional Economics of Foreign Aid.* Cambridge: Cambridge University Press.

Mills, E. 2005. Synergisms Between Climate Change Mitigation and Adaption: An Insurance Perspective. *Mitigation and Adaptation Strategies for Global Change* 12 (5): 809–842.

Moyo, D. 2009. *Dead Aid: Why Aid Is Not Working and How There Is a Better Way for Africa.* New York: Farrar, Straus and Giroux.

Nakhooda, S., J. Sohn, and K. Baumert. 2005. *Mainstreaming Climate Change Considerations at the Multilateral Development Banks.* WRI Issue Brief. Washington, DC: World Resources Institute.

Narayan, D. 1995. *The Contribution of People's Participation: Evidence from 121 Rural Water Supply Projects.* ESD Occasional Paper series. Washington, DC: World Bank.

Neuhoff, K., S. Fankhauser, E. Guerin, J.C. Hourcade, H. Jackson, R. Rajan, and J. Ward. 2009. *Structuring International Financial Support to Support Domestic Climate Change Mitigation in Developing Countries.* London: Climate Strategies.

Norrington-Davies, G. 2011. *Climate Change Financing and Aid Effectiveness: Cameroon Case Study.* Paris: OECD.

Norrington-Davies, G., and N. Thornton. 2011a. *Climate Change Financing and Aid Effectiveness: Kenya Case Study.* Paris: OECD.

———. 2011b. *Climate Change Financing and Aid Effectiveness: Tanzania Case Study.* Paris: OECD.

Ockwell, D., J. Watson, G. MacKerron, P. Pal, F. Yamin, N. Vasudevan, and P. Mohanty. 2007. *UK-India Collaboration to Identify the Barriers to the Transfer of Low Carbon Energy Technology.* Project Report. London: Department for Environment, Food and Rural Affairs.

OECD. 2003. Glossy of Statistical Terms. Paris: OECD. www.stats.oecd.org/glossary/detail.asp?ID=6043

———. 2010a. Measuring Aid for Energy. Paris: OECD. www.oecd.org/dac/stats/45066235.pdf

———. 2010b. Measuring Aid to Agriculture. Paris: OECD. www.oecd.org/investment/aidstatistics/44116307.pdf

———. 2010c. Tracking Aid in Support of Climate Change Mitigation and Adaptation in Developing Countries. Paris: OECD. www.oecd.org/investment/stats/45906157.pdf

———. 2011a. *Statistical Reporting by the Bill and Melinda Gates Foundation to the OECD DAC: Aid to Health Data Now Includes World's Largest Private Foundation.* Paris: OECD.

———. 2011b. *Aid Effectiveness 2005–10: Progress in Implementing the Paris Declaration.* Paris: OECD.

———. 2012. OECD DAC Statistics on Climate-Related Aid. www.oecd.org/dac/stats/FactsheetRio.pdf

Popp, D. 2011. International Technology Transfer, Climate Change, and the Clean Development Mechanism. *Review of Environmental Economics and Policy* 5: 131–152.

Radelet, S. 2006. *A Primer on Foreign Aid.* CGD Working Paper 92. Washington, DC: Center for Global Development.

Rajan, R.G., and A. Subramanian. 2008. Aid and Growth: What Does the Cross-Country Evidence Really Show? *The Review of Economics and Statistics* 90 (4): 643–665.

Riddell, R.C. 2008. *Does Foreign Aid Really Work?* Oxford and New York: Oxford University Press.

Sachs, J.D. 2006. *The End of Poverty: Economic Possibilities for Our Time.* New York: Penguin Press.

Scholl, A. 2009. Aid Effectiveness and Limited Enforceable Conditionality. *Review of Economic Dynamics* 12 (2): 377–391.

Seres, S., E. Haites, and K. Murphy. 2009. Analysis of Technology Transfer in the CDM Projects: An Update. *Energy* 37 (11): 4919–4926.

Stokke, O., ed. 1995. *Aid and Political Conditionality*. London: Frank Cass & Co..

The Geneva Association. 2009. The Insurance Industry and Climate Change-Contribution to the Global Debate. The Geneva Reports: Risk and Insurance Research. Geneva: The Geneva Association.

Thornton, N. 2010. *Realising Development Effectiveness: Making the Most of Climate Change Finance in Asia and the Pacific*. A synthesis report from five country studies in Bangladesh, Cambodia, Indonesia, Philippines and Vietnam. Prepared for the Asia Pacific Climate Change Finance and Aid Effectiveness Dialogue. Manila, Philippines: Capacity Development for Development Effectiveness Facility for Asia and Pacific.

Tsikata, T.M. 1998. *Aid Effectiveness: A Survey of the Recent Empirical Analysis*. Paper on Policy Analysis and Assessment of the International Monetary Fund. Washington, DC: IMF.

UN. 2006. *Definition of Basic Concepts and Terminologies in Governance and Public Administration*. Committee of Experts on Public Administration. New York: United Nations Economic and Social Council.

UNDP-UNEP Poverty-Environment Initiative. 2011. *Mainstreaming Climate Change Adaptation into Development Planning: A Guide for Practitioners*. Nairobi: UNDP/UNEP.

UNFCCC (United Nations Framework Convention on Climate Change). 2007. *Investment and Financial Flows to Address Climate Change*. Bonn: UNFCCC.

———. 2009. *Report of the Conference of the Parties on Its Fifteenth Session, Held in Copenhagen from 7 to 19 December 2009*. Copenhagen: United Nations.

VandeHei, J., and P. Blustein. 2005. African Aid Is Doubled By G-8. *The Washington Post*, 9 July. www.washingtonpost.com/wp-dyn/content/article/2005/07/08/AR2005070801776.html

Victor, D.G. 2011. *Global Warming Gridlock: Creating More Effective Strategies for Protecting the Planet*. Cambridge and New York: Cambridge University Press.

Webster, P.J. 2013. Meteorology: Improving Weather Forecasts for the Developing World. *Nature* 493: 17–19.

World Bank. 1998. *Assessing Aid: What Works, What Doesn't, and Why?* Washington, DC: World Bank.

———. 2011. *Mobilizing Climate Finance: A Paper Prepared at the Request of G20 Finance Ministers*. Washington, DC: World Bank.

———. 2012. *State and Trends of the Carbon Market 2012*. Washington, DC: World Bank.

CHAPTER 3

Lessons Learnt about Foreign Aid for Climate Change Related Capacity-Building

Zexian Chen and Jingjing He

Summary Despite the ongoing efforts of foreign aid to promote capacity building in developing countries, little is known about the effectiveness of foreign aid in terms of developing climate change-related capacity, what lessons and experiences can be drawn from past and present aid projects, what areas of foreign aid can be improved to boost capacity building, and what successful aid experiences can be applied to a wider context.

This chapter follows the structure outlined above, and aims at answering each of the four essential questions covered in subsequent sections. It is worth pointing out that these issues are intertwined and should also be treated as an integral whole—it is for the sake of simplicity that four separate sections are devoted to the discussions. Given the complexity and large scope of these four essential questions, this chapter can by no means provide complete answers but should, rather, be viewed as a starting point to invite further research.

Z. Chen • J. He (✉)
Institute of International Law, the Chinese Academy of Social Sciences, Beijing, China

Y. Huang, U. Pascual (eds.), *Aid Effectiveness for Environmental Sustainability*, https://doi.org/10.1007/978-981-10-5379-5_3

This study begins with a review of the literature on capacity building and foreign aid. The introductory section first underscores the important role of foreign aid in assisting the developing world's shift to a green growth trajectory, as applicable to countries that simply lack the resources and capacity to pursue sustainable development on their own. The aid literature suggests that lack of capacity is one of most crucial problems facing developing nations in their fight against climate change. Something that has become clear in the literature is that the developing countries need not only more financial and technical resources but also institutions, procedures and incentive structures that enable them to make more effective use of resources. This highlights the importance of strengthening capacity building in aid recipient countries. The section then moves on to a detailed discussion of the concept and scope of capacity, as well as of the capacity building/development process. Finally, it examines the 15 key areas where existing climate change capacity building activities are centred, and categorizes the climate change capacity building projects already in existence.

Section 3.2 aims to answer the question of 'what works?' Based on a series of case studies, it identifies successful aid experiences for climate change capacity building, each of which is exemplified by actual projects. To be more specific, this research finds that (1) capacity building programmes are most successful when they are country driven and demand driven; (2) that education improvements can translate directly into heightened climate change mitigation and adaptation capacity; (3) that monitoring of results and evaluation of capacity building activities are important for improving the effectiveness of future aid interventions; (4) that the exchange and sharing of data, information, expertise and financial resources at all levels could help to promote the best practices in responding to the climate change crisis; and (5) that it is important to improve policymakers' decision-making with regard to climate change issues through capacity building.

Section 3.3 attempts to answer the question of 'what could work?' Although some progress has been achieved with existing capacity building assistance projects, many areas call for improvement. On the basis of some case studies, this section tries to explore what areas have the potential to work, and what issues arise in delivering on this potential. Among the lessons to be noted in future aid interventions are: (1) the lack of cooperation between different climate change aid projects that affects the effectiveness of the implementation process; (2) the mismatch between internal and external capacity building resources that limits the effective execution of aid interventions; (3) the lack of initiative on the part of aid-receiving governments; (4) non-existence of a mechanism flexible enough to enable developing nations to apply for specific and timely assistance; (5) the need for capacity building aid to be an integral part of support arrangements in all relevant areas addressing climate change; (6) the need for improving long-term aid commitments to developing countries; and (7) that lack of country ownership can negatively impact on the effectiveness of climate change aid activities.

Section 3.4 deals with the question of 'what is scalable?' It investigates what aspects of foreign aid for capacity building need to be delivered on a larger scale to enhance its positive effects on improving climate-related capacities in the developing world. This section demonstrates that aid efforts need to be scaled up to boost climate change-related capacity building: (1) to raise public awareness of the urgency of tackling climate change at all levels of society; (2) to boost the international negotiation capacity of developing countries, so as to win more support and help from the international community in battling against climate change; (3) to support green technology transfer and development; and (4) to deal with the shortage of climate change professionals in Africa.

Section 3.5 addresses the question of 'what is transferrable?' It investigates what experiences from current aid projects can be transferred across countries and across projects. In fact, almost all successful approaches summarized in Sect. 3.2 (on 'what works?') could be transposed from one related project to another or from one developing country to another, but the following practices in particular should find application in a wider context. First, the country-owned and demand-driven foreign aid approach for climate change capacity building (discussed in Sect. 3.2) should be advocated in a wider context. Second, knowledge, information and experiences gained from past projects could be transferred to other aid programmes in similar situations. Third, experiences generated by successful aid projects on capacity building for the clean development mechanism (CDM) are transferrable to other developing nations that urgently need aid support in order to benefit from CDM.

Section 3.6 concludes the chapter, pointing out that although there is no 'fits-for-all' solution to climate change-related capacity building problems, most of the experiences and lessons discussed here are applicable, with modifications to allow for local features, in a wider context.

3.1 Introduction

Events related to the current climate change crisis have repeatedly alerted humankind to the urgency of tackling this pressing global challenge before it is too late. Developing countries, which have contributed negligibly to the present climate change problem, are, nevertheless, hit the hardest by and most vulnerable to its negative effects. In fact, according to the 2010 'World Development Report' (World Bank 2010), developing countries bear most of the costs of climate change damage, currently around 70–80 %. As Roberts and Parks (2006) point out, developing countries actually suffer 'a double injustice', in the sense that environmental degradation and climate change will impinge on the poor countries hardest and worst, a situation to which they have contributed little, but they are required to be 'part of the solution' by cutting greenhouse gas (GHG) emissions at the expense of economic development. Furthermore, an economic slowdown in these countries can jeopardize their ability to address their pressing development problems. Developing countries are plagued with poverty, lack of adequate healthcare, as well as high unemployment and gender inequality. Climate change can only intensify such existing development problems.[1]

The only way out for developing countries is to pursue a green growth development path that aims to achieve harmony between development and environmental conservation.[2] However, many developing countries simply lack the financial and technological resources and capacity to follow such eco-friendly transformation. This makes a strong case for the developed countries—in other words, the countries mainly responsible for the current environmental crisis—to help the developing world battle climate change and achieve a sustainable development trajectory.

Climate change, one of the most challenging issues facing the world, requires unified and urgent global action. Foreign aid for the development of green growth offers a plausible solution in that it not only helps the developing countries but also supports the interests of developed countries themselves. Aid is needed in particular to promote green technology transfers, to help establish frameworks that foster green growth and to enhance capacity building in low-income communities to raise people's awareness and capacity to pursue a low-carbon growth path.

The lack of capacity is a serious problem confronting developing nations with regard to climate change. Capacity constraints are a major impediment to these countries' shift towards a low-carbon development path.

One recognition emerging in the aid literature is that developing countries need not only more financial and technical resources but also institutions, procedures and incentive structures that enable them to make more effective use of resources. This highlights the importance of strengthening capacity building in recipient countries (Degnbol-Martinussen 2002). *Capacity*, as defined by the United Nations Development Programme (UNDP) (2009) is 'the ability of individuals, institutions and societies to perform functions, solve problems, and set and achieve objectives in a sustainable manner'.[3]

There are broadly three levels of capacity: namely, the individual, the organizational and the system level or the enabling environment (UNDP 2011a), as is shown in Fig. 3.1. Capacity at the individual level refers to the skills, experience and knowledge that allow individuals to perform (UNDP 2009).[4] Access to resources and experiences that can develop individual capacity is significantly influenced by organizational and environmental factors which, in turn, are shaped by the extent of capacity building in each individual. Capacity at the organizational level is the internal structure, policies and procedures affecting an organization's effectiveness (UNDP 2009).[5] An enabling environment or capacity at the system level refers to the political, economic, policy, legal and regulatory systems within which organizations and individuals function (UNDP 2011a).[6] This sets the overall scope for capacity building. Capacity building aid programmes at the system level aim for outcomes in the form of

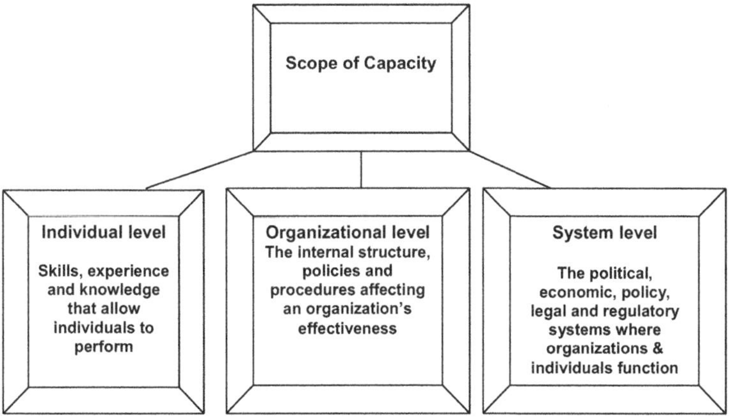

Fig. 3.1 Three levels of capacity (Source: UNDP 2009)

sound governance, effective policymaking and transparent institutions. Among the targeted objectives are high quality of law enforcement, high standard of governance and management, and effective human resource development. It is important to note that these three levels of capacity form an integrated system, in which each level affects the others in a dynamic manner—the strength of each depends on and determines the strength of the other domains (UNDP 2009).

The concepts of 'capacity building' or 'capacity development' are about establishing human, organizational and institutional capacity. Given that the awareness and competence of individuals and households are important determinants of whether a society can undertake responsive actions in the face of the current climate crisis, capacity building is at the root of all effective foreign aid efforts to drive sustainable development.

As suggested in the literature, capacity building is the dedication to the strengthening of economies, governments, institutions and individuals through education, training, mentoring and the infusion of resources. Capacity building aims to develop secure, stable and sustainable structures, systems and organizations, with a particular emphasis on using motivation and inspiration for people to improve their lives. According to UNDP (2009), capacity development is a process through which individuals, organizations and societies generate, strengthen and sustain the competence to set and achieve their own development objectives over time. UNDP has also developed its own approach to the capacity development process, which involves five steps (UNDP 2011b), as depicted in Fig. 3.2.

The climate change capacity building activities of countries party to the UN Framework Convention on Climate Change (UNFCCC) are centred around the following 15 key areas: institutional capacity building; enhancement and/or creation of an enabling environment; national communications; national climate change programmes; GHG inventories emission database management; vulnerability and adaptation assessment; capacity building for implementing adaptation measures; assessment for the implementation of mitigation options; research and systematic observation; development and transfer of technology; improved decision-making, including assistance for participation in international negotiations; clean development mechanism (CDM); capacity building activities to reduce climate change-induced vulnerabilities in the poorest and most vulnerable communities; education, training and public awareness; and information and networking.

It is widely acknowledged among the UNFCCC member countries that proper and adequate capacity is essential to tackling climate change.

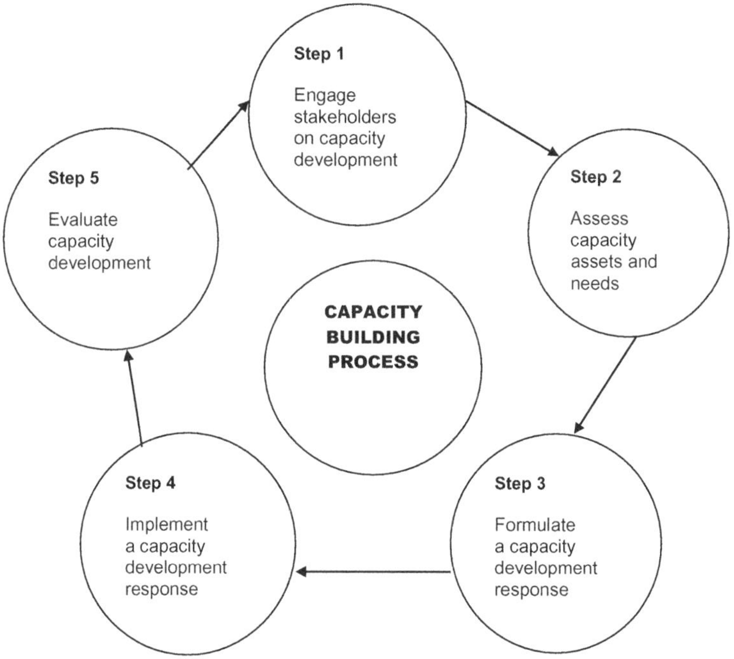

Fig. 3.2 Capacity building process (Source: UNDP 2011b)

According to UNFCCC (2012), adequate capacity needs to cover a variety of issues related to the root causes of climate change, to promote the means for climate change mitigation and adaptation as well as to make sure that climate change considerations are prioritized in development agendas, in the educational curriculum and in people's everyday activities.

Current climate change-related capacity building assistance programmes vary by type, purpose and size, with some aid projects being country specific, some targeting a group of developing countries and others offering general assistance with information and expertise to all developing nations. For example, the Czech Republic has received targeted country-specific aid projects, which helped to build its institutional and regulatory capacity on domestic and international CO_2 emissions trading, and GHG monitoring and reporting. Latvia, Estonia and Lithuania have received assistance with institutional capacity building, while Russia, Ukraine and Belarus were given aid for the development of national GHG inventories.

Generally speaking, capacity building aid projects at present can be categorized according to the funding sources into bilateral assistance, international institutions and multi-lateral assistance, as well as aid projects stemming from business associations and non-governmental organizations (NGOs). Examples of bilateral assistance include aid projects from the US Environmental Protection Agency to Poland, Slovakia and the Czech Republic for developing recommendations on national CO_2 emissions trading schemes; and the Dutch government's financial assistance to Bulgaria and Romania for climate change capacity building initiatives. Among the cases of multi-lateral assistance are the UNDP aid projects to help the transitional economies with capacity building as well as the assistance from the Nordic Regional Organizations (e.g. Nordic Task Force for Climate Issues and Nordic Energy Research Institution) to developing countries in the Baltic region in the area of climate, energy and environment. NGOs and business associations are also an important source of foreign aid to boost capacity building. Examples include the World Resources Institute's support for climate protection aid projects, aimed at assisting developing nations find less emission-intensive development paths and to create effective climate policy and institutional frameworks.

Despite ongoing efforts by aid agencies to promote capacity building in developing countries, little is known about the effectiveness of existing foreign aid in terms of developing climate change-related capacity, what lessons and experiences can be drawn from past and current aid projects, what areas of foreign aid can be improved to boost capacity building and what successful aid experiences could be applied to a wider context. As already outlined, this chapter attempts to find answers to these concerns by investigating what works, what could work, what is scalable and what is transferrable in foreign aid for capacity building. Although each question will be analysed on its own, it should, however, be noted that these issues are intertwined and should be treated as an integral whole.

3.2 What Works

What evidence exists on the effectiveness of aid for climate change-related capacity building? This section discusses the successful experiences of existing foreign aid projects for capacity building through illustrations of actual aid interventions. A summary of capacity building aid activities is given in Table 3.1.

Table 3.1 Examples of capacity building aid activities for 'what works'

Aid activities	*What works*	*Year and location*	*Donors*
Decentralized Service Delivery: a Makerere University Training Pilot Project			
	• Represents a successful example of the new country-driven and demand-driven approach that fosters a strategic long-term relationship between an in-country capacity 'supplier' and local 'demand' for capacity building	2002–6 Uganda	World Bank
Climate and Development Knowledge Network (CDKN)			
	• Example of international assistance to boost knowledge, capacity and awareness of climate impacts in the developing countries. Helps various decision-makers develop climate policies, transfer green technologies and mobilize green growth funding sources	2011–15 worldwide	UK, Netherlands
Cambodia Climate Change Alliance (CCCA)			
	• An example of employing effective outcome indicators to measure and compare institutional capacity development with other similar institutional capacity development projects • The aim is to deepen the understanding of how to enhance the effectiveness of foreign aid for institutional capacity building	2009–14 Cambodia	EU, Sida, Danidaand UNDP
Forest Resource Assessment in Nepal			
	• Successful example of existing foreign aid efforts to establish a climate change database for developing countries • Aims to improve the provision of quality forestry data, and enhance data processing for developing national green forest policies and national decision-making process	2009–14 Nepal	Finland
Strategic Initiative to Address Climate Change in the LDCs [Least Developed Countries]			

(*continued*)

Table 3.1 (continued)

Aid activities	*What works*	*Year and location*	*Donors*
	• Exemplifies donor efforts to improve the decision-making ability of policymakers by offering technical and policy support to develop their capacity to access and implement climate finance, mainstream climate change mitigation and adaptation responses into sustainable national development plans, and effectively engage in international climate negotiations	2010–13 worldwide	UNDP
UNDP's Capacity Development for Policymakers to Address Climate Change			
	• Aid interventions to enhance policymakers' decision-making to coordinate ministerial and stakeholder opinions on climate change, to enhance understanding of the magnitude of the national efforts needed to address climate change and to provide support in long-term climate change planning at the national level	2008–12 worldwide	UNDP

Source: Compiled by authors, based on various sources

3.2.1 *Country-Driven and Demand-Driven Approaches*

According to the Organisation for Economic Co-operation and Development (OECD) (Levina 2002), it has been proven in practice that capacity building programmes have the best chance of succeeding when they are country driven, include a wide range of national stakeholders and involve a high degree of in-country ownership. This approach has been followed in many existing and forthcoming climate-related capacity building projects with encouraging results.

For example, the Decentralized Service Delivery: A Makerere University Training Pilot Project, which took place in Uganda in 2002, is a sustainable capacity building project which adopted the country-driven and demand-driven approach, resulting in progress being achieved ahead of schedule (World Bank 2005). Past capacity building training projects by the World Bank and other donors used to be 'short-term, narrow in coverage, supply driven, uncoordinated, ad hoc, and relied heavily on external technical assistance'. This Uganda aid project is a successful example of the new approach that fosters a strategic long-term relationship between an in-country capacity 'supplier' and local 'demand' for capacity building. Thanks to this promising new approach, progress indicators for the project were met by 2004 or exceeded ahead of schedule.

3.2.2 *Education and Training*

Considerable progress has also been made in terms of education, training and raising public awareness of climate change and its impacts. As suggested by the literature, such capacity building activities as education improvements can translate directly into increased environmental awareness, as well as heightened climate change mitigation and adaptation capacity, which then in turn contribute to a country's low-carbon development. According to UNFCCC (2012), educational programmes on environmental issues and climate change have been implemented in many UNFCCC member nations at all levels of society, from primary schools to universities. Educational initiatives to raise public awareness of climate change are also being introduced by civil society and within targeted communities. These efforts have led to satisfactory results, which imply enhanced environmental awareness and heightened climate change mitigation and adaptation capacity.

The aid project Climate and Development Knowledge Network (CDKN) is one example of international assistance to boost developing countries' knowledge, capacity and awareness of climate impacts (EU 2012). The five-year programme, jointly funded by the UK and the Netherlands, is aimed at assisting 40 developing nations to enhance their awareness, knowledge and competence in the field of climate change. The programme encompasses six private organizations and NGOs covering three continents. Through its professional team of scientists, economists and policy analysts, the project helps public, private and non-governmental decision-makers develop climate policies, transfer to green technologies and mobilize green growth funding sources. For instance, in Rwanda, the CDKN programme provided support for the Ministry of Natural Resources to develop a green growth and climate resilience strategy, which enables direct action to promote climate change mitigation and adaptation actions in the economy, enhances public awareness of the damage resulting from climate change and boosts institutional capacity to tackle climate change. Another CDKN foreign aid project is in Bangladesh, and was launched to empower the government through capacity and expertise building, as well as science and policy support. Thanks to this successful aid intervention, Bangladesh has transformed itself from a vulnerable actor to a leader in international climate negotiations.

3.2.3 Results Monitoring

Monitoring, reporting and sharing of information on capacity building activities, and the experiences and lessons deriving from them are important for improving the effectiveness of future aid interventions. As indicated by the European Union (EU) (2012), in terms of monitoring climate change-related capacity building programmes, the same standards are followed as in other 'normal' development projects. Such monitoring criteria include: relevance; efficiency; development effectiveness; development impact; sustainability; use of a country's own institutions and systems; ownership; management for results; mutual accountability; coordination; complementarity; and coherence. Past aid intervention experiences suggest that it is more effective to monitor climate-related capacity building objectives and results within an aid programme's overall evaluation process. It is worth noting that on top of the general monitoring standards listed, indicators measuring the effectiveness of aid programmes for capacity building also need to be tailored to each specific capacity building activity and defined jointly with the aid-recipient country.

The initiative of Cambodia Climate Change Alliance (CCCA) provides a case in point of outcome indicators to measure development of institutional capacity to tackle climate change (EU 2012). The CCCA project is multi-donor endeavour funded by the EU, Sida, Danida and UNDP to deal with climate change and disaster risks in Cambodia through improved institutional capacity and enhanced resilience to climate change impacts. This project was carried out over the period 2009–14 and was anchored in the Cambodian government's National Climate Change Committee programme. A major anticipated outcome was enhanced institutional capacity to coordinate national climate policymaking, and to monitor implementation of national climate change strategy. The project followed the measurement indicators indicated above to compare outcomes with other similar institutional capacity development projects, with an aim to deepen the understanding of how to enhance the effectiveness of foreign aid for institutional capacity building.

3.2.4 Data and Resources Sharing

The global nature of climate change calls for the exchange and sharing of data, information, expertise and financial resources at all levels to promote best practices in the response to climate change impacts and facilitate climate-related research (UNFCCC 2012). This requires the establishment of information databases in developing nations, where gaps in capacity currently exist in terms of data collection, dissemination and accessibility by the international community. Fortunately, many capacity building aid activities have been devoted to enhancing the quality of data and the dissemination of climate-related information. Furthermore, capacities have been strengthened in some developing countries for international cooperation, collaboration and networking, which facilitate climate information sharing and pave the way for international north–south and south–south cooperative climate research.

One successful example of current foreign aid efforts to establish a climate change database for developing nations is the project Forest Resource Assessment in Nepal, funded by Finland for the period 2009–14 (EU 2012). This programme centred on creating a consistent system for collecting and sharing basic data on forest stock, biomass, soil carbon and biodiversity in Nepal. Its objectives included improving the provision of quality forestry data, and enhancing the processing of data for developing national green forest policies and decision-making at the national level. The programme collected information about forestry activities as pertaining climate change, and has strengthened Nepal's capacity of reducing emissions due to deforestation. In addition to boosting capacity building in Nepal, the project promoted cooperative research and international networking among research institutes in Finland, Nepal and Vietnam. Among its achievements were the development of local partners' capacity for data acquisition and analyses as well as promoting south–south cooperation in terms of mobilizing local and regional resources.

3.2.5 Improved Decision-Making

Aid activities for capacity building play an important part in improving the decision-making capacity of developing country policymakers. As suggested by the literature (e.g. World Bank 2010), human beings are 'myopic decision-makers', who tend to strongly discount future events and give more weight to problems closer in space and time. Thus, policymakers are likely to assign lower priority to the climate change challenge than to other domestic problems. This highlights the importance of developing the capacity of making informed decisions in the policymakers responsible for the design and implementation of sustainable development strategies. Fortunately, considerable aid efforts have been devoted to enhancing policymakers' knowledge and awareness of climate changes and its impacts.

One such donor attempt to improve policymakers' decision-making through capacity building is the Strategic Initiative to Address Climate Change in LDCs, which was funded by the UNDP from 2010 to the end of 2013. This project offered technical and policy support to 26 countries for developing their knowledge in accessing and implementing climate finance, mainstreaming climate change mitigation and adapting responses to national sustainable development plans, and effectively engaging in international climate negotiations. Another good example is the Capacity Development for Policymakers to Address Climate Change (UNFCCC 2012). This programme, which was funded by the UNDP for the period 2008–12 was aimed at building the national capacity of coordinating ministerial and stakeholder opinion on climate change in 19 developing countries, to enhance their understanding of the magnitude of the task ahead and to provide capacity support in their national long-term climate change planning.

3.3 What Could Work

Although progress has been achieved with capacity building assistance projects, there are many areas still needing improvement. This section tries to explore what areas have the potential to work, and what issues arise in delivering on this potential. Relevant cases are also presented to deepen our understanding of ways to improve the effectiveness of capacity building aid activities. Table 3.2 gives a summary of the examples discussed in this section.

3.3.1 Lack of Cooperation

Many of the existing capacity building assistance initiatives have an ad hoc character and there is little cooperation among the various projects, a fact that limits the effectiveness of the implementation process. For instance, it is noted that many climate change adaptation activities are carried out as stand-alone or disconnected projects. Fragmented climate adaptation

Table 3.2 Examples of capacity building aid activities for 'what could work'

Aid activities	*What could work*	*Year and location*	*Donors*
Innovative Insurance Products for Climate Change Adaptation			
	• Highlights the importance of enhancing internal capacity building resources of developing countries • Aimed at enhancing knowledge and awareness of the stakeholders, to foster professionalism through training and to build the capacity of Ghana's insurance sector	2009–13 Ghana	Germany
Water Programme for Environmental Sustainability (WPA)			
	• WPA forms an integral part of the long-term, multi-purpose and multi-sectoral climate aid programmes, where aid activities complement each other within existing national/regional climate adaptation and mitigation programmes • Successful implementation of WPA highlights the advantage of including capacity building as an integrated part of an overall climate aid programme	Effective from 2004 Serbia, North Africa, Vietnam and China	Italy
PAKLIM—Policy Advice on Climate Change and Environment			
	• Efforts by donor community to employ a country-owned and demand-driven approach in foreign aid for capacity building, highlighting the importance of ownership in capacity building aid processes • Climate capacity building measures were jointly designed, carried out and monitored by the donor and recipient government	2009–16 Indonesia	Germany

Source: Compiled by authors, based on various sources

finance hinders the mainstreaming of climate change responses into planning and development processes, and raises transaction costs for both donors and recipients (World Bank 2010).

Yet attention has been paid to this problem and efforts are being made to shift the ad hoc approach to a more systematic level, with more agencies developing comprehensive capacity building programmes to achieve synergy (Levina 2002). Donor agencies now strive to keep each other informed of programmes and actions implemented in the same region and to cooperate when possible. For instance, before the start of any programme, there should be donor coordination to reduce duplication. For similarly targeted projects in the same region, it could save donor time and money to utilize available, earlier studies to avoid duplication of needs assessment. One additional useful action could be to establish an online or actual climate change capacity building coordination centre, or hub, for exchanging information and resources, requesting assistance and undertaking mutual activities.

3.3.2 *Lack of Internal Capacity*

There is a frequent mismatch of internal and external capacity building resources, with many aid initiatives from donor countries unable to be matched to internal resources or competence in recipient countries. Thus, there is a need to encourage developing countries to strengthen relevant resources and capacity in order to be able to match donors' interventions.

The aid programme Innovative Insurance Products for Climate Change-Adaptation in Ghana serves to highlight the importance of enhancing the internal capacity building resources of developing countries (EU 2012). Established to improve Ghana's capacity to manage the socioeconomic costs and risks brought on by climate change, this programme developed and introduced climate change-related agricultural insurance schemes for chosen value chains for implementation during the years 2009–13. But despite the benign intentions of donor countries, Ghana has had little experience with, and only limited capacity for, this kind of insurance scheme, creating a gap between internal and external capacity building resources. Nonetheless, owing to efforts to enhance the knowledge and awareness of the stakeholders, to foster professionalism through training and to build capacity of Ghana's insurance sector, the gap was bridged, paving the way for effective implementation of the project.

3.3.3 *Lack of Initiative*

Another observed obstacle to the effective launch of capacity building aid programmes is the lack of initiative on the part of recipient country governments. The passiveness of stakeholders (e.g. governments, public and industry) in many developing nations discourages the design of effective climate change policies and prevents these nations from reaping benefits from foreign aid. Accordingly, there is a need to encourage the developing nations to take the lead with donors in capacity building efforts, to harmonize donor support around the priorities of the beneficiary countries (World Bank 2005).

3.3.4 *Lack of Flexible Aid Application Mechanism*

There is no mechanism flexible enough to enable developing nations to apply for specific and timely assistance. In a limited number of cases, developing countries can apply for special programmes to request overseas assistance for capacity building. However, in most cases, developing countries are not equipped with the necessary resources to apply for country-driven capacity building assistance.

3.3.5 *Lack of an Integral Approach*

Given its cross-cutting nature, foreign aid for capacity building should be an integral part of support arrangements in all relevant areas, ranging from climate change adaptation and mitigation, response measures, green technology transfer to development of low-carbon market mechanisms (EU 2012). For instance, the Water Programme for Environmental Sustainability (WPA) funded by Italy since 2004 represents a good attempt to adopt an integrated approach to capacity building activities (EU 2012). Conducted in Serbia, North Africa, Vietnam and China to improve integrated water resources management and protection in response to climate change impacts, the programme explores effective adaptation practices for water management in the face of climate variability and climate change impacts. Capacity building is well integrated into the overall programme, as the project centres on capacity building, regional coordination, promotion of effective technological practices and dissemination of sound governance measures. It is important to note that individual climate capacity building activities should not be isolated events but rather an integral part of long-term, multi-purpose and multi-sectoral climate aid programmes, where activities complement each other within existing national/regional climate adaptation and mitigation schemes. The successful implementation of this aid project underscores the advantage of including capacity building as an integrated part of an overall climate aid programme.

3.3.6 *Lack of a Long-Term Approach*

Climate change capacity building programmes should be long-term iterative processes, employing a flexible and adaptive manner to take changing circumstances and emerging challenges into consideration. Shortage of sustainable financial resources has been recognized as one of the barriers impeding the launch of long-term programmes. Accordingly, more aid commitment from the developed countries is needed to secure a sustainable flow of funds for capacity building in developing countries.

3.3.7 *Lack of Country Ownership*

Ownership is essential to the success of foreign aid for capacity building. As van de Walle and Johnston (1996) point out, 'recipient governments can be said to "own" an aid activity when they believe that it empowers them and serves their interests'. Aid programmes for climate change must therefore be designed, carried out and monitored in collaboration with local partners to make sure that the implemented capacity building activities are country owned and demand driven.

The PAKLIM project in Indonesia represents a positive attempt by the donor community to employ a country-owned and demand-driven approach, which deserves wider application in other climate capacity building aid projects. This aid programme, funded by Germany for the period 2009–16, offered policy advice on climate change and strengthened capacity building on climate change mitigation and adaptation at national and local levels of government (EU 2012). In terms of climate change mitigation, PAKLIM was aimed at assisting the Indonesian government to launch a set of climate change strategies to reduce GHG emissions and pursue a sustainable economy. In terms of climate change adaptation, it strove to help the government improve national climate policy frameworks for local-level adaptation. In this process, the German donor designed, carried out and monitored climate change capacity building measures jointly with the Indonesian government, deciding mutually what climate adaption measures to implement, and defining the measurement indicators for monitoring the effectiveness of the adopted climate capacity building measures.

3.4 What Is Scalable

This section investigates what aspects of foreign aid for capacity building need to be delivered on a larger scale to enhance aid's positive effects for improving climate-related capacities in developing countries. As identified by the UNFCCC synthesis report (2012), aid efforts need to be scaled up at least in the following areas to boost climate change-related capacity building. Examples of effective aid interventions are discussed in this section to illustrate where more capacity building aid efforts are required. Table 3.3 gives a summary of the cases discussed.

3.4.1 Enhance Awareness and Knowledge

As reported by many UNFCCC member countries, climate change and its impacts are still not well understood. As the World Bank indicates (2010), misconceptions about the dynamics of climate change lead to complacency,

Table 3.3 Examples of capacity building aid activities for 'what is scalable'

Aid activities	*What is scalable*	*Year and location*	*Donors*
Capacity Building Activities in the Iberoamerican Region			
	• Exemplifies donor efforts to boost awareness and knowledge of climate change impacts in developing countries • Similar efforts still needed to educate governments and the public of the urgency of tackling climate change in other developing countries, many of which still lack adequate understanding of climate change issues	2008–11 Iberoamerican region	Spain
Negociations Climate Toute l'Afrique Renforcee (NECTAR)			
	• An example of donor efforts to boost LDCs' climate negotiation capacity • Aimed at assisting LDCs develop national strategies to enhance capacity to participate in international negotiations and to target obstacles in negotiations facing Africa	2008 African LDCs	France
Regional Gateway for Technology Transfer and Climate Change Action			
	• An example of the efforts from the international aid community to facilitate capacity building for green technology transfers	2011–13 Latin American countries	Spain and Norway

(*continued*)

Table 3.3 (continued)

Aid activities	*What is scalable*	*Year and location*	*Donors*
Prosol Industrial			
	• Aimed at designing an effective financing mechanism to overcome investment obstacles in the diffusion of solar thermal technologies in the industry • Represents a sound example of international donor community efforts to boost capacity to support green technology adoption efforts, and establish market mechanisms for attracting green investment	2010–14 Tunisia	Italy
Capacity Building in Development of Policy Framework for Promotion of Low Carbon Emission Societies			
	• Example of scaled-up foreign aid efforts to enhance institutional capacities with regard to the diffusion of clean technologies • More efforts are needed to build relevant capacities to establish an environment conducive to green technology diffusion and to reduce barriers to green technologies and services	2009–13 Central Asia	South Korea
Intergovernmental Panel on Climate Change Scholarship			
	• Offers young people in developing countries new opportunities to study (PhD level) and contribute climate change research	ongoing LDCs	IPCC
Youth and United Nations Global Alliance Education Programme on Climate Change			
	• Aimed at children and youth to promote their participation in climate change mitigation and adaptation activities • Due to a shortage of similar programmes for climate change education in the LDCs, more aid efforts are called for	2011 worldwide	Food and Agricultural Organisation (FAO), UN and youth agencies

Source: Compiled by authors, based on various sources

and support for emission reduction policies is hampered by people's limited understanding of the dynamics of climate change.[7] For some developing countries, climate change is perceived to pose less risk than other hazards and is given low priority in national development plans. Therefore, more efforts are required to raise public awareness of the urgency of tackling climate change at all levels of society. Effective channels to enhance people's awareness include educational programmes to boost local understanding, and specific scholarship programmes to encourage research on climate change.

Capacity building activities by Spain in the Iberoamerican region reflect the efforts from donor countries to increase awareness and knowledge of climate change impacts in developing countries. In collaboration with other multi-lateral and regional organizations, Spain has launched a number of aid projects in the region to improve understanding of the pertinence of climate change impacts, and to enhance the capacity of the developing countries to deal with these issues. This project has organized a series of climate-related workshops, aimed at enhancing public awareness of climate change, building governments' capacity to design and implement climate strategies, and improving access to climate finance. Despite the progress achieved by this aid programme, much greater foreign aid efforts are needed to educate governments and the public of the urgency of tackling climate change in developing countries, many of which still lack adequate understanding of the climate change crisis.

3.4.2 *Boosting International Negotiation Capacity*

The LDCs are significantly impacted by climate change, but they play a small part in international climate negotiations owing to the lack of relevant capacity. This calls for more aid efforts to scale up their ability to take part in international negotiations so as to win greater support and help in battling climate change. Only through active involvement in the global arena can the voices and national interests of LDCs be better reflected in the international climate regime.

One such example is the NECTAR project (Négociations Climat Toute l'Afrique Renforcée), which is aimed at assisting African LDCs develop national strategies so as to enhance their capacity to participate in international negotiations and to target the obstacles they face during talks (EU 2012).[8] Implemented in 2008 and funded by France, the project organized a series of workshops to improve the capacity of African climate negotiators, to offer them relevant contacts within other countries and to train governments in climate negotiation strategies and skills. As a result, the programme helped to enhance the LDCs' influence and bargaining power in the Durban conference not only because of better negotiating capacity but also through strengthened alliances with other LDCs. Capacity building aid programmes such as the NECTAR project will continue to benefit LDCs by enabling them to make better use of new climate financing mechanisms after the launch of the Green Fund in Durban.

3.4.3 *Green Technology Diffusion*

UNFCCC parties have identified green technology transfer and development as one capacity building area that calls for scaled-up aid efforts. As voiced by many developing nations, there must be more investment from the donors to develop local capacity to support green technology diffusion, both with regard to 'hard issues' and 'soft issues'. 'Hard issues' refers to the ability to access clean technology and having the skills to apply it. 'Soft issues' are the enabling conditions associated with the adoption/non-adoption of such technology in developing countries, including its cost, market failures that hinder its implementation, and issues related to technology design such as poor adaptation to local situations (UNFCCC 2012). Also impeding green technology development and transfer are the developing countries' lack of commitment to clean technology diffusion in national long-term sustainable development plans, and inadequate institutional capacity to support green technology adoption. But these obstacles can be conquered with scaled-up aid from the international community.

The Regional Gateway for Technology Transfer and Climate Change Action aid project exemplifies the efforts of the international aid community to facilitate capacity building for green technology transfer (UNFCCC 2012). Jointly funded by Spain and Norway to the amount of US$7.2 million for the years 2011–13, the project aimed at assisting Latin American countries boost the mobilization and sharing of knowledge on climate change issues covering the entire technology cycle, from green technology development to clean technology transfer and employment, with regard to both hard issues and soft issues. The project helped recipient countries develop the necessary skills to access green technology, to integrate it within their national development strategies and to create an enabling environment for the adoption of the technology.

The Prosol Industrial project in Tunisia, funded by Italy from 2010 to 2014, intended to design an effective financing mechanism to overcome investment obstacles for the diffusion of solar thermal technologies into the industry there (UNFCCC 2012). This aid programme represents a sound attempt by the international donor community to boost capacity to support green technology adoption, especially with regard to establishing an effective market mechanism to attract green investment.

Capacity Building in Development of Policy Framework for Promotion of Low Carbon Emission Societies in Central Asia is another example of scaled-up foreign aid efforts to enhance institutional capacities to overcome barriers to and to promote the diffusion of clean technologies (UNFCCC 2012). Funded by South Korea for the years 2009–13 to provide assistance with institutional capacity building activities in Central Asia, this project aimed at conquering the above-mentioned barriers by establishing an enabling framework for low-carbon technology development and transfer, including improving national institutions, raising public awareness on climate change and developing markets for energy efficient technologies.

In spite of the progress being made in developing capacity among LDCs for adopting clean technology, it is clear that greater efforts are still required to generate the expertise needed to establish an environment that is advantageous to the diffusion of green technology and to reduce the barriers hindering green technologies and services.

3.4.4 Training Climate Change Professionals

Scaled-up capacity building aid support is urgently required to deal with the shortage of climate change professionals in Africa, especially at the PhD and postdoctoral levels. This is viewed as one of the critical impediments to the dissemination of climate-related information and knowledge, and the conducting of scientific climate research. PhDs and postdoctoral researchers could also contribute to furthering the standard of education in the LDCs, which lack sufficient professional teaching staff for undergraduate and postgraduate study on climate change.

Among the existing aid efforts in this regard is the Intergovernmental Panel on Climate Change Scholarship programme, which offers selected young people in developing countries new opportunities to study and contribute to climate change research (UNFCCC 2012) by funding their PhD studies. The Youth and United Nations Global Alliance Education Programme on Climate Change targets children and youth to promote participation in climate change mitigation and adaptation activities, and to encourage their involvement in international climate negotiation processes.[9] Despite the positive effects of these existing educational schemes, there is a shortage of similar aid programmes to meet the educational needs related to climate change in different LDCs. More aid efforts are therefore called for in order bridge this gap.

3.5 What Is Transferrable

This section investigates and discusses what experiences gained from existing aid projects can be transferred across countries or across projects, and what aspects of aid interventions are potentially transferrable. In fact, almost all the successful approaches summarized in Sect. 3.2 ('what works') are transferred, albeit with modifications to cater to local circumstances. The following practices in particular should find application within a wider context (summarized in Table 3.4).

3.5.1 Transfer of Country-Owned and Demand-Driven Approach

The country-owned and demand-driven approach promoted by OECD, UNEP and the World Bank is transferrable to other capacity building programmes by other donors. It highlights the importance of shifting the focus of capacity building aid from technical assistance 'fixes' and supply-driven aid to a demand-driven, outcome-based and country-owned approach (Levina 2002; World Bank 2005; UNEP 2009). Similarly, the EU finds that the principle of a scientific and systematic approach has been effective in its climate change capacity building programmes. This implies that such an approach could be transferred to other aid programmes for wider application within the donor community. The fact that this approach has found successful applications among numerous international donor projects demonstrates that the principle is transferrable to other capacity building aid cases, and should be advocated in a wider context once local features 'based on local ownership, demand driven-processes and responding to expressed needs of partners' are added.

Table 3.4 Examples of capacity building aid activities for 'what is transferrable'

Aid activities	*What is transferrable*	*Year and location*	*Donors*
Support to CDM Establishment and Operationalization			
	• Represents a sound example of plausible aid efforts to boost developing nations' capacity of launching CDM projects • Project experiences are transferrable to other similar projects	2011 Bosnia and Herzegovina	UNDP
Capacity Development for the African, Caribbean and Pacific Capacity Development for the CDM			
	• Countries successful in registering CDM projects have developed the complementary technical capacity to achieve the CDM objectives. This programme helps to remove capacity barriers hindering countries from fostering a robust carbon market and from utilizing the CDM scheme for low-carbon development	Ongoing worldwide	UNEP
Workshop on Enhancing the Regional Distribution of CDM Projects			
	• Organized to disseminate CDM information and knowledge, provide training on project application and share good capacity building practices for CDM project development among developing nations	2011 Nepal	UNFCCC
Stakeholder Consultation Workshop on Standardized Baseline under CDM			
	• Aimed at promoting the transfer of good practices and dissemination of CDM knowledge of the CDM among developing nations, this workshop facilitates the access of under-represented regions to the CDM by enhancing their understanding of standardized baselines	2011 Nepal	UNFCCC

Source: Compiled by authors, based on various sources

3.5.2 *Transfer of Knowledge and Experience*

The Innovative Insurance Products for Climate Change Adaptation aid programme discussed earlier is a good example of knowledge sharing and capacity building across Africa and beyond, particularly in terms of narrowing the gap between internal and external capacity building resources. As specified in the project objectives, experiences are to be systematically collected and distributed to other countries, and will also be transferred to other climate change aid programmes with similar circumstances (EU 2012).

3.5.3 Transfer of CDM Knowledge and Practice

The CDM is a flexible market mechanism under the Kyoto Protocol, offering developing countries an opportunity to benefit from investments in emission reduction development projects.[10] According to the UNFCCC (2012), some developing countries have reported considerable progress in registered CDM project activities while others have been less successful because of limited technical capacity, political understanding or will. In comparison, countries with successful CDM experiences have developed the complementary technical capacity to be able to achieve CDM objectives, thanks to donor support that covered marketing, training, information dissemination and market development (UNFCCC 2012). The positive experiences gained from aid projects earmarked for CDM development are transferrable to any developing nation in urgent need of similar support in order to take advantage of CDM and emissions reduction investments.

Plausible aid efforts to boost developing nations' capacity of launching CDM projects include a UNDP aid project which offered capacity building support for Bosnia and Herzegovina in establishing and adopting national rules and procedures for CDM approval. Training was also provided on the procedures for registering CDM projects. This scheme has generated encouraging results; a pipeline of projects is being created, and another approved and submitted to UNFCCC. Another UNEP aid programme that offers sound lessons is the Capacity Development for the African, Caribbean and Pacific for the CDM. This programme strives to enhance the capacity of aid recipient countries for identifying, designing, approving, financing, implementing and monitoring CDM projects. These activities help remove capacity barriers that hinder developing countries from fostering a robust carbon market and from utilizing the CDM for low-carbon development.

A series of workshops has also been organized by international donors to disseminate CDM information and knowledge, provide training on project application and share good capacity building practices for CDM project development. Examples include the UNFCC workshops on 'Enhancing the

Regional Distribution of CDM Projects in Asia and the Pacific', and the 'Stakeholder Consultation Workshop on Standardized Baseline under CDM' to facilitate the access of under-represented regions by enhancing their understanding of standardized baselines (UNFCCC 2012).

3.6 Conclusion

Capacity building is at the root of all effective foreign aid efforts in driving sustainable development. As highlighted by UNDP (2011b), the successful development of a country depends on sufficient capacity. Without supportive policies, laws, institutions and education in place, a country simply does not have the foundation to pursue long-term and well-rounded development. It is important that donors shift the focus of foreign aid for capacity building from technical assistance 'fixes' and supply-driven aid to a demand-driven, outcome-based and country-owned approach. Given its cross-cutting nature, foreign aid for capacity building should be an integral part of support arrangements in all relevant areas, from climate change adaptation and mitigation, response measures and green technology transfer to development of low-carbon market mechanisms. Moreover, it should be a long-term iterative process, applied in a flexible and adaptive manner so as to take changing circumstances and emerging challenges into consideration. Although aid generally has a positive effect on the availability of finance, it can at times have a negative impact on governance, and thus undermine government actions. Therefore, foreign aid for capacity building should be delivered with care and purpose in order to improve the effectiveness of capacity building in developing countries (Barnett 2008).

Although there is no recipe that 'fits-all' to tackle all climate change-related capacity building problems, most of the experiences and lessons discussed in this chapter are transferrable to a wider context, albeit with modifications to fit local circumstances. However, it is important to note that all the identified barriers that impede capacity building are in fact interrelated, although they are discussed here separately for the convenience of the reader. It is unlikely that fixing one or two problems can make capacity building work, and all these challenges need to be met in an integrated way.

Capacity building is important, yet it is not easy to measure the effectiveness of foreign aid capacity building interventions because of the 'soft' and dynamic nature of capacity. By answering the four essential questions of 'what works?', 'what could work?', 'what is scalable?' and 'what is transferrable?', this chapter investigates how effective is foreign aid at present with regard to capacity building, and how to improve future aid projects by taking advantage of past experiences and lessons, inviting further research to deepen our understanding of this significant issue.

Notes

1. For example, increased maximum temperatures and changes in rainfall patterns are already having a negative impact on agriculture and food security in many low-income communities. Many coastal nations are suffering from damage to their ocean fisheries resulting from problems of ocean acidification (Howes and Wyrwoll 2012).
2. Green growth can be defined as 'fostering economic growth and development, while ensuring that natural assets continue to provide the resources and environmental services on which our wellbeing relies' (Hallegatte et al. 2011). Unlike the traditional pattern of economic growth, which was achieved largely at the expense of the environment, green growth aims to achieve synergy between economic progress and environmental protection that is vital to realizing the goal of sustainable development.
3. Capacity is the essential lubricant of international development. Examples of areas particularly relevant to developing countries include education, training and raising public awareness of climate change. Strengthening government delivery with trained professional staff is a reoccurring theme, as is the establishment of climate change research and policymaking bodies. The term 'adaptive capacity' has entered the language of the fight against climate change.
4. Some aspects of capacity are generated formally through education and training, while others are obtained informally through doing and observing (UNDP 2009).
5. Capacity at the organizational level is where the benefits of an enabling environment are put into action and where a collection of individuals gather together for one purpose (UNDP 2009).
6. An enabling environment covers the rules, laws, policies, power relations and social norms that govern civic engagement (UNDP 2009).
7. Many people actually misunderstand the risks of climate change, believing that simply stabilizing GHG emissions at the current rate would stabilize GHG concentration in the atmosphere and stop further climate change (World Bank 2010).
8. The usual difficulties faced by LDCs include linguistic barriers, lack of scientific climate policy and action plans, and lack of coherent negotiation techniques and methodologies.
9. This aid project is jointly organized by FAO, other UN agencies and youth organizations (UNFCCC 2012).

10. As specified in Article 12 of the Kyoto Protocol, the CDM was set up with two major yet equally important aims: to mitigate GHG emissions in a cost-effective manner and to boost sustainable development in the host countries.

References

Barnett, J. 2008. The Effect of Aid on Capacity to Adapt to Climate Change: Insights from Niue. *Political Science* 60 (1): 31–45.

Degnbol-Martinussen, J. 2002. Development Goals, Governance and Capacity Building: Aid as a Catalyst. *Development and Change* 33 (2): 269–279.

EU (European Union). 2012. Submission by Denmark and the European Commission on Behalf of the European Union and Its Member States. Capacity-building under the Convention and under the Kyoto Protocol: Information on the Activities Undertaken Pursuant to Decisions 2/CP.7, 2/CP.10 and 29/CMP.1.

Hallegatte, S., G. Hela, M. Fay, and D. Treguer. 2011. *From Growth to Green Growth: A Framework.* WB Policy Research Working Paper 5872. Washington, DC: World Bank.

Howes, S., and P. Wyrwoll. 2012. *Climate Change Mitigation and Green Growth in Developing Asia.* ADBI Working Paper 369. Tokyo: Asian Development Bank Institute.

Levina, E. 2002. *Climate Change Capacity Building in Annex I EITs: Issues and Needs.* OECD Report. Paris: OECD.

Roberts, J.T., and B. Parks. 2006. *A Climate of Injustice: Global Inequality, North-South Politics, and Climate Policy.* Cambridge, MA: MIT Press.

UNDP (United Nations Development Programme). 2009. *Capacity Development: A UNDP Primer.* New York: UNDP.

———. 2011a. *Practitioners' Guide: Capacity Development for Environmental Sustainability.* New York: UNDP.

———. 2011b. *Supporting Capacity Development.* New York: UNDP.

UNFCCC (UN Framework Convention on Climate Change). 2012. *Synthesis Report on the Implementation of the Framework for Capacity-Building in Developing Countries.* Bonn: UNFCCC.

Van de Walle, N., and T.A. Johnston. 1996. *Improving Aid to Africa.* Washington, DC: Overseas Development Council.

World Bank. 2005. *Capacity Building in Africa: An OED Evaluation of World Bank Support.* Washington, DC: OEC-World Bank.

———. 2010. *World Development Report: Development and Climate Change.* Washington, DC: World Bank.

CHAPTER 4

The Effectiveness of Foreign Aid for Sustainable Energy and Climate Change Mitigation

H-Holger Rogner and Kei-Kit Leung

4.1 Introduction

Foreign aid and technology transfer are essential means towards meeting the UN's Sustainable Development Goals (SDGs) as well as to facilitate adaptation to and mitigation of climate change. They also constitute fundamental components of sustainable development. Although access to affordable and clean energy services is a key prerequisite for the realization of all SDGs and climate change objectives, generation of energy or access to it are not SDGs. Development aid, therefore, has been preferably allocated to the specific quantitative and time-bound SDGs, rather than to enabling investments in energy access or for clean and sustainable energy services.

Nevertheless, development aid disbursements allocated to renewable energy technologies, infrastructures and efficiency programmes have been far from negligible since 2000s. Quite a number of specialized funds have been established since the 1992 United Nations Conference on Environment and Development (UNCED), also known as the 'Rio

H.-H. Rogner (✉) • K.-K. Leung
International Institute for Applied Systems Analysis (IIASA), Schlossplatz, Austria

Y. Huang, U. Pascual (eds.), *Aid Effectiveness for Environmental Sustainability*, https://doi.org/10.1007/978-981-10-5379-5_4

Summit 1992', in support of facilitating technology transfer and investment in sustainable energy.

The United Nations Secretary General's initiative Sustainable Energy for All (SE4ALL), launched in 2011, is one of the most ambitious campaigns undertaken to date aimed at transforming the world's energy systems. It is based on the premise that current energy systems are unsustainable and in danger of compromising the future in a variety of areas in which they themselves constitute some of the most serious challenges of our century: security, climate change and other environmental impacts, poverty eradication and equality. To address all these challenges, SE4ALL has three goals to reach by 2030: universal energy access, doubling the rate of energy efficiency improvements and doubling the global share of renewable energy in the final energy mix.

Reaching the triple objectives of SE4All by 2030 will require substantial amounts of capital from the public sector, the private sector and the donor community. The private sector will have to provide the bulk of investments and finance for the implementation of SE4All. The international financial institutions will have a key role to play in mobilizing financing, in providing risk coverage and in acting as a catalyst for private-sector involvement and energy-sector reforms (AGECC 2010).

Energy-related development aid during the first decade of the twenty-first century is estimated at close to US$60 billion, of which renewable energy technologies received about US$8 billion (OECD 2012a). Multilateral development banks (MDBs) provided some US$70 billion of energy-related funding, of which energy efficiency projects received US$8.2 billion and renewable projects US$10.1 billion. However, the Secretary General's Advisory Group on Energy and Climate Change (AGECC) estimated the funding requirements for access and energy efficiency at US$205 to US$245 billion annually (AGECC 2010)—a huge funding gap compared with the funds available between 2000 and 2010.

Given this large shortfall, it is crucial that the limited resources for sustainable energy and environmental protection are used most effectively. This chapter, therefore, explores the effectiveness of foreign aid on the advancement of sustainable energy, as well as reviewing enabling conditions and policy prerequisites. This chapter aims to document 'what works', 'what could work', 'what is scalable' and 'what is transferable' in the context of sustainable energy investment. This is fundamental for private-sector participation and the extent to which public and concessional finance is leveraged with venture capital, private equity and/or asset finance.

4.2 Foreign Aid, Energy and Climate Change

The terms 'foreign aid' and 'financial assistance' are not precisely defined in the development literature (Urban and Wolcott 2009). They are generally interpreted as the voluntary transfer of public resources, from a government to another independent government, to a non-governmental organization (NGO) or to an international organization, such as the World Bank (WB) or the United Nations Development Programme (UNDP), with at least a 25% grant element. Foreign aid usually targets the socioeconomic factors underlying poverty, with the objective to improve the living conditions in the recipient country (Lancaster 2007). Foreign aid encompasses various types of financing mechanisms, including official development assistance (ODA),[1] other official flows (OOF),[2] foreign direct investment (FDI), direct budgetary support (DBS), basket funding as well as conditional and unconditional funding. The multi-lateral development banks (MDBs) or international financial institutions (IFIs) such as the World Bank Group (WBG) and regional development banks as well as private–public partnerships (PPPs), via a combination of ODA, private-sector funding and/or funding from foundations and NGOs are transaction channels for transferring foreign aid.

There also exists a variety of funds that support projects targeted at mitigating or adapting to climate change in the context of sustainable development: for instance, the Global Environment Facility (GEF); the Climate Investment Funds (CIFs), which include the Clean Technology Fund (CTF), the Scaling Up Renewable Energy Programme in Low Income Countries (SREP), the Forest Investment Programme (FIP) and the Pilot Programme for Climate Resilience (PPCR). The flexible mechanisms under Kyoto Protocol Clean Development Mechanism (CDM) and joint implementation (JI) are also vehicles for the transfer of sustainable energy technologies and associated financial assistance.

Energy has long been the stepchild of foreign development aid, especially ODA. Although recognized as a necessary but not sufficient ingredient for sustainable development, energy's cost share in the generation of welfare was much lower than the production factors of labour and capital. The adverse consequence of the oil price hikes of the 1970s on the terms of trade of oil-importing developing countries were quickly forgotten after the collapse of global oil market prices in 1986. The small amounts of ODA and FDI were largely targeted at upstream fossil resource and electricity-sector development (thermal generation and transmission and distribution).

This changed in 2004 when oil prices started to climb, and culminated in 2007 and 2008 when they reached their historical peak. High oil prices disproportionally affect the poor, as developing countries spend a significant share of their income on crude or oil products imports. The increase in oil prices (or rather fossil fuel prices in general) has made investment in energy efficiency measures and renewable energy an economically attractive proposition.

Indeed, both ODA and FDI in energy efficiency and renewables began to rise rapidly. But even more effective than high energy prices has been the growing recognition that the potential detrimental effects of climate change could well undermine the gains of development assistance. As the SE4All initiative demonstrates, sustainable energy access and mitigation of climate change but also coping through safeguarding against its adverse impacts are now seen as non-negotiable prerequisites for achieving the SDGs.

An increasing variety of financial instruments for combating climate change has been developed under the umbrella of the UNFCCC, creating a dynamic financial innovation trend that has gained momentum since 1990s. But including the financial flows associated with these instruments in the current ODA reporting system is not allowed due to the additionality condition—that is, climate change-related foreign aid should not be donated at the expense of regular development aid (ODA) for economic development in poor nations. Hence, climate-related funding must be reported separately by the various institutions administering these funds or instruments.

This does not mean that ODA cannot be used for the support of energy efficiency and renewable projects in developing countries motivated by climate mitigation objectives. While there was an extensive debate about 'additionality' before the ratification of the Kyoto Protocol, the Organisation for Economic Co-operation and Development (OECD) Development Assistance Committee (DAC) decided in 2004 to exclude only those CDM activities that governments directly use to purchase certified emission credits (CERs) (Michaelowa and Michaelowa 2007).

The developed countries that signed the Rio Conventions in 1992 (and that were amended in 2010) committed themselves to assist developing countries in the implementation of the conventions. The DAC monitors aid targeting the objectives of the Rio Conventions through its 'creditor reporting system' (CRS) using the so-called 'Rio markers'.[3] In short, there is also a sizeable ODA contribution to funding climate change mitigation (one of the Rio markers) either directly through bilateral support or indirectly through contributions to the MDBs.

Table 4.1 summarizes the development of foreign aid (ODA and OOF) in support of energy and renewables for the period 2002–10. The data in

Table 4.1 Total and energy-related gross ODA and OOF disbursements, 2002–10, in millions US$

	2002	*2003*	*2004*	*2005*	*2006*	*2007*	*2008*	*2009*	*2010*
ODA—All donors	67,367	80,414	92,149	120,771	120,241	122,168	144,423	139,893	148,380
ODA—DAC countries	58,575	69,432	79,854	107,838	104,814	104,206	121,954	119,778	128,465
OOF—All donors	1003	767	−2773	3975	−8090	−816	2831	9482	4940
OOF—DAC countries	119	−420	−5418	1986	−9822	−5491	−55	10,119	5878
Total ODA and OOF	68,370	81,181	89,376	124,746	112,150	121,352	147,253	149,376	153,320
Energy—DAC	889.6	945.1	1632.2	3083.5	3046.4	3476.6	3716.7	3417.9	4910.5
Energy—MDBs	499.0	444.1	647.7	579.3	598.0	899.5	1697.6	1768.5	2385.2
Energy—All donors (DAC)	1388.6	1389.2	2280.0	3662.8	3644.3	4376.0	5414.3	5186.5	7295.7
Energy—OOF	1346.5	708.9	723.6	627.6	669.5	1290.9	1667.8	6867.9	5508.1
Total energy of which	2735.2	2098.1	3003.6	4290.4	4313.8	5667.0	7082.1	12,054.4	12,803.8
Renewables, incl. Hydro	153.7	239.9	358.3	619.4	653.1	1077.0	1044.0	1566.1	2082.2
—Hydro	109.1	176.5	296.0	544.6	488.3	818.4	666.6	792.5	1186.9
—Geothermal	8.4	0.5	2.3	2.4	3.6	11.8	33.8	53.9	46.7
—Solar	8.9	35.6	25.4	27.3	37.0	44.3	130.6	389.5	440.7
—Wind	17.1	22.4	27.6	32.7	112.4	186.0	180.8	291.6	332.3
—Ocean	0.0	0.0	0.0	0.0	0.0	0.0	0.0	0.1	0.1
—Biomass	10.2	4.9	7.0	12.4	11.4	16.6	32.2	38.5	75.5

(*continued*)

Table 4.1 (continued)

	2002	*2003*	*2004*	*2005*	*2006*	*2007*	*2008*	*2009*	*2010*
Share of energy in total ODA and OOF, %	4.7	3.0	3.8	4.0	4.1	5.4	5.8	10.1	10.0
Share of renewables in total energy, %	5.6	11.4	11.9	14.4	15.1	19.0	14.7	13.0	16.3
Share of renewables in total ODA and OOF, %	0.2	0.3	0.4	0.5	0.6	0.9	0.7	1.0	1.4

Source: OECD (2012a)

Note: A disbursement is the actual placement of resources at the disposal of a recipient country or agency. This is different from 'commitment', which is a firm written obligation by a government or official agency, backed by the appropriation or availability of the necessary funds, to provide resources of a specified amount under specified financial terms and conditions and for specified purposes for the benefit of a recipient country or a multi-lateral agency. The conditionality often leads to delays or lower actual disbursements

Table 4.1 show the actual gross disbursements to recipient countries which, for a variety of reasons, can be notably lower than the original commitments made. The energy-related funding in total ODA and OOF disbursements has increased considerably since 2005, that is, from 4% to 10%, while the share of renewables in total energy fluctuated and grew only slightly over the period.

Tables 4.2 and 4.3 summarize the energy and climate-related funding of the WBG and the Asian Development Bank (ADB). Despite the economic and financial crises (or perhaps because of them), the annual outlays for low-carbon projects have increased significantly after 2007.

4.3 Funding Mechanisms

Renewable technology and energy efficiency measures are more upfront investment-heavy than conventional technologies and fuels. In most developing countries, government resources alone are inadequate to meet the large investment requirements of scaling up renewable energy technologies and energy efficiency—in the past, they were rarely sufficient to support conventional supply infrastructures. The bulk of investment funding, therefore, will have to come from the private sector. Without some kind of incentive, private-sector investors find investing in energy efficiency and renewables a risky proposition—long amortization periods, lack of market regulation, poor governance, subsidized tariffs and prices that often do not cover generating costs are some of risk factors. The adoption of energy-efficient technologies, processes and devices suffer from limited access to capital and lack of sufficiently attractive loan conditions that would offer competitive returns as well as uncertain returns.[4] Funding to overcome the investment barrier and to provide access to capital, therefore, is a make or break factor for sustainable energy development in these countries.

Any instrument that makes climate change benefits financially 'visible' to investors or consumers would reduce overall economic and financial risks. Absent such instruments, mobilizing multi-lateral and bilateral financing institutions is vital for ensuring a sustainable development of clean energy markets, especially efficiency and renewables (N'Guessan 2012).

The following list summarizes the principal funding sources available for investments in renewable energy, energy efficiency and climate change mitigation which, except for carbon financing, are standard financing routes for any investment in developing countries:

Table 4.2 Energy and climate rated funding by the World Bank Group

	2000	*2001*	*2002*	*2003*	*2004*	*2005*	*2006*	*2007*	*2008*	*2009*	*2010*	*2011*
Energy efficiency	295	193	67	177	92	217	761	753	1521	1685	1802	1551
Renewable energy	765	26	350	229	221	784	594	840	1471	1678	1905	2977
Thermal generation				599	272	100	511	364	1087	987	4287	290
Other energy				816	370	278	248	717	1015	1702	2019	1783
Transmission & distribution				216	248	906	1465	458	1.650	1204	2208	1397
Upstream oil, gas, coal				333	496	578	1074	729	972	1076	725	182
Total energy financing	2756	2776	2914	2370	1699	2863	4653	3862	7670	8332	12,947	8181
Total low carbon				406	350	1237	1660	1761	3338	3363	5584	5937
Total access				794	537	986	1018	905	1784	2201	1020	1031

Source: Computed by author based on Tirpak and Adams (2008) and WBG Energy Portfolio Data

Note: WBG includes the International Bank for Reconstruction and Development (IBRD), the International Development Association (IDA), the International Finance Corporation (IFC), the Multilateral Investment Guarantee Agency (MIGA) and the International Centre for Settlement of Investment Disputes (ICSID). These institutions include the following energy and climate-related funds: Global Environment Facility (GEF), Programme for Scaling Up Renewable Energy in Low Income Countries (SREP), Clean Technology Fund (CTF), the Strategic Climate Fund (SCF), Climate Investment Funds (CIF), the Special Climate Change Fund (SCCF) and the Least Developed Countries Fund (LDCF)

Table 4.3 Funding for energy efficiency and renewables by the Asian Development Bank, in millions US$

	2002	*2003*	*2004*	*2005*	*2006*	*2007*	*2008*	*2009*	*2010*	*2011*
Energy efficiency	0.0	0.0	0.0	0.0	120.0	0.0	0.0	112.9	400.1	500.0
Renewable energy	297.3	0.0	0.0	0.0	161.6	41.9	426.5	216.6	265.2	214.6
Thermal generation and upstream fossil	0.0	0.0	0.0	0.0	0.0	27.9	0.0	902.9	135.0	309.9
Other energy	0.0	0.0	0.0	0.0	0.0	726.8	0.0	0.0	150.0	42.9
Transmission and distribution	124.0	0.0	400.0	386.6	210.5	310.0	860.5	605.4	651.3	1.827.5

Source: ADB (2012)

- country government sources;
- domestic private-sector sources (equity and quasi-equity);
- conventional lending instruments (commercial banks, insurance companies, pension funds, etc.);
- equity and quasi-equity;
- ODA and OOF;
- multi-lateral institutions such as the WBG, various regional development banks and special funds such as the Global Energy Fund (GEF) and United Nations organizations;
- FDI;
- carbon financing; and
- new and upcoming innovative financing.

The financing of development and climate change agendas is intimately linked (as are climate change mitigation and renewables/efficiency). Development and climate change mitigation (and adaptation, of course) are the two sides of the same coin. Projected climate change appears likely to exacerbate poverty and undermine development, especially in least developed countries. Development can increase or reduce emissions and thus increase or reduce vulnerability to climate change (Rogner et al. 2007). Moreover, climate mitigation can also provide co-benefits, such as improved health outcomes, energy or food security. Mainstreaming climate change mitigation is therefore an integral part of (sustainable) development or green growth.

4.3.1 Official Development Assistance

For effective climate mitigation, the allocation of ODA, therefore, must necessarily accommodate and anticipate the effects of climate change—from traditional development activities that promote economic development and welfare of developing countries to a more proactive and impact-focused structure that fosters green growth, that is, development and climate protection.

Indeed, ODA gained a new focus when UNFCCC was established. After a slow start, ODA support for climate change mitigation picked up momentum after COP-13 (see Sect. 4.3.3) held in 2007 in Bali and then more than tripled between 2007 and 2010. OECD-DAC estimates that bilateral ODA for mitigation-related activities averaged US$9.7 billion (about 7.7% of DAC's total bilateral ODA) per year in 2008–10. These figures include contributions to specific climate funds, such as the Climate Investment Funds. In addition, DAC members' core contributions to multi-lateral organizations contain a climate-related share which in 2010 was US$718 million (OECD 2011).

Three bilateral financial institutions—Agence Française de Développment (AFD), Japan's JICA, Germany's Kreditanstalt fur Wiederaufbau (KfW)—and the European Investment Bank (EIB) together provided US$13 billion concessional and non-concessional assistance for climate action in developing countries in 2009, with over two thirds going towards mitigation. Some 85% of the concessional assistance is disbursed through ODA channels (Atteridge et al. 2009; UNEP 2010).

4.3.2 Multi-Lateral Agencies

Multi-lateral and national development banks continued to be important contributors to renewable energy asset finance in 2011. Provisional data collected by Bloomberg New Energy Finance from projects on its database suggest that these institutions provided US$17 billion of finance for renewable energy in 2011 (UNEP-FI 2012).

The central purpose of multi-lateral development banks and similar agencies is to promote economic and social progress in developing countries by helping to raise productivity so that their people may live a better and fuller life. This includes promoting energy-efficient and renewable energy for both access to affordable energy services to enable socioeconomic development as well as environmental and climate protection. MDBs provide a wide array of support and services to developing countries, ranging from concessional and non-concessional loans to human resource development and capacity building.

MDBs help developing countries take maximum advantage of existing instruments, such as the Global Environment Facility (GEF), various carbon funds associated with the CDM and JI, the Carbon Partnership Facility, the Climate Investment Funds (CIF) including the Clean Technology Fund (CTF) and the Scaling Up Renewable Energy Programme for Low Income Countries (SREP).

These institutions assist in the finance of demonstration, deployment and transfer of low-carbon technologies with a significant potential for long-term avoidance of GHG emissions consistent with the development objectives of the recipient countries. For example, the SREP supports new renewable energy technologies including solar, wind, bioenergy and geothermal, as well as hydropower with capacities normally not exceeding 10 MW per facility.[5] Complementary technical assistance is considered essential for transformative and enduring change and country engagement and ownership. SREP, therefore, supports planning and pre-investment studies, policy development, legal and regulatory reform, business development and capacity building (including for knowledge management and monitoring and evaluation) (CIF 2012).

At the lower end of financial assistance is the Small Grants Programme (SGP) of UNDP/GEF. The SGP grants range from US$7000 to US$20,000 and are provided directly to community-based organizations and NGOs for the implementation of sustainable development programmes at the grass roots level. Since its inception in 1992, the SGP's strategy has evolved to deliver support to low carbon and climate resilient technologies in more than 130 countries. Typically, SGP projects involve community participation and align with national sustainable development priorities.

A new institution established in 2011 is the International Development Finance Club (IDFC), a network of 19 international and national development banks from both developed and developing countries.[6] IDFC supports a wide range of development projects that contribute to poverty alleviation, sustainable development and green growth (Höhne et al. 2012). Climate finance and access to clean sustainable energy service are central IDFC objectives. In 2011, a total of US$74 billion was disbursed in support of green energy development and GHG mitigation (Höhne et al. 2012).

Services provided by MDBs that are equally important as actual loans include packaging of several financial instruments and risk management (interest rate, exchange rate, price and market risks) as well as capacity building in these areas.

4.3.3 Climate Finance and UNFCCC

Developed countries were called to provide new and additional financial resources to mitigate climate change. When the Kyoto Protocol was adopted in 1997, the clean development mechanism (CDM) was introduced, allowing developed countries to invest in climate mitigation and take credits. The MDGs established by the UN in 2000 included mitigation of climate change as part of the development goals. Negotiations in subsequent annual Conference of Parties (COP) of UNFCCC further refined the means, mechanism and conditions for the transfer of funds for the purpose of mitigation and adaptation of climate change. COP-13 at Bali called on developed countries to provide 'financial resources to support action on mitigation' and for developing countries to receive assistance in 'technology, financing and capacity building'. COP-15 at Copenhagen further agreed to 'scale up, new and additional, predictable and adequate funding as well as improved access' (UNFCCC 2009). Close to US$30 billion were committed by the developed countries for the period 2010–12. They also committed to a goal of providing US$100 billion per year by 2020 to address the needs of developing countries. To put this commitment into perspective: during 2006–10, ODA disbursed as aid for energy was, on average, US$3.7 billion per year.

4.3.4 Micro-Finance

Micro-finance and micro-credits are particularly effective mechanisms for deploying renewable energy technology to off-grid poor communities. Typically, these communities cannot afford to switch to modern renewable energy, and the cost of connecting them to the main grid is too expensive. Most of the people also have no access to banking services (bank accounts) or are not 'bankable'.

Micro-finance is the provision of financial services such as loans, savings or insurance to poor people and small business which are not of interest to standard financial institutions. For small business and micro-entrepreneurs, micro-finance is a means to access loans without the high transaction costs and red tape of interacting with traditional banks. Generally, micro-finance does not require collaterals or equity. It is based on relationship building and partnerships using local social networks between civil society, local banks and intermediaries, often NGOs or community cooperatives. The former provide low interest loans or grants to the intermediaries, the latter then make and administer loans to poor families and small businesses. Most micro-finance schemes are revolving funds, that is, loans that are repaid over relatively short timelines and then recycled as new loans, 'keeping the money working and in the hands of borrowers' (Grameen Foundation 2012).

Payback calculations including interests for home photovoltaic (PV) systems are often determined by 'replacement expenditure', in other words, the amount that a family would have spent on kerosene for lighting. Loan amortization collection is usually carried out by the intermediaries or community cooperatives. Micro-finance programmes are funded by loans, grants, guarantees and investments from individuals, philanthropists, social investors, local banks, foundations, governments and international financial institutions.

4.3.5 *Essential Instruments*

4.3.5.1 Market Incentives

A power purchase agreement (PPA) or feed-in-tariff (FIT) can be a vitally important component of a renewable electricity project for recipient/utility and seller/sponsor alike. In essence, PPAs reduce the revenue risk emanating from the uncertainty of future sales volumes and sales prices, especially in deregulated markets. For the utility selling electricity, PPAs/FITs provide the long-term revenue stream needed to secure financing for the project. In the case of a project sponsor not having the up-front capital required to invest in the plant, a PPA/FIT can serve as collateral for a bank loan. For the project sponsor providing project finance, PPAs reduce the risk of financial default of the utility or loan recipient. For the utility customer, PPAs/FITS ensure for the customer the availability of renewables generation applicable towards meeting its renewable portfolio standard (RPS) or climate mitigation commitments. As policies continue advancing renewable energy development, knowing and understanding the intricacies of PPAs is becoming increasingly important.

PPAs/FITs create long-term relationships between parties. Key elements of PPAs/FITs are the terms and conditions regarding price of the electricity, penalty for non-compliance, duration of the agreement which can be up to 20 years and more, periodical PPA/FIT revisions in the light of possible market changes, termination clauses, reporting and verification requirements and so on. In short, PPAs govern the transaction for years into the future. These terms are usually heavily negotiated prior to the signing of PPAs and are meant to fairly allocate risks between participating parties. As the renewable energy community continues to grow, these negotiations are becoming more and more sophisticated.

4.3.5.2 Regulation

However, in the presence of regulatory uncertainty, for example, before or during a shift from regulated to deregulated market structures, large customers or distributors may be reluctant to commit themselves to long-term PPAs/FITs (WB/IMF 2006). Regulatory certainty and predictability are essential for PPAs and finance of energy projects.

MDB support programmes can serve as leverages to encourage private-sector participation both for funding and capacity building. The involvement of MDBs in partner countries can lead to changes in the regulatory environment, resulting in lower market risks and increased competition. Technical risks of projects can be shared and thus reduced via support for pilot projects (WB/IMF 2007).

The adoption of renewable technology and energy efficiency measures has often been linked to technology 'appropriateness', which is generally defined as 'the degree to which a technology fits its specific context of use; to be relatively low cost, locally made and serviced, and well suited to their cultural, material, and ecological contexts' (Nieusma and Riley 2010: 5).

It has been recognized that technology transfer and implementation on the ground alone are usually not sufficient to achieve their goals. For example, while funding access to electricity by itself is desirable, failing to attend to a broader set of factors could limit access to a short-lived affair. Key factors include careful attention to community capacity building: educating electricity consumers on what types of appliances would not be allowed, training plant operators on operations and maintenance, creating organizational procedures for troubleshooting and conflict resolution in advance of system breakdown but also the need for rate payment schemes for the electricity used by consumers. Regardless of social equity and other development goals, these factors are required simply to ensure that the technology remains functional and economically viable over time.

Nieusma and Riley (2010) argue that a combination of renewably sourced electrification, with productive income-generation activities based on the now available access to electricity, improves the acceptance of, and care for, the technology. Here development assistance, for instance by development-oriented NGOs, that incorporates a range of social, organizational and economic goals with electrification has proven successful, especially as such an integrated approach greatly facilitates payment for the electricity used and thus the payback of loans for the project. Yet technical functionality trumped all other aspects. Without it, the income generation and other benefits would be void—precisely what the production of electricity is designed to leverage. Successful renewable technology implementation requires a 'package approach', including support for training and maintenance, project finance, payment collection, development of market incentives as well as opportunities for income generation (Barton 2007).

Capacity building for effective utilization of development aid for clean energy finance and climate change mitigation must also focus on a technology's intangible knowhow and services (Brewer 2008).

Key findings of a case study on six African countries indicate that the effectiveness of foreign aid and carbon finance is hampered by a lack of domestic leadership, with the adverse effect that the national responses are driven by international and donors' priorities and are not necessarily linked to national priorities (Thornton et al. 2011). Apropos access to finance, studies observed that recipients had to conform to donor or funders' requirements and schedules (e.g. budget cycles) rather than funders conforming to recipients' needs and requirements. Policy capacity at the national and local levels of how to define climate change mitigation and climate finance is often lacking (especially the concept of additionality, which is not even harmonized among different funders (Thornton et al. 2011)).

While the Paris Declaration on Aid Effectiveness of 2005 requires donors and recipients to jointly manage the implementation of mitigation programmes, lack of coordination between funders and the recipient government appears to be another area for improvement of the effectiveness of green energy funding.

4.3.5.3 Business Models

Merely supplying renewable energy systems to a few tens of homes in a given village is unlikely to gain climate change mitigation or sustainable development benefits. What is needed are business models that:

- Establish a mechanism that would pave the way for the commercialization of rural household electrification in the developing world. Numerous projects have demonstrated the willingness of rural households to pay for, say, solar home systems as long as they are given access to credit;
- In the case of village-wide energy systems, develop commercially viable distribution and service chains for these renewable energy services using local entrepreneurs and multi-sector collaboration;
- Further integrate technology transfer and access to credit with other poverty alleviating (income generating), health and education improving measures (see Nieusma and Riley 2010). Renewable energy programmes such as the 'Whole Village Development Model' (SELF 2013) or the 'Base of Pyramid (BoP) Model',[7] (Prahalad

> 2004) have been adopted by numerous communities in different countries and cultural settings, in large part because of the real and measurable indirect (non-energy) benefits deriving from the implementation of the business model.

Successful implementation of a business model often depends on the products and services matching local needs and expectations (e.g. charging possibilities for mobile phones in addition to solar lights).

4.4 Evaluation of Effectiveness

The literature on the effectiveness of foreign aid is vast. It focuses primarily on the impact that foreign aid has had in reducing poverty and inequality, advancing economic growth, building capacity and accelerating achievement of the MDGs. Indicators used by the World Bank cover aid received as well as progress in reducing poverty and improving education, health and other measures of human welfare (WB 2012a). As regards effectiveness of foreign aid, especially ODA, there is hardly an aspect of it that has not yet been criticized (de Coninck et al. 2010). Symbols of failed ODA range from 'white elephants' in the form of developing country leaders' Swiss bank accounts to calls to abolish aid altogether based on the rationale that dependence on international aid undermines democracy and governmental accountability (Moyo 2009). Easterly (2005) argues that IMF's structural adjustment loans (SAL) of the 1980s did not generate the intended outcome of improved per capita growth as a function of increased SAL lending.[8]

Undeniably there are instances of aid failing to work. Yet, the accumulation of empirical evidence shows that aid has had broadly positive effects on socioeconomic growth and development (Addison et al. 2011). Arndt et al. (2009) conclude that, on balance, foreign aid's impact 'is positive and conforms to priors from modern growth theory'. World Bank indicators show that human welfare improved through the positive aid impact on HIV, primary education and child mortality (WB 2012a).

The quantified and time-bound MDGs agreed in 2000 made monitoring the effectiveness of aid somewhat more transparent. Annual progress reports on the MDGs assess to what extent goals are met. Although progress has been made since 2000, it is not evenly distributed across the developing world. It seems likely that many targets will be missed in most regions (UNDESA 2012).

Failure of many development aid schemes result from a disregard for the complexity of institutions and incentive systems. MDBs often impose conditions and restrictive policies on recipient countries that disregard engagement with the way the recipients see their problems and which stifle individual country actions (Sen 2006). Bilateral aid driven by geopolitical factors does not have an effect on growth, as it primarily serves donors' global geopolitical interests (Headey 2007). Failure also results from policy advice given by donor country experts in the absence of a proven track record of economic development policies ('no one-size-fits-all'). The Barcelona Development Agenda (2004) concluded that

> there is no single set of policies that can be guaranteed to ignite sustained growth. Nations that have succeeded at this tremendously important task have faced different sets of obstacles and have adopted varying policies regarding regulation, export and industrial promotion, and technological innovation and knowledge acquisition.

Easterly (2012) therefore suggests freeing development aid from the delusion that it can accomplish development and focusing it on financing particular tasks and projects for infrastructure development and agriculture, education and health—in essence, practical steps towards meeting the MDG and SE4All targets. Dedicated funding of energy efficiency programmes and the spread of renewable energy technologies would perfectly fit such a template.

4.4.1 Measuring Effectiveness

The effectiveness of foreign aid for sustainable energy may be assessed along several dimensions: (1) increased *access* to clean energy services, (2) improved *affordability* of energy services, (3) *reduced environmental* impacts (local and climate mitigation) and (4) rates of *efficiency* improvements throughout the energy system (including reducing energy intensities within the economic production process) and deployment and market penetration of renewable technologies. All four dimensions are interrelated, with numerous trade-offs between them.

Access is about bringing modern forms of energy to households, community institutions (schools, health clinics, etc.) and productive applications including micro-enterprises and agriculture (e.g. irrigation and food processing). It is also to do with connecting households in underprivileged parts of large metropolitan areas (the urban poor). It is usually considered synonymous with access to electricity—both grid and off-grid—and clean cooking fuels. Access also means adequate supply, reliability and quality, especially during periods of highest demand, and, in the case of off-grid intermittent sources, adequate back-up and/or storage capacity.[9] At the national level, energy security is yet another aspect of access.

Affordability of modern energy services, that is, the upfront costs for the purchase of electric devices or modern cooking stoves, hook-up costs as well as electricity and fuel charges, are often constraints to access even in the presence of a supply infrastructure. Energy charges and their collection, however, must fully cover supply costs in order to be consistent with the concept of sustainable energy.

Technological deployment of renewable energy technologies (as well as energy-efficient infrastructures and end-use devices) is a key enabler of sustainable energy, while the efficient generation and use of energy services is a central pillar of sustainable energy. Doing more with less generally reduces pollutant emissions and waste, enhances energy security and can lower the costs of energy services, hence improving affordability.

Environmental impacts span a wide range of adverse consequences for human health, air and water quality and agricultural productivity due to anthropogenic interference with the climate system. Pollution and waste from energy resource extraction and energy conversion increasingly stress and overburden the carrying capacities of ecosystems. Viewing these environmental impacts in monetary terms reveals how they progressively undermine the economic gains from the production and use of energy services.

On the supply side, the potential of renewable technologies that tap the abundant energy flows provided by nature to supply an increasing share of the world's energy demand, while delivering substantial sustainability benefits, has long been acknowledged (IPCC 2011). On the demand side, energy efficiency improvements have been confirmed as the low hanging fruit for any pathway to sustainable energy (GEA 2012). Jointly, accessibility and efficiency can curb GHG emissions, cut emissions that cause poor air quality and regional water acidification, enhance energy security, improve the balance of payments through lower energy import bills and safeguard against the price volatility of international fossil fuel markets.

Unlike in the OECD, where, over the last decade, renewably generated electricity has grown at more than twice the rate of total generation, in the developing countries, the market penetration of renewables has barely kept pace with total supply, despite all these benefits.[10] One reason for this is that renewables have long been the stepchild of foreign aid, even though substantial development assistance for energy supplies has been provided by OECD donor countries and multi-lateral development banks (MDBs). Lack of financial support may have been one reason. Another reason could also be that the effectiveness of the foreign aid expended for renewable energy over the last two decades has been wanting (1.4% of ODA for renewable energy in 2010).

This chapter explores the effect of foreign aid for sustainable energy development with a view as to 'what works', 'what could work', 'what is scalable' and 'what is transferable' on the effectiveness of foreign aid in boosting sustainable energy in developing countries.

The effectiveness of foreign aid has been a highly controversial topic, especially with regard to what works, what could work and what does not work. On balance, the empirical evidence on its effectiveness is discouraging (Djankov et al. 2006; Easterly 2006; Moyo 2009; Doucouliagos and Paldam 2009). Given that the purpose of foreign aid of promoting development and alleviating poverty leaves ample room for a vast array of areas for aid application—ranging from economic reform, building institutions and good governance to building roads, pipelines lines and power plants—measuring its effectiveness is not a straightforward affair. It is certainly much easier to measure success related to 'hard' infrastructure projects than to, say, 'soft' institutional reforms. However, having successfully implemented a solar PV system in a rural village, for instance, which previously had no access to electricity, is by itself insufficient as a measure of effectiveness of foreign aid on sustainable energy. Important aspects to consider are the project's effect on the betterment of the social, institutional, economic and/or environmental conditions within the village.

More specifically related to aid effectiveness on climate mitigation, based on an analysis of aggregated data from 80 low- and middle-income countries between 1973 and 2005, Kretschmer et al. (2011: 86) conclude that 'Aid tends to be effective in reducing the energy intensity of GDP in recipient countries.… the carbon intensity of energy use is hardly affected. Scaling up aid efforts would thus be insufficient to fight climate change beyond improving energy efficiency'.

In 2002, the Monterrey consensus was forged when donors recognized that their fragmented efforts added undue cost and reduced their effectiveness. In 2003, OECD convened a high-level forum on harmonization in Rome, during which donor countries agreed to better coordination with the recipient countries and to work towards a more acceptable working relationship. Subsequently, in 2005, the Paris Declaration on Aid Effectiveness was endorsed by 137 countries, all the major multi-lateral funding agencies and several large NGOs. The Paris Declaration established five principles for effective aid (OECD 2008b):

1. *Ownership*: Developing countries set their own strategies for poverty reduction, improve their institutions and tackle corruption;
2. *Alignment*: Donor countries align behind these objectives and use local systems;
3. *Harmonization*: Donor countries coordinate, simplify procedures and share information to avoid duplication;
4. *Results*: Developing countries and donors shift focus to development results and results get measured; and
5. *Mutual accountability*: Donors and partners are accountable for development results.

Subsequent to the Paris Declaration, two high-level fora have taken place at three-year intervals (2008 in Accra, Ghana and 2011 in Busan, Republic of Korea) to review and assess the progress of implementation. While the Paris Declaration is slowly being considered in the actual process of planning, execution and evaluation of foreign aid projects, these five principles provide a benchmark for an assessment of the flow of funds to developing countries over the years.

4.4.2 Review of Effectiveness

Numerous renewables projects funded by foreign aid starting twenty years ago were reviewed to see where they stand now in 2013. Most of the early renewable energy projects served the purpose of providing desperately needed access to electricity and other modern energy services; projects and were rarely motivated by sustainable energy or climate mitigation objectives. This approach changed fundamentally with the coming into force of the CDM in 2005, and climate change benefits are now an integral part of a project's rationale.

Successful implementation of renewable energy technologies in ODA-recipient countries has been used in this chapter as a proxy for action that fosters sustainable energy and generates immediate climate change benefits. Until the enforcement of the Kyoto Protocol in 2005, ODA support for renewables has been marginal compared with ODA disbursement for fossil resource development and combustion technologies. Furthermore, the modest investments in renewables were predominantly driven by the ultimate ODA paradigm of advancing economic development rather than environmental protection and mitigating climate change, albeit that these co-benefits are now duly recognized.

The literature on specific investment projects that received development assistance from DAC member countries, MDBs or any of the other climate finance channels is huge. The majority describe technical aspects of the projects. Unlike studies and reviews of policy relevance and effectiveness, only a small portion of the technical studies goes further and reviews project effectiveness and lessons learned from successful or less successful projects. A notable exception is the project evaluations carried out by MDBs, international and national funding institutions. These evaluations examine the outcome of projects guided by a uniform set of performance indicators.

The following sections summarize our literature review of the effectiveness of foreign aid (in essence, ODA) for climate mitigation using the four categories (1) what worked, (2) what could work, (3) what is transferrable and (4) what is scalable. A selection of projects reviewed is presented in the Appendix to this chapter.

4.4.2.1 What Worked

The following list of factors translates insights, prerequisites and lessons learned from specific local project evaluations into more generic terms of what worked in advancing the deployment of sustainable energy and efficiency measures supported by foreign aid:

- Consistent government policy and strong commitment at all levels—national, regional and local—play a key role for the adoption and use of renewable energy technologies. Implementation of renewable projects predominantly occurs at the local level. National legislation, for example renewable energy targets, and support policies provide the framework conditions for local government action. Local governments assume multiple roles from planning authorities, decision-makers, managing of local infrastructures, service providers and prime interlocutors for their citizens and businesses. Local government or community championing of renewable energy development is an attribute of successful project implementation.

- Regulatory frameworks and policies that incentivize private investment and guide structuring of the financial and ownership arrangements for large renewable projects are indispensable.
- Of equal importance is the presence of enabling laws and national policies such as renewable energy acts, and energy market regulation with provisions and incentives for private sector involvement. This includes, but is not limited to, policies such as those that guide the purchase, grid connection and transmission of electricity generated from renewable energy sources and incentives ranging from feed-in tariffs, power purchase agreements, direct and indirect capital subsidies, to accelerated depreciation or exemption from various taxes and duties.
- Full alignment of the project with local, regional and national development plans is one of the principal components of successful renewable technology deployment.
- Government guaranties for ODA loans are an important sign of government support and concurrence with a renewable project.
- The presence of excellent renewable resource flows alone is not sufficient to attract funding for renewable energy projects. In the past, only countries with an adequate enabling environment and long-term stable comprehensive public policy and strong political commitment have succeeded in developing and maintaining renewable energy supplies. Therefore, grants or funding assistance should be first allocated to creating an enabling environment for renewable projects and to removing barriers. This includes appropriate legislation and regulation, raising awareness and disseminating information, capacity building and so on, as well as the development of renewable energy resource assessments (wind, solar or biomass maps), technology development, grid issues, economic and financial analysis, localization and industrial development opportunities, education and awareness-raising. Knowhow transfer from more advanced countries as an integral part of development assistance is often at the root of successful programme implementation. It reinforces both lender confidence and governments' resolve for further development of renewable energy programmes.
- A legitimate public authority that sets rules and obligations and enforces them in a transparent and equitable manner instils credibility and confidence in investors. Because of their intermittent availability, policies on preferential grid access are particularly important

for wind or solar farms. The minimum requirement is a fair and open grid access.

- Policies giving grid access to renewable energy are the most critical. Feed-in laws—the main instrument used in Europe to promote wind energy—have the advantage of giving developers long-term stability and predictability. In many countries, other forms of public support, such as tax credits, soft and concessional loans, an increase in electricity tariffs or portfolio standards (quotas), are used along with feed-in tariffs.
- Extensive pre-feasibility studies and comprehensive energy planning, taking into account local circumstances plus environmental impact assessments with involvement of stakeholders, especially the local communities directly affected by a project, add credibility and raise funding agency confidence in project viability. Comprehensive energy plans for larger, beyond community-level projects help communicate the rationale of the project and its relative merits against alternative options.
- Community ownership of micro-energy projects through in-kind (labour) or financial contributions to its establishment helps community identification with the project. Team building of the community with consultants, timely availability of funds and compensation to farmers and households affected by rezoning or land acquisition are other elements contributing to the successful implementation and operation of a project.
- Micro-finance schemes with payback arrangements based on avoided costs of the recipient household or business, and arranged and administered by local NGOs or user organizations make access to modern energy services affordable. Micro-finance and foreign aid in the form of SGPs pair up well.
- Capacity building to elevate cooperatives and intermediaries to competent financial and business agencies is essential for sustained success of micro-finance.
- One of the success factors for micro-financing is when women actively involved in the livelihood of the family are incentivized to subscribe to the renewable energy (through means such as biogas cook stoves, solar home systems, micro-hydro or micro-wind turbines) which in turn provides opportunities for income-generating activities for women.
- Availability of rural electricity is a necessary but not sufficient condition for the poor to improve their economic status. They need access

to income-generating activities, seed money for small and medium enterprises and markets for their products, which in turn sustain the operation of the renewable energy technology via revenues from energy sales.

- Providing electricity is not an end in itself. Therefore, as important as income generation is a tripartite package approach of renewable technology, storage dealing with intermittency and end-use equipment (lamps, radios, TV, water pumps, refrigeration, etc.)
- Privatized operation and maintenance with user tariffs that cover costs appear to be a critical element of success. Many failures can be attributed to the absence of user tariffs or insufficient collection of revenues. The business model of community ownership with in-kind contribution from the community is further enhanced when local capacity is developed to manage, operate and maintain the energy system. Such cash-generating activity has functioned best via private-sector participation within the local community. Capacity building of community members in the operation and maintenance of the equipment or micro-system is essential, as is basic business education for small enterprises providing maintenance services and revenue collection. Reliable maintenance service ensures acceptability.
- Good plant operation management and routine maintenance schemes need competent operators, skilled technicians, incidentals and spare parts. The lack of any of these elements can seriously jeopardize an otherwise successful project.
- Supplemental activities, such as training for para-technicians, marketing and knowledge sharing in the productive use of energy, are important for sustainability of micro-finance and use of renewable technologies that maximize benefits.
- Integration of larger renewable energy projects, for instance hydropower, with community infrastructure development, such as local road improvement or improved health and education services, enhances overall acceptance of a project otherwise viewed as 'disruptive' by the local community.
- The simultaneous development of domestic manufacturing capability (supply push) and a domestic market (demand pull) has been extremely successful.
- Distribution and allocation of project risks between different entities or partners to parties best suited to manage such risks is important. For example, there are good track records of public–private partner-

ships (PPP), where governments retain significant responsibilities over certain portions of an asset and entrust private parties with their operation and management.
- Making the climate mitigation benefits financially visible to investors, for example earning income through the sales of emission credits, reduces economic risks and makes attracting funding easier.

Effective use of foreign aid for sustainable energy and climate mitigation is not the result of any individual factor alone but rather of a well-coordinated package of features tailored to the local conditions in the targeted area.

4.4.2.2 What Could Work

The phrase 'what could work?' necessarily includes a fair degree of speculation and uncertainty. It intrinsically suggests that correcting all the things that caused failure in past projects could probably make them work in the future. In any case, knowledge management and information exchange on 'what worked' and lessons learned are prerequisites for 'what could work'—at the minimum, it prevents reinventing the wheel. Foreign aid is an essential ingredient to most, if not all, aspects of 'what could work'.

Mechanisms for the dissemination of 'what worked', lessons learned in specific cases and the application of design practices to inform effective policy designs and implementation strategies in other jurisdictions do not always exist—but a comprehensive data and information system could work in accelerating the deployment of sustainable energy. Sustainable energy and energy efficiency are capital intensive—for local manufacturers of key components, such as PV modules, and for consumers who purchase PV systems. Access to finance, especially longer-term loans and financing vehicles that reach the target consumers, is at the root of 'what could work'.

Strengthening domestic capital markets, providing basic insurance services and risk management instruments would improve access to capital and lower financing costs. For example, an energy efficiency project in China fell by the wayside because of a lack of collaterals for the lending. In India, the ADB covers up to 50% of the payment default risk on commercial bank loans of up to 15 years to private-sector developers of small solar power projects (WB/IMF 2011). From a policy perspective, MDBs can encourage developing countries' institutions in mainstreaming clean energy options into national development plans and investment decisions. Foreign aid for partial risk guarantees, currency hedging or commodity and interest rate risk management lowers borrowing costs and thus the barrier for the investment in renewable energy.

The expected rate of return (ERR) required by public or private investors often cannot be met. Financial and regulatory incentives, as well as concessional financing schemes, can significantly lower the investment and ERR barriers. Here simple tax reductions and exemptions tend to have the lowest impact (not considering economic opportunity cost). By contrast, concessional financing schemes tend to have the highest impact and are likely to be the most cost-effective incentives in terms of their overall impact on generating costs and competitiveness (Kulichenko and Wirth 2011). Cost reductions could also result from scaling up of the level of implementation. But more importantly, the countries involved would benefit from the project economically through a high localization factor of technology components and services. The latter could be accelerated by concessional loans and grants.

Acceptance beyond the demonstration period is enhanced when the households can make income-generating uses of the energy. Helping the local community to develop and implement business plans improves the likelihood of collecting revenues that pay for the energy services. Programmes and grants for community development could be instrumental in developing and maintaining the consumer base for low-carbon energy. This could be accomplished if the financial institutions (and foreign donors) willing to finance solar home systems, micro-hydro or mini-grid solutions, were also funding associated end-use investments. Support of businesses that utilize the electricity would help create a profitable demand load.

Introduction and development of micro-finance programmes by local institutions without requiring conventional banking and consumer credit would increase the potential customer base. Repayment of loans, however, is intimately tied to new income-generating activities.

Typical financing packages offered by commercial banks as well as development financing institutions require feasibility studies, loan collaterals and equity from borrowers, all of which come at a cost. Measures to reduce these costs would encourage a wider participation of commercial banks, lower interest rates and, consequently, a final price that is affordable to poor consumers. An affordable price (calculated as upfront cost or instalment payments) would ideally be similar to the cost that the consumer would bear for kerosene or diesel. Creative ways of collecting payment is also important for the sustainability of the scheme.

Employment opportunities (another form of income-generating activity) in the renewable energy business as technicians, plant operators, bookkeepers, fee collectors and managers also help sustain the consumer community's interest in the utility. Therefore, education and training in all aspects of renewable energy systems could make a renewable technology a sustainable community asset.

A broader 'programmatic' approach is more effective than a narrowly defined 'project' design for the deployment of sustainable energy. Foreign aid could consider a multi-disciplined approach—that is, in addition to funding the design, building and operation of renewable energy systems, ODA could integrate community development and introduction of microfinancing for business as part of the programme. These aspects of development would create the 'pull' effect for the success and effectiveness of low-carbon technologies.

Gender is usually not included in foreign aid projects for sustainable energy. However, in recent years, there has been an increase in research on gender issues with foreign aid support. Gender is slowly being recognized as an issue in energy delivery. Women are actively involved in the livelihood of the family and, therefore, more open to embracing new concepts or new technologies that hold promise of improving their living standards. An accelerated involvement of women is likely to have a positive impact of the deployment of sustainable energy. Mainstreaming gender issues in project design and implementation could create additional impetus to sustainable energy.

Succession planning is essential for a seamless continuation of activities and services provided by aid-supported individuals and institutions or directly by the sponsor organization once the funding period ends.

Maintenance service businesses are often reluctant to service large but low population density areas. Here additional economic incentives could improve the 'serviceability' of such areas and thus the deployment of sustainable energy technologies.

Clear and mutually agreed objectives (outcomes) of a low-carbon energy project increase the probability of success. Strengthening long-term self-reliance—via revenue generation and reinvestment—improves the probability of success but requires that revenue covers costs. It is easier to make a profitable energy system socially beneficial than to make a socially beneficial plant profitable. Low-carbon energy should be promoted for its role in securing livelihoods or enabling small enterprises, rather than as an 'energy programme' or environment protection programme (Khennas and Barnett 2000).

Many renewable energy projects will generally be able to comply with CDM eligibility rules. Capacity building in utilizing CDM for finance could enlarge the financial pool for renewable energy and efficiency programmes. CDM finance is not necessarily sufficient. If combined with other policies such as FITs, it would generate additional revenue and remove non-economic barriers. FITs and CDMs pair up quite well for reducing investment risks and providing stable revenue. The engagement of multiple suppliers for similar technologies or micro-systems forfeits potential cost reductions of 40–50% associated with bulk purchases (Mahama 2012).

Demonstration projects should hold the potential for scaling up to commercially sized and economically viable sustainable energy projects.

Martinot et al. (2000) made an assessment of GEF's loan portfolio for PV projects based on the project cycle of the World Bank solar programme. The WBG (2007) also published a report on the lessons learned from WBG's funding experience of PV projects, valued more than US$600 million, in 30 countries for a total of 62 MW capacity. These two publications offer the following recommendations on 'what could work' pending further investigation, with emphasis on:

- affordability through fee-for-service and consumer credit;
- use of GEF resources for non-recurring costs related to business and market development;
- access to finance and incremental risk-sharing;
- explicit linkages to rural electrification policies and planning;
- commercially feasible business models that are sustainable and can be replicated;
- project finance design must be flexible apropos the local situation;
- support must be for the technology of choice as consumers choose among several options;
- private equity is not always the best finance as the returns are less than what could be obtained in other ventures;
- good government relations and support are necessary;
- quality of product must not be comprised; and
- other financing vehicles must be accessible (explore all options).

4.4.2.3 What Is Transferrable

Transferability of 'what worked' and 'what could work' depends on numerous factors ranging from geographical, weather, climate and environmental conditions to energy resource endowment, stage of economic infrastructure development, governmental structures, social and cultural realities (this an indicative and not comprehensive list). Although one size does not fit all, many of the features listed under 'what worked' above are transferrable if adjusted to account for the different circumstances present in the target area or country.

For example, there is no inherent reason why the rationale for and development of appropriate macro-level framework conditions (legislation, government commitment, private sector participation, incentives, etc.) are not transferrable to other jurisdictions. Likewise, 'participatory approaches' to create, nurture and capacitate communities to build, own and operate micro-energy systems are essentially location independent. Examples of successful south–south cooperation are often replicable.

Government participation or co-ownership enables access to international development assistance and reduces dependence on local financial institutions. Furthermore, activities such as pre-project planning, resource mapping, site and infrastructure evaluation are prerequisites for any successful project. Human resource development and capacity building are always transferable.

A liberalized electricity market and experience with independent power production help attract private-sector participation. The need for tariff structures that cover costs, especially after the end of the funding period, are universally applicable. BOT (build-operate-transfer) schemes require long-term PPAs or FITs plus a supportive and transparent market regulation.

The distribution and allocation of project risks between different entities or partners to parties best suited to manage such risks, for instance PPPs and PPAs, have been successfully applied in different jurisdictions. Clearly defined and delineated business models and financing schemes plus guaranteed market access, and not just on financing the demonstration plant, are generally location independent. Business models where small micro-enterprises are playing a central role in delivering micro-renewable energy system-based services to rural areas have already been transferred to several countries. Incentives such as FITs and associated laws have also been transferred and implemented in several different countries. Micro-credits arrangements for financing renewable energy service packages with local intermediaries and user organizations are rarely location dependent. Especially at the village level, the SGP support of user organizations or intermediaries is transferrable.

Depending on the size of the country, the simultaneous development of domestic manufacturing capability of system components or entire system (supply push) and of a domestic market (demand pull) is transferrable if the domestic market is sufficiently large or if a coordinated regional (several smaller countries) market can be accessed. This may need intergovernmental agreements on cross-border taxation and such like.

4.4.2.4 What Is Scalable

In the context of foreign aid and sustainable energy, scalability is about broadening the project planning, finance, implementation and management approach and its beneficiaries to a larger scale. It may involve changing the project design, approach or focus but it is grounded in the same fundamentals as the original project (UNDP 2006).

Scalability depends on many features and dimensions—political and regulatory, institutional, economic (market structure and size), financial and technological. In many instances, scaling up affects several dimensions simultaneously and, therefore, is a collaborative effort across the dimensions concerned, and so coordination is paramount. Scalability also depends on a society's preference and weighting on issues such as jobs, energy security and environmental protection. Scalability involving capital-intensive technologies is generally easier in an environment with a strong public-sector presence than in a setting where myopic private-sector objectives dominate.

Capacity building and human resource development are probably most suitable for expansion, and generate fast returns. Dispersed individual renewable technologies can be scaled up to energy parks/farms yielding higher returns on lower generating costs through streamlined procurement, management and maintenance. Off-grid local micro-energy systems can be expanded to larger island systems with eventual grid integration.

A more commercial version of user organizations are renewable energy service companies (RESCOs). RESCOs either own the mini-grid infrastructure or lease it from a governmental or non-governmental organization and rent out electricity-generating equipment, such as solar home systems and electric devices, at fixed monthly tariffs, or sell electricity services by the hour (IEA 2011). Like user organizations, RESCOs carry out maintenance and repair services, again utilizing local labour, and manage financial transactions, billing and revenue from user-fee collection.

The sustainability of renewable projects supported by micro-finance schemes often depends on the capability of the technology supplier and the micro-finance intermediary to reach large numbers of clients and a speedy recycling of the funds (UNDP 2012a). Micro-credits for financing renewable energy service packages with local intermediaries and community-based user organizations can be, and have been, replicated in numbers (horizontal scalability) but with growth that will eventually overlap with commercially provided financial services. Fee-for-service systems are scalable to the point of traditional utility type generation, distribution, management and administration. The horizontal scalability of SGP-funded projects has been demonstrated for more than two decades, and has spawned numerous community-based organizations and by now has developed into an overarching network committed to sustainable energy and environment protection (UNDP 2012b).

In fact, SGPs are also vertically scalable, that is, income-generating activities and new local business opportunities eventually make them financially self-sustainable without the need of foreign aid support, and subsequently integrate them in the standard economy.

MDBs and multi-lateral funds are increasingly called upon to contribute to the scalability of affordable renewable energy technologies by 'supporting regional research, the testing of new technologies in selected countries, greater deployment of new technologies through technology transfers, and regional manufacturing of packaged renewable energy products and sub-assemblies' (ADB 2009).

4.4.3 Further Considerations

4.4.3.1 Bilateral versus Multi-lateral Aid

MDBs and other international financial institutions offer several advantages compared with bilateral assistance. They provide a platform for collective action, and help to contain donor competition and minimize conflict among donors (Burall et al. 2006). Moreover, recipient governments generally have a greater say in the aid allocation process. In contrast, bilateral assistance often has a long history of engagement between donor and recipient country institutions; greater coherency with other national policies such as trade and security; and greater flexibility than multi-laterals (Urban and Wolcott 2009).

4.4.3.2 Access and Acceptance

Initiatives to increase energy supply in developing countries have not necessarily reached the poor, and initiatives designed specifically to increase energy access for the poor have not taken full advantage of clean energy technologies (OCI 2011). There are several reasons for this. Despite energy-sector reforms in many developing countries encouraging private-sector involvement, energy supply remains predominantly an affair of large state-owned utilities or energy companies. Even the private-sector entities tend to become relatively large while pursuing 'conservative management philosophies' (Mahama 2012)—incremental rather than fundamental in technology choice or business development—not necessarily what customers and markets or policymakers expect and demand. Initially, the techno-economic performance of new renewable technologies is often inferior (upfront investment costs, functionality, familiarity, convenience, etc.) to their existing fossil fuel alternatives and hence of little attraction for utilities and their affluent customer base (Christensen and Raynor 2003). In contrast, these technologies tend to appeal to those lacking access to modern energy services.

Development aid in whatever form is crucial to accelerate the market penetration of clean energy technologies. Mahama (2012) argues that policy and implementation advice for energy development, especially when linked with development aid, is provided by foreign technical experts who are often ignorant of local conditions and needs. As a consequence, development aid lacks the necessary effectiveness in providing access to clean energy services. Effective climate change mitigation requires concerted action over the long term by many partners in industry, finance, government, academia and multi-lateral organizations. ODA and other forms of assistance remain essential tools for stimulating investment in climate-friendly technologies. If the recent increase of ODA for climate-related projects is an indication, remarkable progress has been accomplished already.

But those achievements will not enter the mainstream unless new approaches—on the policy front and in finance—arise to complement existing initiatives. The problem is this: sustainable development through clean energy is still being addressed through short-term financing and regulatory frameworks that are not aligned to the immense scale of the challenge facing the globe (WB 2012b). Government resources alone are inadequate to meet the large investment requirements of scaling up renewable energy services and energy efficiency while private-sector investors continue to find it too risky to invest in clean energy. Therefore, mobilizing multi-lateral and bilateral financing institutions is vital for ensuring sustainable energy development and climate protection.

Gender is usually not included in most energy or utility projects but it is increasingly being recognized as an issue in energy delivery. Renewable energy is being adopted for many rural off-grid communities, most of them among the poorest populations. Micro-finance has been proven effective in bringing micro-hydro and solar PV to these communities. As intimated already, one of the success factors for micro-finance is when women are incentivized to subscribe to the renewable energy projects that provide them with opportunities for income-generating activities (UNDP 2004, 2011). Projects with gender-related elements achieved their overall objectives in relatively greater proportion than projects similar in sector and year of approval but without gender actions (Murphy 1997). Other studies also found that when gender issues were included in project design, greater end-user acceptance is achieved and more entrepreneurial activities were initiated (ENERGIA/DfID 2006). More recently, in assessing approaches to bring modern lighting to Africa, a study by the IFC found that women are both important beneficiaries and key facilitators of successful energy access programmes (IFC 2011).

Finally, the following list summarizes the key elements for a successful market penetration of sustainable energy technologies:

- high level of government engagement and responsibility;
- institutional capacity for managing technical and policy components;
- critical mass of human resources with technical and policy expertise;
- investment and financial support driven by ‘need’ or ‘demand’;
- technology solution contributes to wellbeing (development, alleviation of poverty);
- energy technology (hardware and software) is appropriate for the selected site;
- financial and economic viability of projects;
- governments are market enablers based on supportive policies and regulation;
- private-sector participation;
- ODA an enabling tether between governments and the private sector;
- climate benefits ‘visible’ for investors and improve the bottom line;

- investment in plant and equipment includes committing resources for training and capacity building (maintenance, management, business development);
- overall integration in local energy and economic infrastructure;
- innovation in technology, business models and financing; and
- good intentions alone are not sufficient, money alone will not bring change.

4.5 Conclusions

ODA and other forms of development assistance have played and will continue to play an essential role in transferring climate-friendly technologies to developing countries. The effectiveness of the assistance depends on a variety of factors: political, economic, financial, geography/location, infrastructure, social and cultural. Because of the wide differences of these factors across countries, regions and continents, the literature review undertaken for this chapter leads to a first conclusion: one size does not fit all and potential generalization, hence transferability and scalability, of the specific micro-features associated with examples of effective use of foreign aid is limited. What is transferable are the macro-level framework conditions that formed the basis for the success of 'what worked': solid government commitment at all levels, legislation in support of energy efficiency and renewable energy development (mainstreaming of climate change mitigation), incentives for private-sector participation, participatory involvement of local communities (creation of local champions), communicating benefits and risks, linking energy access with income-generating opportunities and access to concessional loans and credits. In short, benefits exceed real and perceived risks or revenues cover costs.

A second conclusion concerns 'what could work'. There are no inherent reasons why effective use of ODA would not work as long as the macro-level framework conditions are fully implemented and fundamental technical project prerequisites exist: sufficient resource base, technology adaptable to local conditions, sufficiently skilled labour and infrastructure availability. ODA might be more effectively spent on filling gaps in the macro-level framework than directly supporting on-site technology implementation.

The literature review revealed little information on scalability. Again, scalability is a matter of the existence and further development of macro-level framework conditions, access to finance and the recognition that the 'low hanging fruit' is harvested first. Scaling up of successful projects, therefore, necessitates due diligence of site conditions, infrastructure readiness and economic project feasibility under varying political and institutional framework conditions.

Finally, the literature review confirmed that effective use of foreign aid for sustainable energy and climate protection is possible and has taken place, yet there is lots of room for improvement. The review also confirmed that without ODA and other forms of development assistance, many renewable and efficiency projects would have not been implemented in developing countries with resultant higher GHG emissions. Even with the rapid decline of renewable technology costs observed in recent years, effective and efficient climate change mitigation in developing countries will hinge on stepped-up development assistance and technology transfer, climate finance and, above all, a comprehensive and binding global environmental agreement for climate protection reflecting the UNFCCC principle of 'our common but differentiated responsibilities'. Future levels of ODA are the mirror image of the world community's resolve to avoid dangerous anthropogenic interference with the climate system.

4.6 Appendix: Selected Studies

Micro-Hydropower in Nepal

Please contact the author for details

Renewable Technologies in Sri Lanka

Please contact the author for details

PV in China

Please contact the author for details

The Ghana Energy Development and Access Project

Please contact the author for details

Concentrating Solar Power in Morocco

Please contact the author for details

Funding Micro-Finance of PV Home Systems for Rural Electrification: Dominican Republic

Please contact the author for details

Pre-project Finance: Wind Mapping in Morocco

Please contact the author for details

Wind in Ethiopia

Please contact the author for details

Wind in Kazakhstan

Please contact the author for details

Wind in China

Please contact the author for details

Non-recurring Costs Related to Business and Market Development

Please contact the author for details

Creating a Level Playing Field: The Philippines

Please contact the author for details

Hydropower: Gansu Province of China

Please contact the author for details

Bagasse in Mauritius

Please contact the author for details

Bagasse in India

Please contact the author for details

Electrification in Bhutan

Please contact the author for details

NOTES

1. Grants or loans to countries and territories on the list of the OECD Development Assistance Committee of ODA recipients (developing countries) and to multi-lateral agencies which are: (1) undertaken by the official sector; (2) with promotion of economic development and welfare as the main objective; (3) at concessional financial terms (if a loan, having a grant element of at least 25%). In addition to financial flows, technical cooperation is included in aid (OECD 2008a, 2012b).
2. Transactions by the official sector with countries on the DAC list of ODA recipients which do not meet the conditions for eligibility ODA, either because they are not primarily aimed at development, or because they have a grant element of less than 25% (OECD 2008a, 2012b).
3. There are four Rio markers, covering: biodiversity, desertification, climate change mitigation and climate change adaptation. Every aid activity reported to the CRS should be screened and marked as either (1) targeting the conventions as a 'principal objective' or a 'significant objective', or (2) not targeting the objective.
4. There are many more barriers to investments in efficiency and renewables, which are extensively covered in the literature (e.g. Reddy 1990; Kostka et al. 2011; Verbruggen et al. 2010; Beck and Martinot 2004).
5. By March 2012, US$37 million (elicited from the WB and the CIF) for 21 projects in seven countries had been approved, 90% of which are projects are in sub-Saharan Africa (CIF 2012). The target is to use the existing commitment of the WB and the GEF to leverage US$1.7 billion from other sources.
6. Agence Française de Développement, AFD (France); Banco Estado, BE (Chile); Bancoldex S.A. (Colombia); Banco Nacional de Desenvolvimento Econômico e Social, BNDES (Brazil); Black Sea Trade and Development Bank, BSTDB (Greece); Caisse de Dépôt et de Gestion, CDG (Morocco); Central American Bank for Economic Integration, BCIE/CABEI (Honduras); China Development Bank, CDB (China); CAF-Development Bank of Latin America/Croatian Bank for Reconstruction and Development, HBOR (Croatia); Development Bank of Southern Africa, DBSA (South Africa); Indonesia Exim Bank (Indonesia); Industrial Development Bank of Turkey, TSKB (Turkey); Japan International Cooperation Agency, JICA (Japan); KfW Bankengruppe (Germany); Korea Finance Corporation, KoFC (South Korea); Nacional Financiera,

NAFIN (Mexico); Small Industries Development Bank of India, SIDBI (India); Vnesheconombank, VEB (Russia).

7. The BoP project is about market development: enhancement of the economic situation by simultaneously providing affordable energy services and income opportunities for the local population as well as profitable business for large corporations (Prahalad 2004).
8. SAL provided finance over a period of several years in return for reforms in trade protection and price incentives for efficient resource use with the aim to assist countries in reducing their current account deficits, strengthening their balance of payments, while maintaining their growth and developmental momentum.
9. While industrialized countries are used to a grid availability of >99.9%, developing countries often experience supply disruptions, voltage variation, etc. on a daily basis. This lack of reliability adversely affects economic productivity and socioeconomic development and, therefore, is an area for immediate improvement. However, for a household currently without access to electricity, availability of electricity even a few hours per day (or rather evening) represents enormous progress.
10. In terms of TWh, renewably generated electricity in developing countries grew by almost 900 TWh over the period 2000–10, more than twice the 405 TWh observed in the OECD (IEA 2012).

References

ADB (Asian Development Bank). 2009. *Evaluation Knowledge Brief on Greenhouse Gas Implications of ADB's Energy Sector Operations*. Reference Number: EKB: REG 2009-38E. Manila: ADB.

———. 2012. *Loan and Grant Financial Information Services Database*. Manila: ADB. Accessed 25 October 2012.

Addison, T., C. Arndt, and F. Tarp. 2011. The Triple Crisis and the Global Aid Architecture. *African Development Review* 23 (4): 461–478.

AGECC (Advisory Group on Energy and Climate Change). 2010. *Energy for a Sustainable Future: Summary Report and Recommendations*. 28 April. New York: AGECC. www.unido.org/fileadmin/user_media/Publications/download/AGECCsummaryreport.pdf

Arndt, C., S. Jones, and F. Tarp. 2009. *Aid and Growth: Have We Come Full Circle?* WIDER Working Paper 2009/05. Helsinki: UNU-WIDER.

Atteridge, A., C. Kehler Siebert, R.J.T. Klein, C. Butler, and P. Tella. 2009. *Bilateral Finance Institutions and Climate Change: A Mapping of Climate Portfolios*. SEI Working Paper. Stockholm: Environment Institute.

Barcelona Development Agenda. 2004. *Forum Barcelona 2004: The Legacy*. www.barcelona2004.org/esp/banco_del_conocimiento/docs/CO_47_EN.pdf??

Barton, J.H. 2007. *Intellectual Property and Access to Clean Energy Technologies in Developing Countries: An Analysis of Solar Photovoltaic, Biofuels and Wind Technologies.* ICTSD Issue Paper Issue 2. Geneva: International Centre for Trade and Sustainable Development.

Beck, F., and E. Martinot. 2004. Renewable Energy Policies and Barriers. In *Encyclopedia of Energy*, ed. C.J. Cleveland. London: Academic Press/Elsevier Science.

Brewer, T.L. 2008. *International Energy Technology Transfers for Climate Change Mitigation: What, Who, How, Why, When, Where, How Much ... and the Implications for International Institutional Architecture.* CESIFO Working Paper 2408. Munich: CESifo.

Burall, S., S. Maxwell, and A.R. Menocal. 2006. *Reforming the International Aid Architecture: Options and Ways Forward.* ODI Working Paper 278. London: Overseas Development Institute.

Christensen, C.M., and M. Raynor. 2003. *The Innovator's Solution: Creating and Sustaining Successful Growth.* Boston, MA: Harvard Business School Press.

CIF (Climate Investment Fund). 2012. *Climate Funds Update.* www.climatefundsupdate.org/listing/scaling-up-renewable-energy-programme. Accessed 15 January 2013.

de Coninck, H., R. Saidi, L. Wuertenberger, and X. van Tilburg. 2010. *Development and Climate: Merging Policy Agendas?* ECN Discussion Paper. LE Petten: Energy Research Centre of the Netherlands.

Djankov, S., J.G. Montalvo, and M. Reynal-Querol. 2006. Does Foreign Aid Help? *Cato Journal* 26 (1): 1–28.

Doucouliagos, H., and M. Paldam. 2009. The Aid Effectiveness Literature: The Sad Results of 40 Years of Research. *Journal of Economic Surveys* 23 (3): 433–461.

Easterly, W. 2005. What Did Structural Adjustment Adjust? The Association of Policies and Growth with Repeated IMF and World Bank Adjustment Loans. *Journal of Development Economics* 76 (1): 1–22.

———. 2006. *The White Man's Burden: Why the West's Efforts to Aid the Rest Have Done So Much Ill and So Little Good.* New York: The Penguin Press.

———. 2012. *Was Development Assistance a Mistake.* New York: Department of Economics, New York University. www.ssrn.com/abstract=2070232

ENERGIA/DfID. 2006. *From the Millennium Development Goals towards a Gender-Sensitive Energy Policy Research and Practice: Empirical Evidence and Case Studies.* CRGGE Synthesis Report. Ab Leusden: ENERGIA/DfID Collaborative Research Group on Gender and Energy. www.dfid.gov.uk/r4d/PDF/Outputs/Energy/R8346-dfid_synthesis.pdf

GEA (Global Energy Assessment). 2012. *Toward a Sustainable Future.* New York and Laxenburg: Cambridge University Press and the International Institute for Applied Systems Analysis.

Grameen Foundation. 2012. *Microfinance Basics*. www.grameen foundation.org/what-we-do/microfinance-basics

Headey, D. 2007. Geopolitics and the Effect of Foreign Aid on Economic Growth: 1970–2001. *Journal of International Development* 20 (2): 161–180.

Höhne, N., S. Khosla, H. Fekete, and A. Gilbert. 2012. *Mapping of Green Finance Delivered by IDFC Members in 2011*. Project number: CLIDE12246. Köln: ECOFYS Germany GmbH.

IEA (International Energy Agency). 2011. *Renewable Energy: Policy Considerations for Deploying Renewables*. IEA Information Paper. Paris: OECD/IEA.

———. 2012. *IEA On-line Data Service*. Paris: OECD/IEA. www.wds.iea.org/wds/ReportFolders/ReportFolders.aspx

IFC (International Finance Corporation). 2011. *IFC Advisory Services in Sustainable Business: 2011 Report*. Washington, DC: IFC.

IPCC (Intergovernmental Panel on Climate Change). 2011. *Special Report of Renewable Energy Sources and Climate Change Mitigation*. Prepared by O. Edennofer et al. (eds.). Cambridge: Cambridge University Press. www.srren.ipcc-wg3.de/report

Khennas, S., and A. Barnett. 2000. *Best Practices for Sustainable Development of Micro Hydro Power in Developing Countries*. Final Synthesis Report. London and Washington, DC: DfID and World Bank.

Kostka, G., U. Moslener, and J.G. Andreas. 2011. *Barriers to Energy Efficiency Improvement: Empirical Evidence from Small- and Medium-Sized Enterprises in China*. Frankfurt School Working Paper 178. Frankfurt: Frankfurt School of Finance and Management.

Kretschmer, B., M. Huebler, and P. Nunnenkamp. 2011. Does Foreign Aid Reduce Energy and Carbon Intensities of Developing Economies? *Journal of International Development*. https://doi.org/10.1002/jid 1788.

Kulichenko, N., and J. Wirth. 2011. *Concentrating Solar Power in Developing Countries—Regulatory and Financial Incentives for Scaling Up*. WB Discussion Paper 24. Washington, DC: The Energy and Mining Sector Board, World Bank Group.

Lancaster, C. 2007. *Foreign Aid: Diplomacy, Development, Domestic Politics*. Chicago: University of Chicago Press.

Mahama, A. 2012. 2012 International Year for Sustainable Energy for All: African Frontrunnership in Rural Electrification. *Energy Policy* 48: 76–82.

Martinot, E., R. Ramankutty, and F. Rittner. 2000. *The GEF Solar PV Portfolio: Emerging Experience and Lessons*. Monitoring and Evaluation Working Paper 2. Washington, DC: GEF. www.thegef.org/gef/sites/thegef.org/files/Docs/Climate%2520Change%2520-%2520The%2520GEF%2520Solar%2520PV%2520Portfolio-Emerging% 2520 Experience%2520and%2520Lessons.pdf

Michaelowa, A., and K. Michaelowa. 2007. Climate or Development: Is ODA Diverted from Its Original Purpose? *Climatic Change* 84: 5–21.

Moyo, D. 2009. *Dead Aid: Why Aid Is Not Working and How There Is a Better Way for Africa*. London: Allen Lane.

Murphy, J. 1997. *Mainstreaming Gender in World Bank Lending: An Update. WB Operations Evaluation Study 16409*. Washington, DC: World Bank. www-wds.worldbank.org/external/default/WDSContentServer/WDSP/IB/1997/03/27/000009265_3970818102953/Rendered/PDF/multi0page.pdf

N'Guessan, M. 2012. *Financing Renewables in Africa: A Review of Funding Sources Available for Promoting Renewable Energy in Africa*. Econoler Presentation, September. Napoli. www.irena.org/DocumentDownloads/events/NaplesSeptember2012/M'Gbra_N'Guessan.pdf

Nieusma, D., and D. Riley. 2010. Designs on Development: Engineering, Globalization, and Social Justice. *Engineering Studies* 2 (1): 29–59.

OCI (Oil Change International). 2011. *Access to Energy for the Poor: The Clean Energy Option*. Washington, DC: OCI. priceofoil.org/wp-content/uploads/2011/06/Access-to-Energy-for-the-Poor-June-2011.pdf

OECD (Organisation for Economic Co-operation and Development). 2008a. *Is It ODA?* Factsheet, November. Paris: OECD. www.oecd.org/dac/stats

———. 2008b. *The Paris Declaration on Aid Effectiveness and the Accra Agenda for Action*. Paris: OECD. www.oecd.org/dac/aideffectiveness/34428351.pdf

———. 2011. *Development Co-operation Report 2011*, 50th Anniversary Edition. Paris: OECD Publishing. www.dx.doi.org/10.1787/dcr-2011-en

———. 2012a. *Statistics on Resource Flows to Developing Countries*. Paris: OECD. www.oecd.org/dac/aidstatistics/statisticsonresourceflowstodeveloping countries.htm

———. 2012b. *DAC Glossary of Key Terms and Concepts*. Paris: OECD. www.oecd.org/dac/aidstatistics/dacglossaryofkeytermsandconcepts.htm

Prahalad, C.K. 2004. *The Fortune at the Bottom of the Pyramid: Eradicating Poverty through Profits*. Philadelphia: Knowledge@Wharton.

Reddy, A.K.N. 1990. *Barriers to Improvements in Energy Efficiency*. Paper presented at the Second International Workshop on Energy and Global Climate Change, 4–6 October. Berkeley, CA.

Rogner, H-H., D. Zhou, R. Bradley, P. Crabbé, O. Edenhofer, B. Hare, L. Kuijpers, and M. Yamaguchi. 2007. Introduction. In *Climate Change 2007: Mitigation*, ed. B. Metz et al., Contribution of Working Group III to the Fourth Assessment Report of the Intergovernmental Panel on Climate Change. Cambridge and New York: Cambridge University Press.

SELF (Solar Electric Light Fund). 2013. *Whole Village Development*. Washington, DC: SELF. self.org/whole-village-development/

Sen, A. 2006. The Man without a Plan. *Foreign Affairs* 85 (2): 171–177.

Thornton, N., C. Cameron, P. Grant, G. Norrington-Davies, and J. Zingel. 2011. Realising the Potential: Making the Most of Climate Change Finance in Africa:

A Synthesis Report from Six Country Studies: Cameroon, Ghana, Kenya, Morocco, South Africa and Tanzania. London: Agulhas. www.agulhas.co.uk

Tirpak, D., and H. Adams. 2008. Bilateral and Multilateral Financial Assistance for the Energy Sector of Developing Countries. *Climate Policy* 8: 135–151.

UNDESA (United Nations Department of Social and Economic Affairs). 2012. *Millennium Development Goals Report 2012*. New York: DESA. www.un.org/en/development/desa/publications/mdg-report-2012.html

UNDP (United Nations Development Programme). 2004. *Gender and Energy for Sustainable Development: A Toolkit and Resource Guide*. New York: UNDP. www.esmap.org/sites/esmap.org/files/Gender%20and%20Energy%20for%20Sustainable%20Development_A%20Toolkit%20and%20Resource%20Guide.pdf

———. 2006. *Expanding Access to Modern Energy Services: Replicating, Scaling Up and Mainstreaming at the Local Level*. New York: UNDP.

———. 2011. Entrepreneurship at Grassroots: Diffusion of Biomass Devices in Informal Industries in India. New York: UNDP. www.snap-undp.org/elibrary/Publications/Case15-TIDE_India.pdf

———. 2012a. *Micro-Credit for Energy Services: Transforming Basic Needs for Poor Households into Effective Market Demand'. Toward an "Energy Plus" Approach for the Poor*. Case Study 3. New York: UNDP. www.snap-undp.org/elibrary/Publications/Case3-Best_of_Pren.pdf

———. 2012b. *20 Years Community Action for the Global Environment, GEF Small Grants Programme*. New York: UNDP. www.sgp.undp.org/index.php?option=com_docman&task=doc_download&gid=281&Itemid=188

UNEP (United Nations Environment Programme). 2010. *Bilateral Finance Institutions and Climate Change—A Mapping of 2009 Climate Financial Flows to Developing Countries*. Nairobi: UNEP.

UNEP-FI (United Nations Environment Programme Finance Initiative). 2012. *Financing Renewable Energy in Developing Countries: Drivers and Barriers for Private Finance in Sub-Saharan Africa*. Geneva: UNEP-FI.

UNFCCC (United Nations Framework Convention on Climate Change). 2009. Copenhagen Accord. Decision-/CP.15, 18 December. Bonn: UNFCCC. www.unfccc.int/files/meetings/cop_15/application/pdf/cop15_cph_auv.pdf

Urban, F., and S. Wolcott. 2009. *Learning for Climate Change Mitigation Finance from Development Assistance*. Working Paper. London: Climate Strategies. www.ids.ac.uk/files/dmfile/ISDALearningfromDevelopmentAssistanceSept2009Report1.pdf

Verbruggen, A., M. Fischedick, W. Moomaw, T. Weird, A. Nadaï, L.J. Nilsson, et al. 2010. Renewable Energy Costs, Potentials, Barriers: Conceptual Issues. *Energy Policy* 38 (2): 850–861.

WB (World Bank). 2012a. *World Development Indicators 2012*. Washington, DC: International Bank for Reconstruction and Development/World Bank.

———. 2012b. *Turn Down the Heat: Why a 4°C Warmer World Must Be Avoided.* Washington, DC: International Bank for Reconstruction and Development/ World Bank.

WB/IMF (World Bank and International Monetary Fund). 2006. *An Investment Framework for Clean Energy and Development: A Progress Report.* Development Committee Report DC2006-0012.Washington, DC: World Bank/IMF.

———. 2007. *Clean Energy for Development Investment Framework: The World Bank Group Action Plan.* Development Committee Report DC2007-0002. Washington, DC: World Bank/IMF.

———. 2011. *Mobilizing Climate Finance.* Paper Prepared at the request of G20 finance ministers. Washington, DC: WB/IMF. www.imf.org/external/np/g20/pdf/110411c.pdf

WBG (World Bank Group). 2007. Selling Solar: Lessons from a Decade of IFC and World Bank Experience. Washington, DC: World Bank Group. www.ifc.org/ifcext/enviro.nsf/AttachmentsByTitle/p_CatalyzingPrivateInvestment_SellingSolar/$FILE/SellingSolar.pdf

CHAPTER 5

A Review of the Nature of Foreign Aid to the Energy Sector over the Last Two Decades

Luis Gomez-Echeverri

5.1 Introduction

This chapter is a desk review of the nature of foreign aid to the energy sector over the last two and a half decades, with an exploratory emphasis on how this aid has been adjusted, or failed to adjust, to the changing needs and circumstances of developing nations. The review is by no means comprehensive. Some of the data that would be relevant for analysis—much of it related to the significant and growing amount of aid from the emerging economies and non-governmental organizations (NGOs)—is not widely available. Moreover, the scarce data published are often not compatible with the data which are readily available from Organisation for Economic Co-operation and Development-Development Assistance Committee (OECD-DAC),[1] which report on aid by the large traditional donors.[2]

In terms of the literature review, and for the purposes of providing a view of ongoing aid, this chapter focuses largely on reports by donors and international organizations themselves and their independent

L. Gomez-Echeverri (✉)
International Institute for Applied Systems Analysis, Laxenburg, Austria

Y. Huang, U. Pascual (eds.), *Aid Effectiveness for Environmental Sustainability*, https://doi.org/10.1007/978-981-10-5379-5_5

evaluation units, which exist in almost every one of these institutions. This material falls mostly into the category of 'grey' literature (literature which has not been "peer reviewed"). The academic literature on foreign aid in general is massive and has been accumulated over the past three or four decades but only a small proportion is directly focused on energy. Needless to say, a comprehensive review of the general literature on foreign aid is outside of the scope of this report.

Information on foreign aid resource flows to developing countries has been recorded since the early 1960s by OECD-DAC. In 1969, this same body defined official development assistance (ODA) to mean the official and concessional part of these flows. Since then, this definition has been broadly used by the international community dealing with foreign aid. But recently, there have been some important shifts. Given this new reality, the original definition by OECD-DAC is outdated.

Another term that is increasingly used is 'development cooperation'. The proponents of this term argue that it is important to use one that denotes a relationship that is based on mutual benefits rather than a one-way relationship of givers and takers. The term development cooperation places the emphasis on collaboration and makes it clear that the objective is to provide resources in order to work together on development outcomes based on an open policy dialogue. With these new developments and with the entry of new and powerful actors (e.g. emerging economies), the term 'foreign aid', many argue, has become outmoded and is no longer considered appropriate. For the sake of simplicity, however, the term 'foreign aid' is used in this chapter and it includes all of the following: aid from the public sector—both from within the OECD and from emerging economies—to developing countries, private sector and NGOs. The recent trend has also led to integrated approaches to foreign aid that involve a variety of stakeholders, often resulting in public–private partnerships for action in the energy sector. Again, the traditional definition of foreign aid hardly captures these new realities. Development cooperation is a more appropriate term to describe the complex web of relationships and collaboration that are required.

5.2 Background

5.2.1 *The Importance of Foreign Aid to the Energy Sector*

The special excerpt publication of the 'World Energy Outlook 2010' (WEO) of the International Energy Agency started with the following statement in its foreword:

> It is an alarming fact that today, in the 21st century, there are still billions of people without access to electricity or clean cooking facilities. The ambitious goals that have been set to eradicate extreme poverty can never be fully realized without acknowledging and confronting this fact.[3]

Six years later, the WEO 2016 shows that these alarming facts still persist. Today, billions of people continue to lack access to the most basic energy services required for subsistence. According to this latest report, 1.2 billion people lack access to electricity, and more than 2.7 billion people still use dirty fuels for their cooking needs. The use of these dirty fuels for cooking is the cause of some 3.5 million deaths a year from indoor air pollution.[4]

Foreign aid to the energy sector is of unquestionable importance, given the poor situation of energy systems in many developing countries and the magnitude of resources that are required to transform them so that they might deliver necessary outcomes for poverty eradication and climate change. A similar message appears in the 'Global Energy Assessment' (GEA 2012: xiii), with some additional warnings: 'Without question a radical transformation of the present energy system will be required over the coming decades',[5] the authors write in their preface, and they note that energy access, climate change, global security and other major challenges of our century are all interrelated—meaning they cannot be properly tackled if not addressed simultaneously, in an integrated manner. The messages of both of these reports, by the GEA and the WEO, hold major implications for understanding foreign aid, the way that it is provided and the focus and magnitude of the aid that is needed.

5.2.2 *The Challenges that Foreign Aid Needs to Tackle*

In sounding the alarm on the magnitude of the problem and on the required scale of the response, both the WEO and the GEA also provide a message of hope. According to the GEA, the magnitude of the challenge of transforming current energy systems to make them more responsive to the needs of our times is immense but the solutions are implementable and affordable. The types of transformations required include: making radical improvements in energy end-use efficiency, achieving greater shares of renewable energy in the final global energy mix, and introducing advanced energy systems for utilizing both fossil and biomass fuels.[6] The GEA also concludes that achieving universal access to modern energy and cleaner cooking by 2030 is possible. The levels of investments required, coupled with targeted policies and subsidies amount to some US$36 to 42 billion per year,[7] a small fraction of the energy investments made annually to respond to the increase in energy demand. The GEA goes on to suggest that in pursuing the transformation required for energy systems to deliver across all fronts, there are various ways that could be taken to get there and that achievable energy portfolio options do exist. However, the GEA also points out that, for a number of reasons, immediate action and early and sustained investments are required, coupled with supporting policy and institutional frameworks to help delivery and implementation. The total amount of investment necessary is estimated to be in the range of 2% of global GDP.[8] While these figures appear large, the message is that not taking action would force us to incur greater costs in the long run.

A more recent report confirms and refines these figures. According to the Global Tracking Framework, some $1.5 billion annually is needed until 2030 in order to reach universal access at Tier 1 (which the report defines as enough energy to charge a mobile telephone and turn on a few light bulbs). For universal access at Tier 5 (defined as full and continuing grid access), $50 billion annually would be required.[9]

Immediate action is necessary for a number of reasons, among which are: (1) to avoid 'lock-in' of long lifetime energy systems and infrastructure that are not compatible with sustainable development[10]; (2) to ensure a stabilization of GHG emissions within a reasonable period of time to improve the chances of staying close to or below the 2 °C global temperature increase agreed to by the UNFCCC Conference of the Parties in the Paris Agreement[11]; and (3) to improve the chances of reaching the targets of the Sustainable Energy for All (SE4All) initiative launched by the UN Secretary General, which is strongly supported globally and endorsed at the Rio+ 20 Summit.[12]

The magnitude of the energy challenges that developing countries face is one of the main justifications for sustained foreign aid and for increasing the efforts to make this aid efficient and effective. Most developing countries will not be able to reach the targets of energy access, renewable composition or general efficiency without international aid and major investments, mostly by the private sector, given the order of magnitude of funding required. The biggest challenges for energy access today are in sub-Saharan Africa (SSA), where there has been slow progress in electrification and the general provision of energy. The facts shown in Fig. 5.1, from the latest Global Tracking Framework 2017, speak for themselves[13]:

Both the IEA 'Special Report on Energy Poverty' and the 'Global Energy Assessment' make the point that without extraordinary efforts, scaled-up investments and policy packages to accompany these investments, real progress in expanding global energy access will be difficult if not impossible to achieve. As the figures of the Global Tracking Framework show, the remaining challenges are immense.

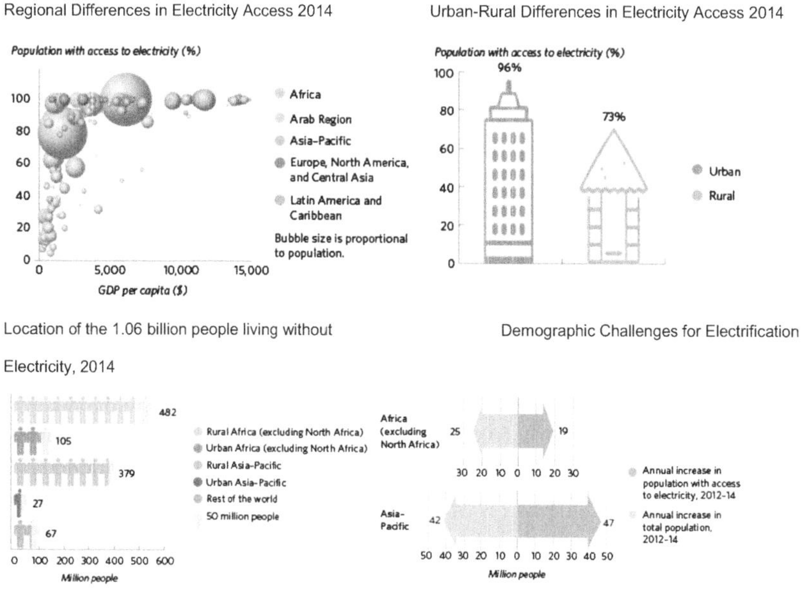

Fig. 5.1 Global Tracking Framework, 2017 World Bank

The implications and results of these studies point to the need for integrated packages of foreign aid in the following areas to help developing countries, particularly those in the less developed category where capacities are weak:

- establishment of sustainable energy baselines;
- identification of needs and opportunities;
- formulation of strategies and plans;
- preparation of investment portfolios to support efforts to scale up investments, mostly by the private sector;
- formulation of proper policy and regulatory frameworks;
- establishment of mechanisms of coordination and promotion of linkages across sectors;
- strengthening of institutions and institutional arrangements;
- enhancement of technological and process innovation across the various sectors of the economy, and building capacity to develop and deploy products of innovation;
- building up of capacities and skills at all levels (individual, institutional and systemic) to support the transformation of the energy systems; and
- forging partnerships, joint ventures and alliances, nationally, regionally, as well as internationally.

5.2.2.1 Changing Perspectives on Energy for Development and Implications for Foreign Aid

Energy has always been central to the global development agenda and in the headlines of the global media. What is new, as of the last few years, is the global effort to address energy issues and their impacts in a more coordinated way. In 1973, the so-called 'energy crisis' led a group of the larger world economies to experiment with a new construct of 'global governance', the G5, in order to address the impacts of that crisis. This new construct is what later became the G7, then the G8 (when Russia was added) and more recently to the G20 (when other large as well as emerging economies were incorporated in an effort to become more inclusive). From that time forward, energy security has been at the centre of the global agenda.

Furthermore, because of the disappointing progress made in many countries to achieve the Millennium Development Goals (MDGs), and a general disappointment with the absence of an energy MDG as many had suggested, and to which many attribute the lack of overall progress, energy has gained a particular prominence. The recently endorsed Sustainable Development Goals (SDGs) prominently feature energy as SDG 7: Affordable and Clean Energy.[14] Many view the fact that energy was not tackled directly in one of the MDGs as one a key reason that countries are doing so poorly in achieving those goals. This, and the persistent levels of poverty and lack of progress in the global negotiations to address climate change prior to the landmark Paris Agreement, led to several bottom-up initiatives with a central focus on energy. Examples of these include the SE4All initiative of the UN Secretary General, which addresses both energy access and poverty as well as climate change objectives (and was described earlier in this chapter), the World Bank Climate Investment Funds, which mostly target climate change objectives, and several other bilateral initiatives by donors, such as the Energy+ Initiative from the government of Norway,[15] which address both energy access as well as climate change objectives, and the Global Climate Change Alliance of the European Union, the UK's International Climate Fund, and Japan's Fast Start Finance programme, which funds a number of climate change mitigation activities. Many UN agencies, including the United Nations Development Program (UNDP), UNEP and others scaled up their work in preparation to COP 21, the Conference of the Parties where the landmark Paris Agreement was reached. Much of the work of these agencies focused on helping countries prepare their INDCs.[16]

Today, there is better science and knowledge, and greater awareness and acceptance of the central role that energy plays in issues such as individual, national and global security, food security, health, gender equality and employment to name just a few. All of these aspects of development have contributed to a recent shift in perceptions on how best to tackle energy challenges (BIAC 2009). While in the past, much of the emphasis of aid resources had been placed on the supply side of the equation—on creating or improving the infrastructure for electricity expansion—recent emphasis has shifted to a broader narrative about energy. The importance of clean energy for cooking and the need to reduce or eliminate the immense dependence on traditional biomass for fuel have gained more prominence, for example, as have concerns surrounding health issues connected with indoor air pollution from the use of such material in dirty and inefficient stoves.[17] Diversification of energy sources and issues of energy security are also now given much greater attention across regions.[18]

5.2.2.2 Changing Composition of Actors in Foreign Aid in Energy: Dangers of Fragmentation

The landscape of foreign aid has also gone through some major changes—and in many areas a complete transformation—in the last 25 years. During this period, there has been a prominent effort by donors to make foreign aid more efficient and effective. Although this effort has been driven by OECD-DAC, the agreements that have emerged from many years of reflection and assessment have also had an impact on the whole business of foreign aid in general.[19] The United Nations has also established a forum for discussing issues related to development cooperation and to improve coordination. Unfortunately, these processes do not include the development cooperation programmes of the emerging economies and other institutions.

The entry of new actors has made foreign aid a more complex system with a large variety of processes and procedures, motives and, in some cases, divergent vested interests. The public sector no longer has a monopoly on the business of foreign aid. Private funding, in the form of philanthropy as well as private foundations,[20] is significant and growing and, in some cases, larger than public funding for some sectors (e.g. climate mitigation and adaptation finance).[21] Many of these institutions and initiatives mobilize both public and private resources and promote public–private partnerships, such as the Clinton Global Initiative. The entry of these various new actors, both public and private, has led to some fragmentation, which in turn has made coordination more difficult and the management of aid by those receiving it more challenging, costly and arriving with heavier burdens on scarce national financial and human resources.[22] The number of official donors alone has grown considerably; in 1960, developing countries each received, on average, aid from two donors—today, the figure is 28. And these are only the large official donors; this estimate does not include the hundreds of other institutions (public, private and non-governmental) which disburse significant amounts of aid today.

In many instances, foreign direct investment (FDI) is combined with capacity and institution development, blurring the lines about what is foreign aid and what is pure FDI. In the area of energy particularly, these FDIs are large, some ten times the size of foreign aid.[23]

The disappointments with the waves of privatization in the 1980s and 1990s, particularly in the energy sector, have given way to more orderly public and private partnerships. In these new partnerships, the essential roles of the public and the private sectors are better recognized and accepted.[24]

5.2.2.3 Changing Trends in Foreign Aid in Energy

Foreign aid to the energy sector has evolved over the years, as a result of a number of factors. First, the energy needs and priorities of countries have changed. Second, foreign aid in general has changed, and foreign aid to the energy sector, a subset of this larger system, has followed this general evolution.

The OECD-DAC, whose members were for a long time the main providers of foreign aid, identifies some clear growth trends in foreign aid to energy that it attributes to a broader evolution in foreign aid as a whole.[25] OECD-DAC traces the steady growth of aid to energy up to the mid-1980s, when the trend began to fall, until the early years of the 2000s. The drop is attributed to the changes introduced to reduce or eliminate tied aid.[26] The 'Helsinki package', which resulted in the 2001 DAC Recommendation to untie aid to the least developed countries, had an impact on the aid to the energy sector, mainly by reducing the presence of foreign aid in the production side of energy (i.e. in supporting large energy production projects). This led to a shift of attention from large infrastructure projects to capacity development and projects that focused on helping countries formulate policies, strategies and institutional infrastructure and governance of the energy sector. The magnitude of resources needed for these types of interventions was by definition much smaller. Thus, falling trends were seen in terms of total resources but not necessarily in the presence of donor support.

This downward trend began to change in the early 2000s, as climate change negotiations increased their momentum and particularly with the formal adoption of the Kyoto Protocol. With this, a trend of increased support for clean energy, renewable energy and energy efficiency began in earnest. This trend has been maintained and has even accelerated during recent years (see Fig. 5.2 below for recent trends[27]), while these recent trends have also been affected by the entry of the new actors into the field of foreign aid.

The increasing priority being given by most countries to energy in the fight against poverty, health impacts, environment and security worries also has an impact on the foreign aid to energy trends over the recent past.

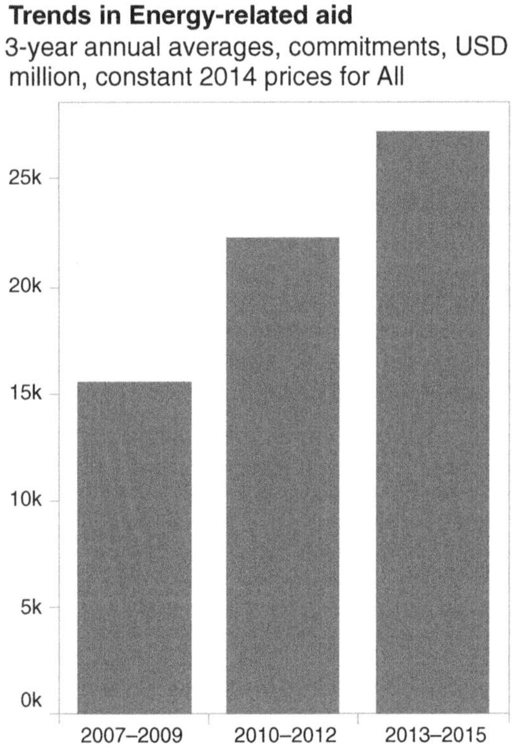

Fig. 5.2 Trends in energy-related aid (Source: OECD DAC CRS http://stats.oecd.org (Downloaded, 15 May 2017))

The declaration of the UN General Assembly to make 2012 the 'Year of Sustainable Energy for All' and the ongoing efforts in making the decade that follows the 'Decade of Sustainable Energy for All' is evidence of this new global concern. These concerns have led to a greater awareness by countries on: (1) the central role that energy plays in addressing these challenges, (2) the urgency of action both in terms of poverty eradication, where keeping up with energy demand alone is challenging, and in climate change, where the window of opportunity to stabilize GHG emissions to a level that will help limit global warming to less than 2 °C is getting significantly smaller as time passes, and (3) the recognition that incremental changes and business as usual will not help address the challenges of our century adequately.[28]

5.3 Literature Review

5.3.1 History and Evolution of Foreign Aid

5.3.1.1 Foreign Aid

The academic literature on foreign aid is massive but it focuses mostly on aid in general rather than on aid to specific sectors. It is not within the scope of this chapter to review that larger literature but it is useful, however, to provide a brief summary of the issues that are more commonly found in the broader literature, as this is considered relevant background to the literature review that is more directly related to the energy sector.

The literature on aid has a long history and one that is as old as the history of foreign aid itself. It dates back to immediately after the Second World War with the reconstruction efforts designed to address the war impacts on Europe and the decision to establish mechanisms that would help avoid a repeat of the Great Depression of the 1930s. Prior to that, aid efforts had focused on military aid and defence and were of a political nature. But immediately after the Second World War, with the creation of the Bretton Woods institutions and the Marshall Plan (also known as the European Recovery Programme), the era of government-to-government assistance and concessional loans for the purposes of reconstruction and development began.[29] The expansion of 'aid for development' to developing countries started at that time.

Foreign aid gained further prominence when Harry Truman publicly announced in his inauguration speech that foreign aid would be an important component of United States' foreign policy.[30] An article by the prominent US political scientist, Hans Morgenthau, then at the University of Chicago, highlighted some of the controversies of the time, mostly related to the concerns that people had about foreign aid being linked to foreign policy.[31] Not surprisingly, much of that early literature was in fact influenced by the question of the political nature of foreign aid. In his article, Morgenthau labelled all types of foreign aid as political. The only aid considered not political was humanitarian assistance.

Since these early days, there have been equal camps of supporters and detractors of foreign aid, as is certainly the case today. Polarization regarding perspectives on the purpose and usefulness of aid has influenced the literature of the last few decades. Prominent critics of foreign aid, such as Bauer,[32] Easterly,[33] and Friedman in his earlier days,[34] argue that aid has contributed to bigger and less efficient governments, larger bureaucracies, enrichment of the elites and considerable waste of money. Those who are more supportive of aid accept some of its failures but argue that the blame for failure lies on both sides of the aid flow, with the donors as well as the recipients, but that aid nevertheless has been a positive influence in developing countries. Some of those in this group include Sachs et al.,[35] Stiglitz[36] and Stern.[37]

The issues that come up most frequently as measures and indicators of success or failure of foreign aid include the impact of foreign aid on savings and growth, as in literature by Boone,[38] Burnside and Dollar.[39] Radelet, Clemens and Bhavnani efficiently summarize the most populous camps of the debate on aid and growth,[40] as: (1) those who believe that aid has no effect on growth but possibly instead contributes to undermining it; (2) those who believe that aid has a positive effect on growth across countries on average, and (3) those who think that aid has a conditional relationship with growth, where it is only helpful under certain circumstances. The literature on these topics is large, ongoing and very relevant to the topic discussion of energy-related foreign aid.

Aid's effectiveness in the fight against poverty and for enhancing countries' chances of reaching the MDGs is another important strand of the debate. Radelet focuses on how donors can improve aid effectiveness in helping countries achieve the MDGs.[41] The three recommendations that he provides apply to all aid, including aid in the energy sector. They include: making aid more goal and results oriented, being less selective and including countries with weak governance and institutions and design programmes to help them make improvements, and tailoring the way that aid is disbursed according to the conditions of individual countries.

5.3.1.2 Foreign Aid and Energy

Some of the best and most up-to-date information on foreign aid and energy from traditional donors comes from the bilateral, multi-lateral and international institutions involved in this work. Their reports on case studies and evaluation units provide a rich background on what works and what can or should be replicated in this arena. Foreign aid in the energy sector is most often provided to specific subsectors. The list of the subsectors that are most commonly used by OECD-DAC includes:

- energy policy and administrative management;
- power generation/non-renewable sources;
- power generation/renewable sources;

- electrical transmission/distribution;
- gas distribution;
- oil-fired power plants;
- gas-fired power plants;
- coal-fired power plants;
- nuclear power plants;
- hydroelectric power plans;
- geothermal energy;
- solar energy;
- wind power;
- ocean power;
- biomass; and
- energy education/training/research.
  (see Fig. 5.3 for recent energy-related aid by sector (OECD/DAC)[42]

Information and data on energy projects by multi-lateral financial institutions are generally provided only in aggregate sums.[43] Most of the literature

Fig. 5.3 Energy-related aid by sector (Source: OECD DAC CRS http://stats.oecd.org (Downloaded, 15 May 2017))

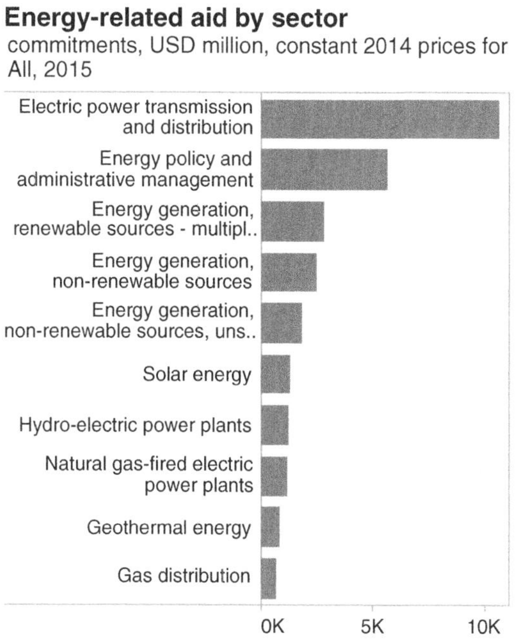

that exists on the activities of these institutions is provided by the institutions themselves, typically on their websites. There are independent efforts to track information on funding for climate change activities (where the energy sector is often a principal beneficiary), but these are not independently evaluated for their accuracy.[44]

Both bilateral as well as multi-lateral institutions have independent evaluation units. All have independent departments that carry out evaluations on an ongoing basis and their reports are publicly available. These evaluations are both global and national and focus on specific aspects of the aid programmes. Literature on these evaluations is comprehensive and provides a great source not only for learning about each donor's policies and strategies on aid but also as a compendium of lessons learned. The volume of these reports has increased in the last two decades, along with the growth of evaluation activities and focus on aid effectiveness. The 23 members of OECD-DAC and seven multi-lateral development banks produce more than 600 evaluation reports a year.[45]

The UN Energy report of 2010, the last time that this tracking was done,[46] makes reference to 130 reports issued by UN Energy members during the period of 2008–9. Most of these are reports that are designed to contribute to the knowledge base and to share the experiences of dealing with the complexities of energy systems in developing countries. They also provide a good store of lessons learned and best practices.

One such rich source in the energy sector is the Energy Sector Management Assistance Programme (ESMAP) hosted at the World Bank.[47] ESMAP was established in 1983 to help countries strengthen their institutional capacity and their planning and policy formulation in the area of energy and through 'upstream' programmes or programmes addressing overall strategies and policies for the sector. It is funded by 15 official bilateral donors and by a trust fund managed by the World Bank. Its website lists over 700 publications of material published or supported by ESMAP and includes technical reports, ESMAP's Knowledge Series (knowledge tools apropos how to guide decision-making about climate change mitigation and low-carbon growth, policy notes, briefing notes and other reports). Similarly, through its Open Knowledge Repository, the World Bank offers hundreds, if not thousands, of specialized reports on almost every aspect of the energy spectrum. Many are free to be downloaded at anyone's convenience.

The reports and publications of the United Nations Development Programme (UNDP) cover climate change finance (institutional issues and leveraging mechanism) and energy access and rural electrification. The UNDP (www.undp.org) also publishes a number of case studies, mostly in the area of energy access, which provide rich information about what works and other lessons born of experience. Other UN Agencies publish reports in their areas of specialization: UNIDO reports on industrial energy efficiency,[48] FAO in energy and agriculture,[49] IAEA in nuclear energy,[50] and others in their respective fields and with regards to their linkages to energy—as, for example, health and energy by the World Health Organization.

Similarly rich libraries of publications are provided by each of the regional development banks. These offer a wealth of reports and publications with a regional focus. Other intergovernmental organizations such as the International Energy Agency (IEA) often publish specialized reports in addition to its annual publication the 'World Energy Outlook'. A relatively new entrant into the field of energy organizations is IRENA (the International Renewable Energy Agency), with headquarters in Abu Dhabi and an office in Bonn, Germany.[51]

There are also NGOs and networks specializing in specific areas of the energy sector that publish reports, case studies and briefs useful to energy practitioners. The Renewable Energy and Energy Efficiency Partnership (REEP) publishes case studies and compendia of best practices and serves as a web-based resource hub for practitioners in both renewable energy as well as energy efficiency sectors. Other institutions, such as the International Chamber of Commerce, the World Energy Council and the World Business Council on Sustainable Development, issue publications of case studies on policies and regulations, energy efficiency, renewable energy development and specific sectoral reports. These reports are often produced by both practitioners and academics, so they combine the rigour of academic method with practical information from on-the-ground work; they are rich with lessons learned and best practices.

5.3.2 *Who Gives Foreign Aid to Energy and in What Areas?*

Most of the literature that is readily available deals with foreign aid by the traditional donors and the data that are largely gathered and kept by

OECD-DAC. Despite growing diversity and the entry of many new actors, the literature and data on these are incomplete, often not reliable nor readily available and, when they are available, not comparable to OECD-DAC data.

5.3.2.1 Aid by OECD-DAC

OECD-DAC keeps track of aid flows and their destinations by sector (see Fig. 5.4). As can be seen in the following tables and figures, a large portion of ODA from OECD went to power generation and transmission (Tables 5.1 and 5.2). The special report by the WEO on energy poverty points out that less attention and resources are allocated to areas such as clean cooking facilities, despite the fact that indoor air pollution resulting from incomplete combustion of dirty fuels is projected to cause some 1.5 million premature deaths a year by 2030 if the situation continues.[52]

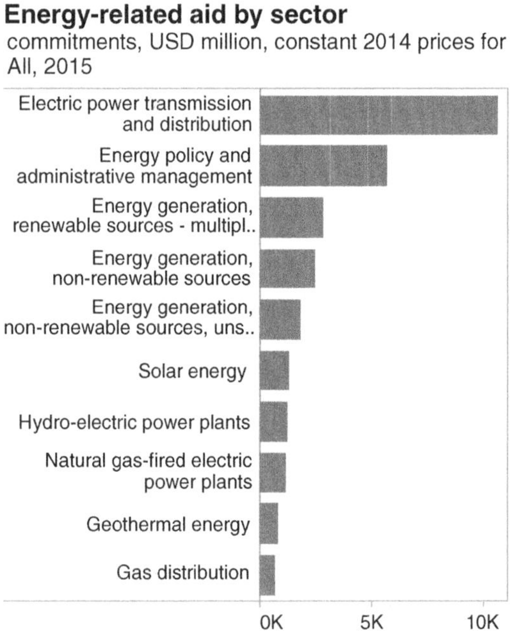

Fig. 5.4 Regional breakdown of aid to energy, 2003 to 2008 ('Regional Breakdown of Aid to Energy by All Donors; Development Cooperation Report 2012: Lesson in Linking Sustainability and Development', OECD 2012)

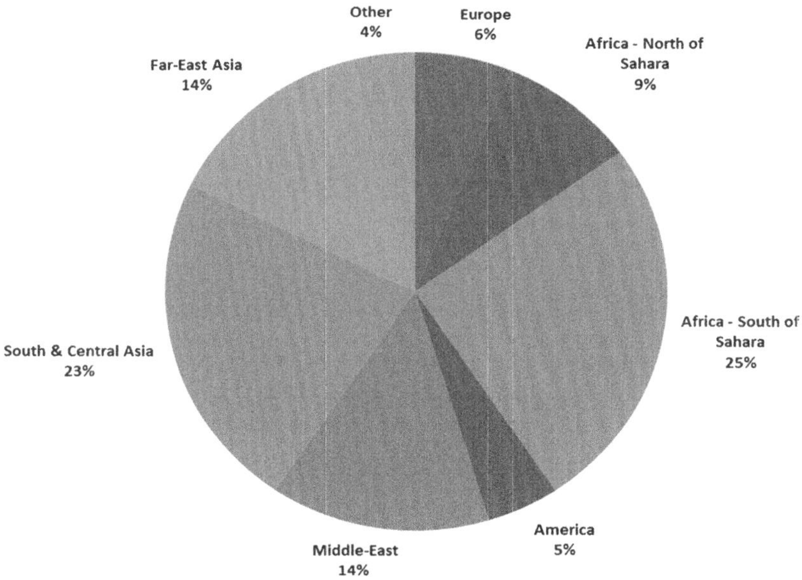

5.3.2.2 Aid By Emerging Economies

Some of the emerging economies, particularly China, have become major foreign aid players, mostly delivering in Africa. Their relationships are broad and cover foreign direct investment, trade and aid. Information and data on aid from these new actors are difficult to obtain for several reasons. One is that there are currently no formal mechanisms or standard formats for regular reporting on aid from these actors. Connected to this issue are allegations of lack of transparency from donor countries.[53] In addition, the relationships with the countries benefiting are often combined into packages that include multiple financial components—investment, trade and aid—which make it difficult to identify where one component begins and another ends.

An example of this synergy is illustrated by the Chinese aid-trade-FDI package signed in 2007 and 2008 with the government of the Democratic Republic of Congo (DRC). This package emerged as a new path to exploit Congo's extensive mineral deposits after an in-depth review of some 61 mining contracts awarded in previous years. Consisting of two large and related investment deals, the two sizeable loans were securitized by providing China with access to cobalt and copper reserves. The loans in turn were tied to an investment package that involved the exploitation of

Table 5.1 ODA in energy by sector: trends over 2006–10 (constant prices, 2011 US$ millions)

Time period	*2005*	*2006*	*2007*	*2008*	*2009*	*2010*	*2011*
Sector(s)							
Energy policy and management	577.7	1016.5	1988.8	2200.2	1019.7	1733.2	1468.7
Power generation/non-renewable sources	574.7	896.8	914.4	441.7	612.9	761.5	921.3
Power generation/renewable sources	416.7	324.3	398.0	732.3	1222.6	1882.1	2121.3
Electrical transmission/distribution	1855.8	1895.8	2172.3	3123.4	3150.1	4331.9	2533.4
Gas distribution	4.2	0.6	48.7	164.1	10.7	2.7	17.2
Oil-fired power plants	20.1	20.6	54.7	441.3	106.1	7.6	84.3
Gas-fired power plants	688.5	3.9	215.1	205.1	37.7	853.8	84.8
Coal-fired power plants	236.3	344.8	0.7	81.4	401.2	0.2	1235.8
Nuclear power plants	36.5	142.8	127.8	294.5	379.1	551.8	206.2
Hydroelectric power plants	590.2	961.3	1376.5	452.0	237.2	743.5	510.0
Geothermal energy	284.3	14.0	8.3	2.6	45.7	721.2	397.3
Solar energy	80.1	62.8	27.7	176.6	347.7	260.4	103.5
Wind power	151.6	109.4	155.6	315.5	222.4	1056.7	8.1
Ocean power		0.5		0.0	0.1	0.1	
Biomass	18.0	24.2	37.3	105.0	138.5	56.8	25.6
Energy education/training	20.8	19.4	17.5	56.2	21.7	33.3	19.6
Energy research	6.9	18.2	2.4	31.0	19.8	5.3	6.6

Source: OECD

Table 5.2 ODA to the energy sector by donor

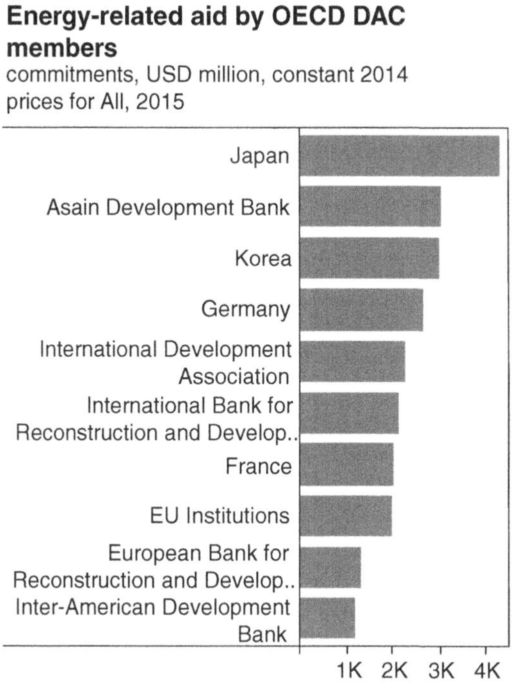

Source: OECD

mineral resources by a joint venture company between China and DRC. In addition to the investments, China is committed to providing support to investments in water, energy, education, transport and health, the key development areas for DRC.

This is a good illustration of China's strategic approach to integrate trade, FDI and aid, but it is also evident in some other donors and emerging economies. In the case of China, the strategy is driven by the need to continue to supply materials to support its growing economy and increasing demand for resources, but is also likely to be promoted by the desire to establish closer relations with an important ally in another continent.

Cooperation with other developing countries is central to China's international relations policy. Its south–south cooperation activities, particularly with Africa, have been growing significantly over recent years and cover many areas (see Fig. 5.5 for period 2010–12). These areas include trade and investment, debt cancellation, training, technical cooperation and capacity building activities. Several of its line ministries are involved.

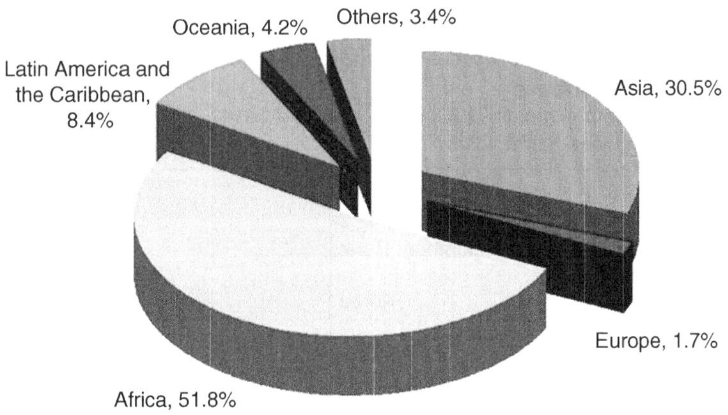

Fig. 5.5 China's Foreign Aid, 2014

China also provides assistance to regional organization such as the African Union (AU).

The most active emerging development partners in Africa are Brazil, China, India, Malaysia, the Russian Federation, the Republic of Korea and Turkey. The fact that oil and gas comprise a large bulk of the exports from Africa to some emerging economies (Brazil, India and China) points to the importance and, perhaps, the increasing presence of energy aid from these countries to Africa. Figures 5.6 and 5.7 show the nature and the magnitude of some of these trade relationships.

From the data available, it is evident that compared to the overall aid provided by OECD-DAC countries, these figures are still relatively small but growing. However, the increasing growth of relationships of these economies with developing countries is bound to push the levels of aid provided to larger shares. It is also possible that these shares are probably larger than what is being captured in data provided in current reports, mostly because the packaging of aid, trade and investment veils the independent levels of each. The box on China describes the challenges of clearly identifying the aid component from the other two frequent components in the China–Africa relationship.

All of the emerging economies are signatories of the Paris Declaration on Aid Effectiveness.[54] Because of the lack of data or proper monitoring of their activities, it is difficult to assess how the aid provided by these countries measures against the five principles of the Paris Declaration.

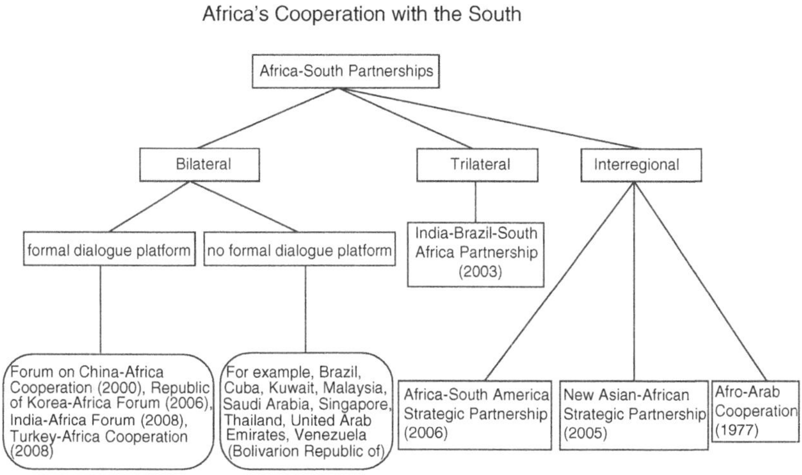

Fig. 5.6 Africa's cooperation with the South (Source: UNCTAD 2010. 'South South Cooperation: Africa and the new forms of development partnerships', Economic Development of Africa report, New York)

There is an increasing pressure on emerging economies to adhere to these five principles. There is also mounting pressure for greater transparency, so that there is more coordination and alignment.

Brazil, China, India and South Africa total economic engagement with developing countries, 2011

	Brazil	*China*	*India*	*South Africa*	*Total*	
	Billions of $	*Billions of $*	*Billions of $*	*Billions of $*	*Billions of $*	*%*
Official Development Assistance	$.36	$2.47	$.73	$.095	$3.66	3
Private Philanthropy	$.02	$.001	$.25	$.096	$.37	<1
Remittances	$.40	$3.60	$9.50	$.70	$14.20	13
Private Capital Flows	$11.00	$48.00	$14.00	$15.00	$88.00	83
Total Economic Engagement	**$11.80**	**$54.10**	**$24.50**	**$15.90**	**$106.20**	**100**

Source: The State of Play of Private Financial Flows, presented to OECD by Dr Carol Adelman, Director, Hudson Institute, Center for Global Prosperity, 15 June 2015

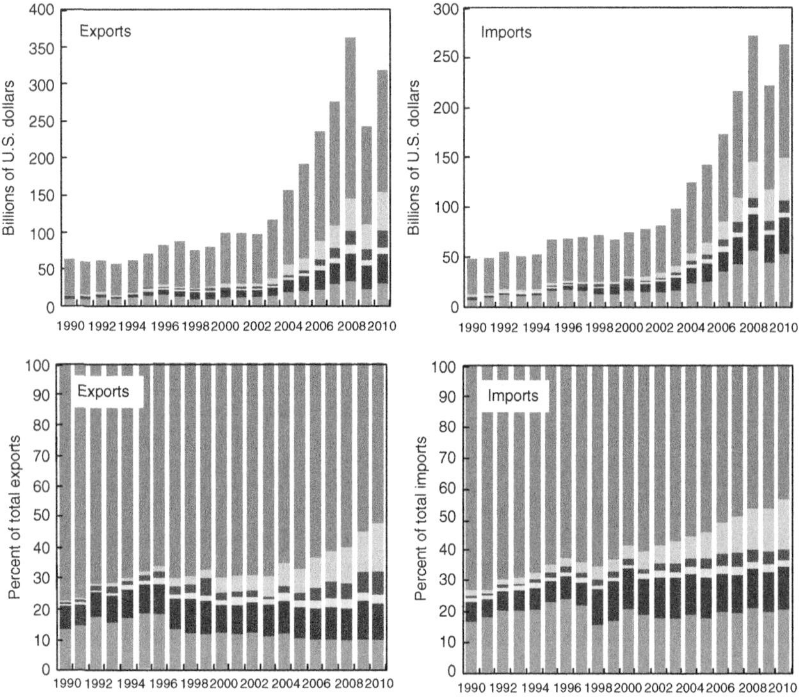

Fig. 5.7 Sub-Saharan Africa: total exports and imports by partner (Source: IMF 2011 'Sub Saharan Africa Sustaining the Expansion', Regional Economic Outlook, World Economic and Financial Survey. Washington, DC)

5.3.2.3 Aid by Private and Non-Governmental Organizations

Development cooperation in general is no longer the monopoly of governments. The private sector and NGOs have become increasingly important actors in development assistance and finance, including in the energy sector. Much of this assistance comes from private philanthropy, private remittances and NGOs. Many of these flows are difficult to track and there is less data as to where this aid goes and how much of it is directed at the energy sector, particularly to projects dedicated to increase energy access in rural communities, poor urban areas and for the betterment of energy services in general.[55] What is evident is that the size of these flows is large and growing and, thus, significant in the review of foreign aid in general.

Much of the data on private sources of development assistance cover only this type of aid coming from OECD countries, where most of it originates. Not many data are available from other regions. Little traces the increasing presence and influence of these new actors, concluding that international development is quietly being revolutionized by this 'private development assistance (PDA)'.[56] She defines PDA as 'cross-border transfers of cash, grants, loans, in-kind contributions, or volunteer time to individuals, NGOs, and governments' and limits PDA to refer to development aid that is:

> (1) undertaken by private actors including individuals, foundations, corporations, private voluntary organizations, universities and collegés, or religious organizations, (2) with promotion of economic development and humanitarian need as the objective; and (3) at concessional financial terms where commodities and loans are concerned.[57]

Little goes on to suggest, reasonably, that the motivations, methods and selection of areas and partners by private donors often differ from those of traditional and better understood public donors. There is a lot to learn from the experiences of these new partnerships and networks as more research is done and a greater wealth of literature becomes available.

Non-governmental organizations per se have also become increasingly important actors. Many are involved in helping to establish codes of conduct in aid provided to difficult industries, including the extractive industries, some of which fall in the energy sector. But many are also involved in supporting countries mainly in the area of energy access to poor sectors of the urban and rural populations. NGOs such as Practical Action are part of an increasing number of NGOS working to define energy access for the poor and to promote the increase of access throughout the developing world through activities that range from outreach and advocacy to project work, capacity building, brokering partnerships and mobilizing resources, and research and publication of literature, guides and policy briefs.[58] Their role, particularly in the energy access area, is crucial in many countries and areas of countries and local communities where the public and private sectors are not so active.

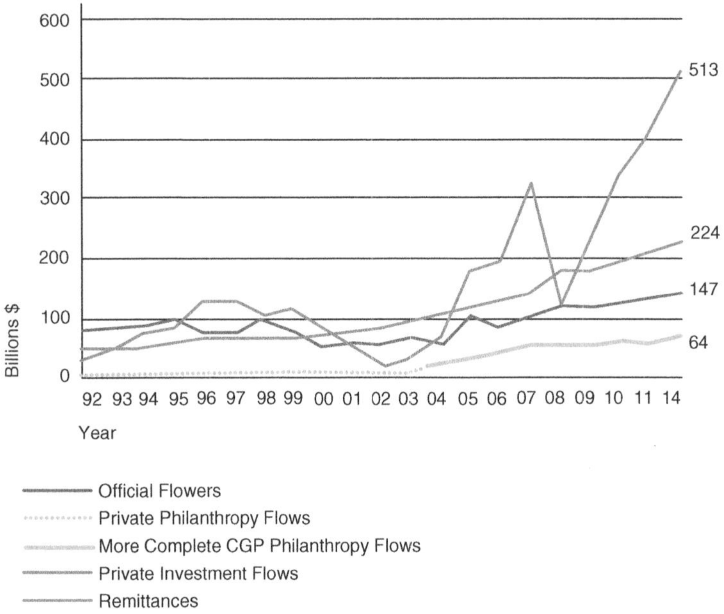

5.3.3 *History of Efforts to Enhance Impact and Effectiveness*

Much of the recent literature on aid is focused on the global effort to improve 'aid effectiveness'—in other words, the ability of aid to achieve its intended aims. These efforts are linked to both substance (how effective is aid in advancing the MDGs?) and processes (in building capacities and efficient governance).[59] They have been led mostly by OECD-DAC, whose efforts have influenced not only their own development cooperation programmes over the years but also those of the development community as a whole, including international organizations such as the World Bank and the UNDP. These efforts culminated in the Paris Declaration on Aid Effectiveness and the identification of its five performance principles.

The major changes that have emerged from these efforts include, among others, increasing the focus on the environment after the United Nations Conference on Sustainable Development in 1992, reducing the level of tied aid as per agreements reached through the 'Helsinki Package', shifting to direct budget support where doing so appears to be more appropriate for delivering on certain outcomes, changing the concept of 'assistance to recipients' to 'cooperation with development partners', with a concomitant effort to align aid to partners' priorities, having the private sector play a greater role, promoting the global development agenda such as that of the MDGs, placing a greater focus on gender issues, linking global development to global security, coming to a consensus on efforts to tackle global development through better and more effective aid and development cooperation, and focusing more aid to LDCs and fragile states which are considered those in most need.[60] The United Nations has also established its own forum 'to review progress in international development cooperation efforts and promote greater coherence among the development activities of different development partners'. Additionally, several UN agencies have undertaken reform efforts oriented towards making their development cooperation more effective and relevant. The recent global endorsements of the SDGs and the Development Agenda 2030 as well as the Paris Agreement are already triggering a new set of reforms that are yet to be assessed for their impact. The new UN Secretary General, António Guterres, has also announced his intention to undertake a major UN reform to make the organization better prepared for the challenges ahead.

In the area of energy, there have also been activities specifically designed to coordinate and foster collaboration among donor groups for greater aid effectiveness. Responding to a call for greater coherence and coordination of the UN system in the energy sector at the 1992 World Summit on Sustainable Development in Johannesburg (South Africa), UN Energy was established. Initially a weak institutional set-up dedicated to sharing information, UN Energy eventually became a central mechanism for UN coordination by more than 29 UN agencies and the World Bank. In 2007, UN Energy elected the head of one of the UN agencies (UNIDO) and reformed its work programme around three thematic clusters: energy access, renewable energy and energy efficiency. These clusters have not only developed programmes of collaboration but also led to the establishment of the Sustainable Energy for All initiative launched by the then UN Secretary General Ban Ki-Moon. SE4All is underpinned by three interrelated goals in support of poverty eradication and climate change, each with aspirations to be reached by 2030: achieving universal access for all to modern energy, doubling the improvements in energy efficiency and doubling the share of renewable energy in the global final energy mix. UN Energy continues to serve as support to the global energy agenda and in support of greater effectiveness of UN development cooperation in the area of energy.[61]

The various efforts to improve aid effectiveness and donor coordination have not been as successful at addressing donor fragmentation, or the proliferation donors making many disjointed payments.[62] Several authors, such as Knack and Rahman, have written on the impacts on donor fragmentation and report that this has a negative effect on government bureaucracies and on effectiveness.[63] In an another paper using Kenya as a case study, Mwega examines the various effects of donor fragmentation on aid effectiveness and reports two categories of negative effects:

1. increased transaction costs related to a diverse set of rules and procedures of donors for managing their aid projects;
2. a limited ability to tap economies of scale, and strains on the already scarce financial and human resources of countries receiving aid.[64]

Lancaster cites cases where lack of donor coordination has led to formulation of projects that require counterpart institutions and funding that is not really available.[65] Collier argues that, in a good policy environment, aid may be beneficial but only up to a point, after which it creates a negative effect due to its sheer size.[66]

5.3.4 What Works

5.3.4.1 General: Making Foreign Aid More Effective and Responsive to Our Century's Challenges

To do full justice to the question of what works in terms of foreign aid in the energy sector would require a massive undertaking that assesses how foreign aid has helped or contributed to eradicating poverty and tackling the main challenges of climate change (through mitigation) in each particular setting and the other challenges of our century in which energy is central. Such comprehensive assessment is outside the scope of this chapter. Much of the literature on foreign aid and poverty eradication is partly linked to the debate about the effects of foreign aid on the overall economic performance of countries—on growth, trade, promotion of private investment and public-sector management, to name just a few. For energy projects, these indicators of success are not as useful. One interesting study on a programme to promote energy efficiency in Mexico in the 1990s concluded that there is a need to reassess the indicators of success for foreign aid in energy and particularly for small-scale projects that promote energy access in general.[67] These are projects that are more knowledge intensive than capital intensive and, therefore, the focus on success criteria should be on institutional determinants and capacity building of individuals, businesses and institutions that create the enabling environment for energy services to be provided.

Because energy was not part of the MDGs, there is little literature assessing how the lack of energy contributed, or not, to the performance in efforts to reach the MDGs. A publication of UNDP and World Bank-ESMAP briefly alludes to the link between lack of energy access and lack of progress on the MDGs. Their preliminary results suggest the need to do more research in this area. Moreover, new and emerging literature is becoming available on the effects of foreign aid in helping countries achieve lower carbon economies and societies. According to the International Energy Agency, to achieve 50% reduction in carbon dioxide emissions by 2050 will require investments of some US$316 trillion, or 17% over the business-as-usual scenario (Modi et al. 2005). A large portion of these investments will be required in developing countries, where the capacity to attract and absorb investments is hampered by the lack of capacity to formulate policy and regulatory frameworks, as well as weak institutions and skills.[68] Foreign aid can and is playing an important role in helping countries build their capacities for attracting investments in the area of climate change.[69]

5.3.4.2 Being Sensitive to National and Local Needs and Priorities

For greatest effectiveness and relevance, foreign aid to the energy sector would need to take into account the very different energy challenges of the various regions of the world. Some of these different challenges, taken from Fontaine et al.[70] are summarized now:

- Europe and Central Asia: Energy security is a major concern for this region. The recent economic crisis has decreased the level of economic activity across the area, lessening the acuteness of concerns around energy security for now but this we can expect to be only temporarily the case.

- East Asia and the Pacific: Energy access and clean cooking needs are the central concerns for this region. The need for massive scaling up of investments for ensuring universal access to modern forms of energy and clean energy for cooking will be a major challenge and one that foreign aid has been supporting for some time.
- Latin America and the Caribbean: Electricity demand and carbon dioxide emissions are expected to increase significantly in this region between now and 2030. Are the institutions and enabling environment and mechanisms in place to attract and sustain the necessary, large amount of resources that will be required in order to put in place the adequate infrastructure to meet these demands?
- Middle East and North Africa: The region has a combination of resource-endowed countries and those that are dependent on foreign resources to satisfy local energy needs. Even those that are rich in hydrocarbon resources have urgent demands for economic diversification and maintenance of infrastructure. Others are still in urgent need of increased levels of energy access.
- South Asia: In this region, the demand for infrastructure and particularly for electricity is growing exponentially. Half of the population still has no access to electricity and most countries require an urgent building up of infrastructure to increase the levels of energy to satisfy high economic growth. This region also has a high ratio of its population as dependent on traditional biomass for basic energy needs.
- Sub-Saharan Africa: The energy challenges in this region are some of the most urgent and complex, combining low rates of energy access, low levels of capacity and skills, poor reliability of existing systems and high costs. The urgency of the energy challenge here is directly related to the region's capacity and prospects for development and economic growth.

5.3.4.3 Illustrative Cases: Helping Countries Achieve Energy Systems Transformation with Well-Timed Strategic Foreign Aid Intervention—the Case of Brazil's Ethanol Programme

Today, Brazil is the largest producer of sugar cane-based ethanol and the second largest producer of ethanol worldwide (second only to the United States, which produces corn-based ethanol).[71] Ethanol sources, naturally, contribute a critical percentage of the fuel used for transportation within Brazil—approximately 40% in 2006.[72] The history of the Brazilian ethanol programme contains many lessons for countries aiming to transform their energy systems as Brazil did, dramatically and in the course of a few decades. The most important lesson learned is that consistent and continuous government policy is critical to engineer and sustain a system transformation.

The Brazilian ethanol programme also offers lessons on the possible strategic use of foreign aid. In the early stages of the industry, when Brazil needed to develop markets by maintaining a competitive edge over petroleum, it turned to the World Bank for support in funding and technical assistance. With a major commitment to research and development for the sector which it considered essential, Brazil built a number of centres of excellence surrounding ethanol technology. To support this effort, Brazil turned to a number of UN institutions, including the UNDP, UNESCO and others. For market development, including the internationalization of its domestic industry, Brazil sought the help of the Inter-American Development Bank.

Brazil is now itself a foreign aid provider, with international cooperation programmes comprised of a mix of FDI, technical assistance and commercial and research joint ventures in several countries that include Angola, Mozambique, Ghana, Zimbabwe, Cuba, Sudan and Senegal.[73] Well-targeted foreign aid arriving at a critical time of need, technical support for institution-building and market development, and support across various sectors critical to the industry made a big difference for Brazilian ethanol. In retrospect, all these factors did not work as part of a coordinated plan of support by the international community but rather because of the *national* commitment and investments on the part of the Brazilian government.

5.3.4.4 Dealing with Market Barriers Through Well-Designed Policy and Institutional Interventions: The GEF Approach to Promoting Energy Efficiency and Renewable Energy Deployment

The operational programmes of the Global Environment Facility (GEF) focus on helping countries remove 'barriers to large-scale application, implementation, and dissemination of least-economic cost energy-efficient technologies (whether commercially established or recently developed)',[74] and also to remove 'barriers to the use of commercial or near-commercial renewable energy technologies (RETs), reduce any additional implementation costs for RETs that result from a lack of practical experience, initial low volume markets, or from the dispersed nature of applications, such that economically profitable "win-win" transactions and activities increase the deployment of RETs'.[75] The support of the GEF ranges from removing institutional and capacity-related barriers, helping establish ESCOs (energy service companies), as in the case of Tunisia, helping develop viable and sustainable markets for renewable energy technologies through the formulation of appropriate policy and regulatory frameworks, supporting demonstration projects to show the viability of the introduction of renewable energy in remote areas, as in the case of Uganda, and promoting lifestyle changes, as in the case of a Manila project that promotes the use of bicycles as an alternative mode of transport.[76] GEF support is aimed at addressing five potential barriers to efficient market-driven dissemination of environmentally sound technologies (ESTs) in developing countries: policy frameworks and the essential role that governments must play in setting policies that are conducive to the adoption of ESTs, mature technologies and the need to ensure that they are robust and operational, making them easier to transfer; awareness and information and the need to have all stakeholders and particularly market participants aware of the technologies, their costs, uses and markets, business and delivery models with businesses and institutions able to deliver and service those markets; and finally, availability of finance for technology dissemination as one of several prerequisites.[77] The more recent programme directions of GEF focus more on energy system transformation, urbanization and sustainable cities, and green infrastructure, to name just a few.[78]

5.3.4.5 Promoting Widespread Adoption of Renewable Energy to Provide Difficult-to-Reach Pockets of Population with Energy Access: China's Success Through a Comprehensive Capacity Building Programme

China has been one of most successful developing countries at increasing the rates of energy access for its population. Despite its outstanding success, there are still significant areas that lack energy access: as of 2014, large pockets of the population (some 57%) lacked access to clean fuels and technology for cooking.[79] Much of the success of access expansion in China has resulted from a strong commitment on the part of the government, policies and measures to support this commitment, and large amounts of domestic investment. But, as is always the case, reaching the last remaining pockets of low access has proved the most difficult. Helping bring energy to these remaining areas is where foreign aid has been helpful, even in a country with the size and resources of China.

A project co-financed by the GEF, the governments of the Netherlands, Australia and China was initiated in 1999 (ending in 2008) with the aim of promoting the adoption of renewable energy technologies in China. The project utilized a mix of comprehensive capacity building measures (which targeted not only key public but also private organizations), support in the formulation of new policies and regulatory measures (such as policies on biogas, wind and village power sectors as well as the Renewable Energy Law of 2005), overall technical assistance and co-financing support of demonstration projects. The overall funding available for the project was approximately US$25 million, a relatively small amount of money in relation to the magnitude of the challenge and the size of the country. Nevertheless, the project achieved some transformative results. It helped to build capacities in the renewable energy industries and to leverage several government programmes in support of energy for the poor. And its demonstration projects of market-based systems helped transform the village power sector. The multi-faceted programmatic approach to the problem helped achieve impressive results in a relatively short period of time.[80]

5.3.4.6 *Energy System Changes that Achieve Universal Energy Access: The Success of Vietnam, which Combined Strong Political Commitment and Sound Policy Frameworks with Well-Targeted Foreign Aid*

The case of Vietnam's success in increasing access to energy through its rural electrification programme provides valuable lessons. In 1975, Vietnam's rate of electrification among poor households in the country was about 2.5%. In 2009, the rate was 96% despite its very low level of average national income. Well-targeted international assistance and foreign aid have played a big role but not the main one. It is true that many international institutions provided much foreign aid, and these included:

- Asian Development Bank and the World Bank (some of it together with the GEF) supporting the electrification of many communes throughout the country, helping in the construction and rehabilitation of small hydropower facilities and distribution networks, and supporting grid extension, roads upgrade and loss reduction;
- UNDP and the Swedish International Development Cooperation Agency supporting the acceleration of electrification in rural and mountainous areas;
- France targeting the electrification of a number of communes, among others (ADB 2011); and

- Assistance has been provided in infrastructure development, such as when IFC provided support for small photovoltaics businesses.[81,82]

Thus, many factors contributed to Vietnam's success in increasing electricity access in a relatively short period of time, including local conditions (ample hydroelectric resources), local customs (which put a premium on electrification and willingness to pay), multiple funding sources and technical assistance from various international foreign aid sources, consistent and persistent policies within a good policy, institutional and regulatory framework context, and a strong political commitment.

5.3.5 *What Could Work*

5.3.5.1 Some Illustrative Cases: Market Transformation for Energy Efficiency or Demand Side Management (DSM)

Several foreign aid programmes support market transformation strategies that facilitate the adoption of energy-efficient products, services and/or practices. In carrying out an evaluation of market transformation projects by the GEF, Birner and Martinot list a number of principles on which these programmes rest: (1) interventions are targeted to specifically identified market barriers, (2) entire markets are changed permanently so the benefits are sustained over time, (3) new products and services or practices appear in the existing markets, (4) private capital and knowhow and competitive market forces push energy efficiency gains, and (5) partnerships between public, private, NGO consumers and other stakeholders come in, to influence market structure and function.[83]

The menu of policies and measures available for these interventions is broad and varied. They range from fiscal incentives to public finance and regulations, some of them are price driven, such as fixed payment and premium payment feed-in tariffs,[84] or quality driven, such as green energy procurement or green labelling. The most effective foreign aid is that which provides support, when needed, to the entire spectrum of the policy framework for market transformation.

That being said, evidence also shows that DSM can be an effective tool and, moreover, that foreign aid can assist in making it successful. The origins of the concept of market transformation came from utility demand-side management programmes in the USA and Sweden. These programmes showed that some of the DSM efforts were creating permanent market transformation, leading many to focus on achieving broader objectives from such transformations. This in turn has led to some bias or less attraction to DSM projects because of their more limited objectives. Evidence suggests, however, that DSM can often also lead to market transformation in the short and medium term. In a post-implementation impact assessment of energy efficiency projects, the World Bank/GEF found DSM to be a useful complementary measure to the longer-term intervention of market transformation. This conclusion was drawn from the analyses of various large energy-efficiency programmes supported by the GEF in Poland, Mexico, Thailand and Jamaica.[85] This evaluation indicated that DSM alone can be quite effective but foreign aid can assist by focusing on helping countries identify those projects where DSM can make long-lasting effects on market transformation. Such foreign aid programmes should be accompanied, whenever necessary, with capacity development, education and outreach programmes and solid monitoring and evaluation to provide feedback for necessary adjustments along the way.

5.3.5.2 Combining Finance and Capacity Building Schemes to Expand Access to Energy Services: The Case of the Use of Micro-Finance in the Philippines

Despite its economic growth in recent years, the Philippines has a large number of people living below the poverty line (25% in 2014) and some 21% without access to electricity.[86] Some 66% of its power is generated from fossil fuels and the rest by hydropower and geothermal-based energy. A project initiated in 2006 had the aim of promoting renewable energy nationwide through a micro-finance project that aspires to reach households and community organizations in remote and off-grid areas in the Philippines. While it is still early to gauge the large-scale, lasting impact of the programme, the model of small-scale financing, multi-stakeholder implementation and engagement, and community involvement is showing promise in terms of large-scale adoption of renewable energy technologies. The lessons learned thus far have been that the sustainability of the model depends very much on the capacity of the suppliers of renewable energy technologies, the capacity of the micro-finance institutions to reach large numbers of clients in remote areas and the ability to revolve the fund to allow new loans to be granted. A close review of the project to date revealed a number of important factors for expanding energy access through micro-financing for households and community-level renewable energy technologies. In all of these, foreign aid has played an important role, and they include support in the following areas: the establishment of an enabling policy environment; market-based approaches in the development and delivery of energy products combined with capacity development and product development; establishment of adequate, accessible and appropriate financing windows; adequate and accessible knowledge support; and finally, capacity development for micro-finance institutions in technology, financing, marketing and risk management.[87]

5.3.5.3 *Enhancing Integration and Collaboration Among Stakeholders Through Integrated Packages that Include, Among Others, Foreign Aid: The Ambitious Model of 'Sustainable Energy for All' to Transform Foreign Aid to Energy*

Sustainable Energy for All (SE4All), an initiative launched by then UN Secretary General Ban Ki-Moon in 2012 and described in Sect. 3.3, promotes the worldwide achievement of universal access to modern energy, doubling the improvements of energy efficiency and doubling the contribution of renewable energy in the global share of the final energy mix. It brings together international organizations, the donor community, the private sector, and non-government and governmental actors to work together in designing strategies and action plans that can fulfil the three target goals. These entities also collaborate on specific foreign aid interventions 'on the ground' to help countries individually contribute to the attainment of the global goal.

In its initial stages, SE4All interventions combined foreign aid, FDI and policy dialogue to promote development outcomes to be reached through programmes of action that promote synergies between the three goals, seek joint approaches that bring several sectors together into action, and involve stakeholders in an inclusive manner. SE4All has now entered a new phase that focuses on implementation and the promotion of partnerships for accelerated action. The first stage consisted of developing rapid assessment to establish baselines on the three goals and identify opportunities for action to promote progress on the goals. Given its infancy, it does not yet have a significant track record. However, it was included in this chapter's review because of its high ambition and innovative approach. The most significant message from this project is that a large number of stakeholders ranging from governments, international organizations, businesses and civil society organizations can—and have agreed to—work together in a large partnership to address world energy issues with the purpose of achieving wide-ranging outcomes in employment, health, environment, security, climate change and other important objectives.

5.3.5.4 Direct Budget Support Versus Traditional Support

Direct budget support—direct assistance to government budgets, often earmarked for particular purposes—has gained more prominence in recent years as an assistance tool, particularly in efforts to support poverty reduction/eradication strategies where increasing the level of energy access is often a main pillar. Under direct budget support the recipient country has the responsibility 'to spend using its own financial management procurement and accountability systems', through either general budget support or earmarked funds for specific sector support.[88]

The advantage that some see in this type of approach is that funding to projects is more predictable, aligned with the budgetary cycle of the recipient countries, and transferred with lower transaction costs. Some donors, such as the United Kingdom and Norway, are strong advocates of this approach. Some argue that it helps to build ownership and capacities, given that most funds are tied to rigorous accountability mechanisms, procedures for monitoring performance and programmes to build capacity in the area of management of public funds.[89] The critics of direct budget support argue that the weakness of institutions in many developing countries may result in fund mismanagement. But these arguments are more about not investing enough in capacity development in foreign aid than they are about criticisms of the recipient countries themselves. Those that promote direct budget assistance are more interested in assessing the conditions for budget support to work effectively. These include:

- government strategies that cover relatively long periods of time with targets and goals;
- good and credible links between these strategies and the government budget (for example, do the energy access strategies have sufficient and credible budgets allocated to make these happen?);
- good efforts at establishing capacities, both institutional and at the individual level, to assign responsibilities with accountability measures;
- adequate counterpart efforts to raise funds, either through taxes or other measures, to increase tax revenue for the implementation of the strategy; and

strong coordination mechanisms among sectors and donors.[90] Some donors, such as the European Union, see direct budget support as a way of promoting countries' ownership of development policies and reforms. In 2015, EU budget support amounted to 1.59 billion euros and constituted 20% of the EU development aid. To be eligible for EU budget support, countries need to have well-defined strategies and policies, a stable macro-economic framework, good public financial management or plans to improve it if necessary, and transparency and oversight of the budget.[91]

5.3.6 *What Is Scalable?*

5.3.6.1 Performance-Based Financing Versus Traditional Financial Assistance

Performance or results-based financing (including for and with technical assistance) has been most commonly used in the transport and health sectors, and there is expansive literature on the use and the experiences gained by practitioners in those fields over the years.[92] More recently, there has been an increasing interest in using this tool in the energy sector. Additionally, there is also more acceptance of performance-based financing as the 'conditionalities' components are linked to objective indicators designed to enhance performance and thus positive project outcomes. The latest example of such efforts is the International Energy Initiative, or Energy+, a partnership launched by the Norwegian government in late 2011.[93] There is increasing interest on the part of the World Bank and other donors to use results-based financing. The Scaling-Up Renewable Energy Programme, which is part of the Climate Investment Funds managed by the World Bank, is also encouraging the use of this approach. Figure 5.8 illustrates how this could be helpful in the energy sector and the funding possibilities.

From general experience with results-based aid in other sectors, and particular reviews by the World Bank's ESMAP, some lessons can be drawn indicating that this type of aid programming offers several possible benefits, such as clear objectives and indicators (which are crucial for interested investors), guarantee of longer-term support rather than project-based and project-bound timelines, increasing local ownership (as it encourages recipients to seek appropriate and local solutions) and capacity develop-

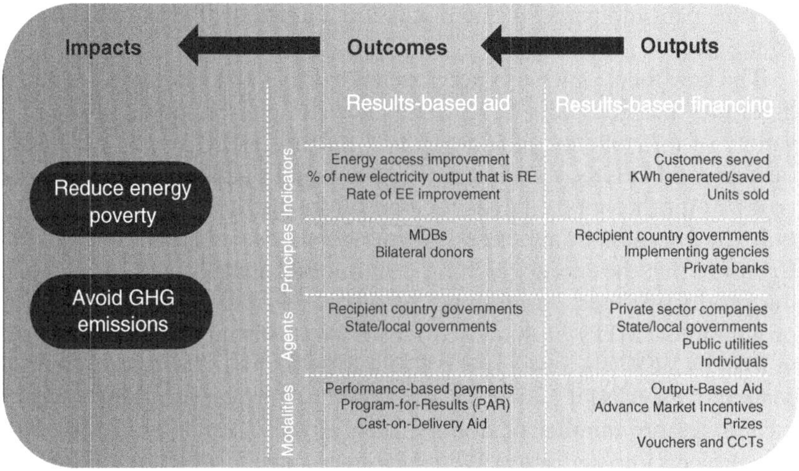

Fig. 5.8 Results-based funding possibilities for the energy sector (Source: Reproduced with permission from ESMAP n.d.)

ment outcomes, as the requirements of local monitoring and management are more demanding. The ESMAP review, however, cautions that this approach is not always adequate or appropriate, particularly with projects offering high risk tactics.[94]

5.3.6.2 Energy Supply Versus Energy Services: The Case of the Multi-Function Platforms

The recent shift—from focusing only on supply (transformers and transmission lines) when addressing energy needs to a vision that also considers the services that energy provides—has enormous implications for foreign aid to energy. This was justified given the dearth of infrastructure. But as these new facilities have been gradually built and improved over the decades, attention has now shifted to whether the energy that is being supplied or produced is really getting to the whole population at a price that is affordable, and for the purposes for which the population requires it. Issues such as health, productive activities for employment and livelihood and agricultural production (with linkages to food security), direct lighting, heating, cooling, cooking, water pumping, refrigeration and communication, not to mention issues of gender and equity, are now part and parcel of the discussion and focus of foreign aid to the energy sector.

The new focus on energy services has led to a rich body of foreign aid literature that considers how to enhance the impacts of aid through a variety of instruments, including micro-finance to help poor businesses operate, build assets, stabilize consumption and shield themselves against crises.[95] Another modality has focused on foreign aid that promotes not only the provision of and access to energy but also extending foreign aid to ensure that the energy reaching communities actually ends up promoting productive uses. An example of this is the multi-function platform programmes (MFPs) now active in many West African countries. Created in Mali in 1994 (the result of a collaboration between the government of Mali and the UNDP), support for MFPs has grown in recent years to include a large number of donors, such as the Gates Foundation, with new projects initiated across West African countries. MFPs combine technical support with social and organizational support to seek solutions that go beyond simply supplying energy. An example of such a programme would be one that supplies a community with a small diesel engine (in many places now being substituted by biofuels) mounted on a chassis with interchangeable attachments that can be used for many tasks including milling, grinding, water pumping, electricity generation, battery charging and communications; thus, energy is paired with production in one project.

The success of MFPs supports the notion that the simple supply of energy is not enough to declare success. The uses of that energy are equally important to consider and, in places, to be provided. In the case of the MFPs, such uses are typically productive ones that engage the community and provide employment, particularly to women, in the area of agroprocessing and promoting economic and social development.

Understandably, MFPs have been the subject of many evaluations to measure their full impact. Results have largely been positive. An article by Nygaard in *Energy Policy* presents a particularly nuanced assessment of MFPS, one that should be useful for those wanting to replicate the model. In his conclusion, Nygaard argues that the very features that attract most donors to the MFP (e.g. its multiplicity of technical functions, ideal and preconceived organizational set-ups and local fuel production) at the same time are allowing a multiplication of criteria of success. This, the author argues, is veiling some shortcomings being detected in areas that are critical to the philosophy of the MFPs, such as environmental aspects, the degree of multi-functionality, vulnerability of the system and other important features.[96] Should single purpose implements be necessarily more successful than complex and ambitious programmes such as the MFPs? As in every foreign aid case, the answer probably lies in both avoiding standardized solutions and the need to test what works for a particular environment before it is scaled up.

5.3.7 *What Is Transferable*

5.3.7.1 Blended Mechanisms (Loans, Grants and Technical Assistance) Versus Stand-Alone Technical Assistance or Finance

Blending facilities are new financial instruments established by the European Union for the purposes of leveraging funding and effectiveness of financial support to developing countries. Since their establishment in 2007, various loan, grants and technical assistance blending facilities have been established, several of which make large investments in the energy sector (e.g. ITFA, Infrastructure Trust Fund for Africa, which funds many energy projects in the region, operates in 47 countries in Africa and works in collaboration with many other bilateral financial institutions including the World Bank and the African Development Bank).[97] Blending loans, grants and technical assistance is not new in the field of development cooperation. What is new is the broadening of the practice to the EU and involving several institutions in a coordinated manner.[98] There are multiple advantages to this type of mechanism as opposed to stand-alone technical assistance or finance. In the energy sector particularly, mixes of technical assistance, finance and, often, targeted subsidies, as in the case of some energy access projects, is the most effective course.

Some of the advantages of blending facilities or mechanisms include:

- possibility of leveraging larger amount of resources as these pools attract investors and donors to participate in programmes and projects that are well embedded in national development strategies and plans;
- greater ownership and responsibility by the recipient country;
- greater donor coordination rather than fragmentation;

- more possibility for integrated and sectoral approaches and alignment with national strategies; and
- possibilities for capacity development.

These new mechanisms, as applied broadly by large groups of donors from the EU, are new but are bound to have a growing influence on foreign aid that would need to be observed for lessons and good practices. What is evident, especially in some regions such as Africa (the ITF) is that it has a considerable portion of its resources going to the energy sector.[99]

5.4 Discussion

The previous section provided samples of the various new or expanding approaches and foreign aid interventions now being delivered to many developing countries. There are a lot of others. Those discussed are a sample but are nevertheless illustrative of some of the most common programmes in operation. Additionally, those few were selected because of the lessons they offer on how foreign aid can or should interact with energy systems; they are indicative of the types of lessons that are becoming available. A number of common issues and the themes that emerge from these could prove helpful in the efforts to improve the performance of foreign aid to the energy sector in developing countries. The following is a brief summary of such.

5.4.1 Foreign Aid in Support of Transformative Rather Than Incremental Change

Given the centrality of energy, the urgency for timely action and the need for transformative change, incremental changes to the energy system will not be sufficient to succeed in addressing adequately the challenges of our century. Can foreign aid help countries pursue more radical transformation? Is foreign aid today already helping to do so? The evidence indicates that, for the most part, foreign aid is project focused rather than programme or system oriented. This, coupled with the fragmentation of donors, does not ultimately pave the way to the type of action that is required for a transformation of the global energy system. Nevertheless, it is evident that foreign aid in energy has indeed helped a large number of countries in many respects. However, those which have achieved impressive transformations, such as Vietnam, did not succeed solely because of foreign aid but rather because well-targeted aid was combined with national persistence, political commitment and coordination with helpful domestic policies and regulatory frameworks. Similarly, in the case of China, where foreign aid played a less important role, transformation was effected through a combination of strong political commitment, specific sector reforms, targeted programmes for electrification, clean cooking and renewable energy, and generous funding. According to the Global tracking Framework of the World Bank, through these means China was able to expand energy access to hundreds of millions of its population, with a now seemingly attainable goal of reaching almost universal access by 2015.[100]

5.4.2 Foreign Aid That Promotes Integration Rather Than Fragmentation

The literature suggests that the most effective strategies for addressing the challenges of today are those that promote integrated approaches to energy system design.[101] Approaches that allow energy policies to be coordinated with policies in other key sectors such as industry, buildings, urbanization, transport, food security and climate change, among others, are the most effective. As new demands in the energy sector emerge that require more integrated approaches and an integrated energy system strategy involving many stakeholders, the fragmented approaches that often prevail in foreign aid will not contribute to helping countries implement strategies that cut across sectors (as is urgently needed) nor will they promote policy dialogue (which evidence suggests has more effect). But fragmentation also works perversely in other ways. It strains human and financial resources of recipient countries as they seek to satisfy the procedures and demands of each donor. It often leads to duplication or, worse, to certain areas of importance being left unattended as countries are busy managing the fragmented aid that they receive. Partly to address problems of fragmentation but also to ensure that aid received is fully mainstreamed and in line with national priorities, some countries are creating national institutions specifically to help manage donor support and funding in the area of climate change and energy.[102] These new institutions provide best practices for emulation by other countries interested to ensure that their foreign aid is fully integrated into their development strategies, needs and priorities.

5.4.3 *Programme- Rather Than Project-Based Approaches to Foreign Aid*

Research shows that much of the foreign aid for energy is project based which, in turn, leads to discrete and punctual action in the energy sector rather than systems-oriented action, which evidence indicates would be the most effective. There have been many efforts to shift away from the project-based approaches. To be truly effective, energy strategies, policies, measures and programmes would ideally need to be coordinated with policies in other key sectors in the economy such as agriculture, health, transport, industry and buildings. Fragmented project-based foreign aid does not lend itself easily to this type of coordination and is not strong on long-term sustainability and country ownership of projects.

Design of programme-based approaches, rather than project-based ones, can more readily be aligned with the strategies of recipient countries. Such approaches can also support sectoral strategies that many countries already have in place and that can be more appropriately supported through sector-wide approaches. Programme-based approaches also fit better into the policy dialogue agenda and give support to policies with targeted goals and results, linked to a budget framework.

5.4.4 *Foreign Aid in Support of Technology Development and Innovation Rather Than Equipment Supply*

Achieving high science and technology capacities is fundamental for countries to advance on economic development goals. Furthermore, technological development, innovation and transfer are vital to the energy transformation that is required for addressing the challenges of our century. Technological development, in its whole spectrum, is therefore necessary for developing countries as they endeavour to make transformations in their energy systems. Ideally, to be effective, foreign aid would need to address the many stages of the innovation process that include starting from research through to incubation, demonstration, market creation (at times, for niche markets), and ultimately, widespread diffusion.[103]

Unfortunately, very little foreign aid has gone to long-term support for innovation and technological development.[104] The foreign aid that is most effective is that which would support countries over a long period of time throughout the various stages of the complex system of energy technology innovation. Unfortunately, many foreign aid projects in the past, with large equipment components but without much support for technology development or transfer, have failed because of a lack of attention to this complexity of innovation and technological development, instead only focusing on part of this complex system.

5.4.5 *Foreign Aid in Support of Public Finance Mechanisms to Mobilize, Catalyse and Leverage Private Investment Rather Than Stand-Alone Project Funding*

Previous sections of this chapter have referred to the immense amount of investment resources that will be required to finance the development, deployment, diffusion and transfer of clean technologies in developing countries, mostly for renewable energy and energy efficiency. It has also been mentioned that a large portion of these resources will need to come from the private sector. Does foreign aid have a place in this large-scale financing operation? Foreign aid has traditionally played an important role in providing loan financing for large infrastructure projects and for reforms and technology development in important sectors. There is no doubt that these will continue to be of importance as they fill a gap in financing needs in a number of countries. But foreign aid can also effectively help countries set up or strengthen what are referred to as public finance mechanisms, which can be crucial in leveraging financing for energy projects and which appear in the list that follows.[105] These include:

- credit lines to local commercial financial institutions (CFIs) for both senior and mezzanine debt;
- guarantees to share with local CFIs the commercial credit risks;
- debt financing of projects by entities other than CFIs;
- private equity funds investing risk capital in technology innovations;
- carbon finance facilities to monetize the advanced sale of emissions reduction to finance project investment costs;
- grants and contingent grants to share project development costs;
- loan softening programmes to mobilize domestic sources of capital;
- inducement prizes to stimulate R&D or technology development; and
- technical assistance to build the capacity of all actors along the financing chain.

By supporting the establishment and strengthening of these mechanisms, foreign aid can help promote investments in clean technologies, particularly those that are in the later stages of the technology innovation pathway but are still facing significant market barriers. Foreign aid can support the strengthening of the management of the private investment flows to bring down market barriers, bridge gaps and share risks with the private sector.[106] Foreign aid can help countries assess these market barriers, target the market segments with the greatest economic prospects, take a more programmatic approach to financing, and define roles and responsibilities for the various actors.

5.5 Conclusion

The importance of foreign aid to the energy sector is linked to the all-encompassing importance of energy in every aspect of modern life and economic activity. It is also connected to the immense and growing energy needs of developing countries—and the quality of the services that this energy can provide.

The nature of foreign aid to the energy sector has been changing over the last few years, along with greater shifts in the needs and priorities of developing countries. Many factors have contributed to these changes and they include, among others, technological changes that have made it easier to reach more the population in a more affordable and reliable way, better knowledge and science (and more acceptance of this science) about the link between energy and other pressing issues such as climate change, health and national security, more flexible mechanisms for delivering aid that combine technical assistance, trade and investments, and lastly, but equally important, the large increase in numbers and diversification of donors that now include many emerging economies and non-government actors.

The most complete data on foreign aid have been collected and reported by OECD-DAC. The group of donors that are participants in this group regularly report in standard and compatible formats and this information is regularly collated and reported upon further. Such systematic and historical data collection and reporting allow for the examination of trends occurring over the last 30 years. These recorded trends include the performance of donors in terms of the quantity of aid provided, to which sectors they provided, and, more recently, on the effectiveness of this aid assessed on criteria agreed upon by this group of traditional donors. The same cannot be said about the data provided by non-traditional donors, such as the emerging economies and non-government actors. In this category of donors, there is less transparency, and when data is available, it is not always comparable to that which is reported by OECD-DAC.

More recently, traditional donors have made great efforts to improve the performance and effectiveness of the foreign aid that they provide. The Paris Declaration and the Accra High Level Forum on Aid Effectiveness constitute efforts by traditional donors to make foreign aid more effective and more in line with the interests and needs of developing countries. The effects of these efforts have spilled over into international organizations, namely the UN and the World Bank, which in turn have introduced measures to make their own foreign aid more effective and coordinated. But these efforts are not always matched by the small but growing group of emerging donors who are expected to continue growing significantly in the foreseeable future.

Assessing the level of effectiveness of improvements over the years is difficult, given the relative scarcity of academic literature on the subject. Much of the literature that does exist focuses on effects at the macro level and aggregate outcomes of growth and poverty eradication. The evidence that exists on concrete interventions with specific technologies targeted to achieve concrete goals and objectives is scarce. This chapter has therefore relied on reports by donors' independent evaluation units. Based on these reports, a brief sample was provided about what appears to work, what could work, what is scalable and what is transferable. The most salient feature that emerges from this survey is that the success of foreign aid is closely linked to how robust the enabling environment (i.e. the policy and regulatory frameworks) is in the receiving countries. This is so much the case that there are efforts by many donors to concentrate their aid only in those countries where there is good governance and strong policymaking capacity. Understandably, this has met some resistance because, as the argument goes, by concentrating aid in strong countries, donors would only exacerbate the existing problems in a large majority of developing countries throughout the world—namely, weak governance capacities and management skills. We must be careful not to penalize the states most requiring assistance. There is an argument to be made, instead, on the need to invest more heavily in capacity development programmes across the board, in conjunction with aid.

Most saliently, success of foreign aid is closely linked to how well that aid is aligned to the needs and priorities of the receiving countries. Work to align aid with domestic goals has made foreign aid more complex and, consequently, higher in transaction costs. Yet ultimately, such shifts are necessary to make aid more effective (if not necessarily efficient). Efforts to coordinate foreign aid have not always been successful. One concern that has been highlighted by this chapter is the increase in number of actors in the foreign aid sector and, along with them, a new proliferation of donors, programmes, funds and projects. Although more funds and projects offer the potential for greater effect, this proliferation and fragmentation has also placed heavier burdens on developing countries.

This chapter has repeatedly alluded to the need for a full transformation of the global energy system—and the importance of foreign aid in helping countries achieve this transformation. The need for the former is irrefutable and the need for the latter should be recognized as such as well. Importantly, this review also emphasizes that this transformation cannot be carried out in a piecemeal manner but rather must occur through integrated approaches designed to address systemic failures and deficiencies. This in turn reveals a number of definite requirements in order for foreign aid to be effective. Aid that promotes coordination across sectors has a better chance of success than aid that is scattered and project based. Aid that is focused on achieving concrete goals which can be monitored and measured has a better chance for success and in achieving buy-in from a large and diverse group of stakeholders. The Sustainable Energy for All initiative launched by the UN Secretary General is a good example of what such an effort can look like. By bringing a large number of countries onboard early, to support very concrete goals, a campaign like this one can enlist the broad support of stakeholders who have space and means clearly to declare which goals and with which actions they are committed to contribute. With consistent coordinated efforts like this, we will be better equipped to address the large problems facing the world today and realize the ambitious goals of our time.

Acknowledgement The author would like to thank Aaron Reuben at the Yale Center for Environmental Law and Policy, Yale University, for his research assistance.

Notes

1. Organisation for Economic Co-operation and Development-Development Assistance Committee (OECD-DAC) is the forum of selected members of the OECD established as a platform for discussing issues on aid to developing countries. The forum includes the World Bank, the IMF and UNDP as observers.
2. Some of this data belongs to some countries (e.g. China) and some NGOs which do not have efficient mechanisms for sharing their data, and some businesses (which in practice do not share certain data due to fears of competition).
3. OECD-IEA (OECD-International Energy Agency) (2010: 3), 'Energy Poverty: How to Make Modern Energy Access Universal'. Special early excerpt of the World Energy Outlook 2010 for the UN General Assembly on the Millennium Development Goals. Paris: International Energy Agency.
4. WEO, Modern Energy for All, http://www.worldenergyoutlook.org/resources/energydevelopment/. Accessed on April 24, 2017.
5. GEA (Global Energy Assessment) (2012: xiii). Global Energy Assessment: Toward a Sustainable Future. Cambridge (UK) and Laxenburg (Austria): Cambridge University Press and International Institute for Applied Systems Analysis.
6. GEA (Global Energy Assessment) (2012). Global Energy Assessment: Toward a Sustainable Future. Cambridge (UK) and Laxenburg (Austria): Cambridge University Press and International Institute for Applied Systems Analysis.
7. The lower and higher figures are provided by the World Energy Outlook, Special Excerpt on Energy Poverty and the Global Energy Assessment respectively.
8. Stern, N. (2007). The Economics of Climate Change: The Stern Review. Cambridge: Cambridge University Press.
9. International Energy Agency (IEA) and the World Bank. 2017. "Sustainable Energy for All 2017—Progress toward sustainable energy" Summary World Bank, Washington, DC.
10. IEA (International Energy Agency) (2011). World Energy Outlook 2011. Paris: IEA.
11. The Paris Agreement was reached at the 21st Conference of the Parties of the UNFCCC. It requires all Parties to declare their efforts to address climate change through their nationally determined contributions and to continue to strengthen these efforts in years ahead. It is the first time that all nations agreed to commit to such targets.
12. The Sustainable Energy for All initiative (www.sustainableenergyforall.org/) was launched by the UN Secretary General at the GA session of

2011. The initiative seeks to promote three goals, all to be reached by 2030: energy access for all, a doubling of energy efficiency improvements in the developing world, and a doubling of the share of renewable energy in global final energy mixes.

13. International Energy Agency (IEA) and the World Bank. 2017. "Sustainable Energy for All 2017—Progress toward sustainable energy" Summary World Bank, Washington, DC.
14. On September 25, 2015, countries adopted a set of goals to end poverty, protect the planet, and ensure prosperity for all as part of a new sustainable development agenda. www.un.org
15. Energy+ is an initiative of the government of Norway to promote access to energy and low-carbon development in developing nations.
16. These are the Intended Nationally Determined Contributions identified the actions that countries intended to take under the Paris Agreement. In their INDCs (to be converted to NDCs after the Paris Agreement came into force), countries outlined the actions they are taking to reduce GHG emissions and other climate-related action, including adaptation measures.
17. *New York Times* (2010). Editorial, 'Cleaner, Healthier Cookstoves'. 23 September. Available at: www.nytimes.com/2010/09/24/opinion/24fri4.html?_r=0 (accessed 10 April 2013); Jolly, D. (2010). 'Energy Access Seen as Vital to Abolishing Worst Poverty'. The New York Times, 21 September. Available at: www.nytimes.com/2010/09/22/business/global/22energy.html?ref=david_jolly (accessed 10 April 2013).
18. See the reports of the Global Alliance for Clean Cookstoves for examples of this attention, available at: www.cleancookstoves.org/resources/reports-and-research/
19. Juselius, K., N. F. Møller, and F. Tarp (2011). 'The Long-Run Impact of Foreign Aid in 36 African Countries: Insights from Multivariate Time Series Analysis'. WIDER Working Paper 2011/51. Helsinki: UNU-WIDER.
20. Since its inception, The Bill and Melinda Gates Foundation made total grant payments totalling US$36.7 billion through 2015. These include programs in global development, health, policy support and US programmes. Bill & Melinda Gates Foundation. (n.d.). Accessed May 16, 2017, from http://www.gatesfoundation.org/
21. Buchner, B., A. Falconer, M. Herve-Mignucci, and C. Trabacchi (2012). 'The Landscape of Climate Finance 2012'. CPI Report. San Francisco, CA: Climate Policy Initiative.; Juergens, I., H. Amecke, R. Boyd, B. Buchner, A. Novikova, A. Rosenberg, K. Stelmakh, and A. Vasa (2012). 'The Landscape of Climate Finance in Germany'. CPI Report. San Francisco, CA: Climate Policy Initiative.

22. Frot, E., and J. Santiso (2008). 'Development Aid and Portfolio Funds: Trends, Volatility and Fragmentation'. OECD Working Paper 275. Paris: OECD Development Centre.
23. Little, H. M. (2010). 'The Role of Private Assistance in International Development'. International Law and Politics, 42: 101–19.
24. Roland, G. (2008). Privatization: Successes and Failures. New York, NY: Columbia University Press.
25. OECD-DAC (2010). 'Measuring Aid for Energy'. Paris: OECD. Available at: www.oecd.org/dac/stats/45066235.pdf (accessed 10 April 2013).
26. 'Tied' aid refers to the type of aid which is directly linked to an obligation to purchase goods and services from the country providing the foreign aid.
27. http://www.oecd.org/dac/stats/energy, downloaded 15 May 2017.
28. Johansson, T., N. Nakicenovic, A. Patwardhan, and L. Gomez-Echeverri (2012). Global Energy Assessment: Toward a Sustainable Future. Cambridge (UK) and Laxenburg (Austria): Cambridge University Press and International Institute for Applied Systems Analysis.
29. ENAN (Encyclopaedia of the New American Nation) (n.d.) 'Foreign Aid: The Origins of Foreign Aid'. Available at: www.americanforeignrelations.com/E-N/Foreign-Aid-The-origins-of-foreign-aid.html (accessed 10 April 2012).
30. Truman, H. S. (1949). Inaugural Address, 20 January. Inaugural Addresses of the Presidents of the United States. Available at: www.bartleby.com/124/pres53.html (accessed 10 April 2013).
31. Morgenthau, H. (1962). 'A Political Theory of Foreign Aid'. The American Political Science Review, 56(2): 301–09.
32. Bauer, P. (1972). Dissent on Development. Cambridge, MA: Harvard University Press.
33. Easterly, W. (2001). The Elusive Quest for Growth: Economists Adventures and Misadventures in the Tropics. Cambridge, MA: Cambridge University Press; Easterly, W. (2006). A White Man's Burden: Why the West's Efforts to Aid the Rest Have Done So Much Ill and So Little Good. New York, NY: The Penguin Press; Easterly, W., R. Levine, and D. Roodman (2004). 'New Data, New Doubts: A Comment on Burnside and Dollar's Aid, Policies and Growth'. American Economic Review, 94 (3): 774–80.
34. Friedman, M. (1958). 'Foreign Economic Aid'. Yale Review, 47(4): 501–16.
35. Sachs, J., J. McArthur, G. Schmidt-Traub, M. Kruk, C. Bahadur, M. Faye, and G. McCord (2004). 'Ending Africa's Poverty Trap'. Brookings Papers on Economic Activity, 1: 117–240.
36. Stiglitz, J. (2002). 'Overseas Aid is Money Well Spent'. Financial Times, 14 April.

37. Stern, N. (2002). 'Making the Case for Aid, Savings and Growth in LDCs: Some Recent Evidence'. In World Bank (ed.), A Case for Aid: Building a Consensus for Development Assistance. Washington, DC. World Bank.
38. Boone, P. (1994). 'The Impact of Foreign Aid on Savings and Growth'. Centre for Economic Performance Working Paper 677. London: London School of Economics.
39. Burnside, C., and D. Dollar (2000). 'Aid Policies and Growth'. American Economic Review, 90(4): 847–68
40. Radelet, S., M. Clemens, and R. Bhavnani (2006). 'Aid and Growth. The Current Debate and Some New Evidence'. In P. Isard, L. Lipschitz, A. Mourmuras and P. Heller (eds), Macroeconomic Management of Foreign Aid: Opportunities and Pitfall. Washington, DC: International Monetary Fund.
41. Radelet, S. (2004). 'Aid Effectiveness and the Millennium Development Goals'. CGD Working Paper 39. Washington, DC: Center for Global Development.
42. http://www.oecd.org/dac/stats/energy, downloaded 15 May 2017.
43. Multilateral financial institutions in this reference include the World Bank and the regional development banks: African Development Bank, Asian Development Bank and the Inter-American Development Bank.
44. One example of such an independent information source is www.climate-fundsupdate.org, which regularly tracks climate change financing.
45. OECD (The Organisation for Economic Co-operation and Development) (2010). Better Aid: Evaluation in Development Agencies. Paris: OECD Publishing.
46. The UN Energy website (www.unenergy.org) provides a full table with a list of activities by agency in each of these areas.
47. ESMAP (www.esmap.org) has operated in over 100 countries through more than 800 activities covering a broad range of energy issues.
48. United Nations Industrial Development Organization.
49. Food and Agriculture Organization of the United Nations.
50. International Atomic Energy Agency.
51. IRENA (International Renewable Energy Agency) (n.d.). About IRENA. Abu Dhabi, United Arab Emirates: International Renewable Energy Agency. Available at: from www.irena.org/Menu/index.aspx?PriMenuID=13&mnu=Pri (accessed 11 April 2013).
52. OECD-IEA (OECD-International Energy Agency) (2010). 'Energy Poverty: How to Make Modern Energy Access Universal'. Special early excerpt of the World Energy Outlook 2010 for the UN General Assembly on the Millennium Development Goals. Paris: International Energy Agency.

53. Woods, N. (2008). 'Whose Aid? Whose Influence? China, Emerging Donors and the Silent Revolution in Development Assistance'. International Affairs, 84(6): 1–18.
54. The Paris Declaration was agreed in 2005 by the OECD-DAC group of donors and is based on five principles to make aid more effective: ownership by recipient countries, alignment of aid to the recipient-developed strategies, harmonization among donors, focus on results, and mutual accountability.
55. Woods, N. (2008). 'Whose Aid? Whose Influence? China, Emerging Donors and the Silent Revolution in Development Assistance'. International Affairs, 84(6): 1–18.
56. Little, H. M. (2010). 'The Role of Private Assistance in International Development'. International Law and Politics, 42: 101–19.
57. Ibid.
58. Practical Action, based in the UK, is one of the many development charity institutions working in the energy-for-the-poor sector.
59. Stern, E., L. Altinger, O. Feinstein, M. Maranon, D. Ruenbergerg, N. S. Schulz, and N. S. Nielsen (2008). 'Thematic Study on the Paris Declaration, Aid Effectiveness, and Development Effectiveness'. Copenhagen: DARA, Ministry of Foreign Affairs.
60. See the OECD-DAC resources hub on aid effectiveness for more information, retrieved from www.oecd.org/dac/effectiveness/
61. UN Energy Knowledge Network (n.d.). 'What is UN-Energy? The United Nations' Inter-Agency Mechanism on Energy'. Available at: www.unn-energy.org (accessed 11 April 2013).
62. Frot, E., and J. Santiso (2010). 'Crushed Aid: Fragmentation in Sectoral Aid'. OECD Working Paper 284. Paris: OECD Development Centre.
63. Knack, S., and A. Rahman (2007). 'Donor Fragmentation and Bureaucratic Quality in Aid Recipients'. Journal of Development Economics, 83(1): 176–97.
64. Mwega, F. M. (2009). 'A Case Study of Aid Effectiveness in Kenya: Volatility and Fragmentation of Foreign Aid, with a Focus on Health'. Wolfenshon Center for Development Working Paper 8. Washington, DC: Brookings Institution.
65. Lancaster, C. (1999). 'Aid Effectiveness in Africa: The Unfinished Agenda'. Journal of African Economies, 5(8): 487–503.
66. Collier, P. (1999). 'Aid Dependency: A Critique'. Journal of African Economies, 8(4): 528–45.
67. Auer, M. (2005). 'Foreign Aid to Promote Energy Efficiency in Mexico: An Institutional Analysis'. The Journal of Energy and Development, 31(1): 85–100.

68. Ballesteros, A., S. Nakhooda, J. Werksman, and K. Hurlburt (2010). 'Power, Responsibility, and Accountability: Re-Thinking the Legitimacy of institutions for Climate Finance'. Final Report. Washington, DC: World Resources Institute.
69. Adaptation Fund (2012). The Adaptation Fund and Direct Access: Supporting Developing Countries in Undertaking Concrete Measures to Adapt to the Adverse Effects of Climate Change'. Washington, DC: The Adaptation Fund. Available at: www.adaptation-fund.org/sites/default/files/DirectAccessMemo29_Oct_2012_0.pdf; Buchner, B., A. Falconer, M. Herve-Mignucci, and C. Trabacchi (2012). 'The Landscape of Climate Finance 2012'. CPI Report. San Francisco, CA: Climate Policy Initiative.
70. Fontaine, A. L., M. Spearman, I. Househam, R. Gunning, and O. Beucher (2012). External Evaluation of ESMAP 2007-2011. Final Evaluation Report. Gatineau, QC: Le Group - Conseil Baastel Itee.
71. Ethanol Producers Magazine (2010). 'International Ethanol Report, 2010'. Available at: from www.ethanolproducer.com/articles/6696/international-ethanol-report-2010 (accessed on 11 April 2013).
72. Sandalow, D. B. (2006). 'Ethanol: Lessons from Brazil'. Brookings Paper. Available at: www.brookings.edu/research/articles/2006/05/energy-sandalow (accessed 13 April 2013).
73. Schelesinger, S. (2012). 'Brazilian International Cooperation and Investments: The Internationalization of Ethanol and Biodiesel'. FASE: Solidariedade e Educaco. Available at: http://issuu.com/ongfase/docs/sergio_schlesinger_eng
74. GEF (n.d. [a]) 5-1. 'Operational Program Number 5: Removal of Barriers to energy efficiency and energy conservation. Washington, DC: Global Environment Facility. Available at: www.thegef.org/gef/sites/thegef.org/files/documents/document/OP_5_English.pdf (accessed 13 April 2013).
75. GEF (n.d. [b]) 6-1. 'Operational Program Number 6: Promoting the Adoption of Renewable Energy by Removing Barriers and Reducing Implementation Costs'. Washington, DC: Global Environment Facility. Available at: www.thegef.org/gef/sites/thegef.org/ files/documents/document/OP_6_English.pdf (accessed 13 April 2013).
76. GEF (n.d. [c]). 'Mitigation: Energy Efficiency. Good Practices and Lessons'. Washington, DC: Global Environment Facility. Available at: www.thegef.org/gef/node/2202 (accessed 13 April 2013).
77. GEF (Global Environment Facility) (2008). 'Transfer of Environmentally Sound Technologies: The GEF Experience'. Washington, DC: Global Environment Facility. Available at: http://www.thegef.org/gef/node/1549

78. GEF 7 Programming Directions and Policy Agenda, Global Environment Facility March 7, 2017.
79. IEA and World Bank, 2027, Global Tracking Framework.
80. UNDP (United Nations Development Programme) (2011). 'Going Renewable: China's Success Story in Capacity Development—Towards an 'Energy Plus' Approach for the Poor: A Review of Good Practices and Lessons Learned from Asia and the Pacific'. Case Study 6. New York, New York: UNDP.
81. A method of generating electrical power from solar radiation.
82. Reiche, K., A. Covarrubias, and E. Martinot (2000). 'Expanding Electricity Access to Remote Areas: Off-Grid Rural Electrification in Developing Countries. World Power 2000. London: Isherwood Production, 62–60.
83. Birner, S., and E. Martinot (2003). 'Market Transformation for Energy Efficient Products: Lessons from Programmes in Developing Countries'. Washington, DC: Global Environment Facility. Available at: www.keion-line.org/misc-docs/Birner_Martinot_EP.pdf (accessed 13 April 2013).
84. Feed-in tariffs are mechanisms that provide renewable energy generators with a fixed long-term price for power and allow them guaranteed access to the electricity grid.
85. World Bank/GEF (2006). 'World Bank GEF Energy Efficiency Projects: Synthesis Report'. Washington, DC: World Bank, Global Environment Fund.
86. IEA and World Bank, Global Tracking Framework 2017.
87. UNDP (2012). 'Micro-credit for Energy Services: Transforming Basic Needs of Poor Households into Effective Market Demand. Towards an "Energy Plus" Approach for the Poor: A Review of Good Practices and Lessons Learned from Asia and the Pacific'. Case Study 3. New York, NY: UNDP.
88. UK DFID (UK Department of International Development) (2004):3. 'Poverty Reduction Budget Support'. DFID Policy Paper. London: DFID.
89. Goody, A. (2009). 'Direct Budget Support'. In A. Goody (ed.), International Development: The Aid Effectiveness Debate. Ottawa: Library of Parliament Research Publications, Parliament of Canada.
90. Norad (Norwegian Agency for Development Cooperation) (2002). 'Direct Budget Support'. Norad Position Paper 16. Oslo: Norwegian Ministry of Foreign Affairs.
91. https://ec.europa.eu/europeaid/node/13967, accessed 16 May 2017.
92. Canavan, A., J. Toonen, and R. Elovainio (2008). 'Performance Based Financing: An International Review of the Literature'. Amsterdam: KIT Development Policy & Practice.

93. The aims of E+ are to support efforts to achieve universal sustainable energy access and the reduction of GHG emissions in developing countries.
94. ESMAP (Energy Sector Management Assistance Programme) (n.d.). 'Results-Based Approaches for Energy Sector Development'. Washington, D.C: World Bank. Available at: www.esmap.org/Results_Based_Approaches (accessed 13 April 2013).
95. Morris, E., J. Winiecki, S. Chowdhary, and K. Cortiglia (2007). 'Using Microfinance to Expand Access to Energy Services: Summary of Findings'. SEEP Network Report. Washington, DC: The SEEP Network.
96. Nygaard, I. (2010). 'Institutional Options for Rural Energy Access: Exploring the Concept of the Multifunctional Platform in West Africa'. Energy Policy, 38(2): 1192–201.
97. Ferrer, N. J., and A. Behrens (2011). 'Innovative Approaches to EU Blending Mechanisms for Development Finance'. CEPI's Special Report for the German Federal Ministry for Economic Cooperation and Development. Brussels: The Centre for European Policy Studies.
98. UK DFID (2011). 'EU Blending Facilities: Implications for Future Governance Options'. London: DFID.
99. Ferrer, N. J., and A. Behrens (2011). 'Innovative Approaches to EU Blending Mechanisms for Development Finance'. CEPI's Special Report for the German Federal Ministry for Economic Cooperation and Development. Brussels: The Centre for European Policy Studies.
100. China (The People's Republic of China) (2012). 'China's Energy Policy 2012'. Beijing: Information Office of the State Council. Available at: www.news. xinhuanet.com/english/china/2012-10/24/c_131927649.htm (accessed 13 April 2013).
101. GEA (Global Energy Assessment) (2012). Global Energy Assessment: Toward a Sustainable Future. Cambridge (UK) and Laxenburg (Austria): Cambridge University Press and International Institute for Applied Systems Analysis.
102. Gomez-Echeverri, L. (2010). 'National Funding Strategies: Their Role in the Transition to a New Paradigm of Global Cooperation in Climate Change'. Ebci Policy Report. Available at: www.eurocapacity.org/downloads/NFEsPolicyReport.pdf (accessed 13 April 2013).
103. Grubler, A., F. Aguayo, K. Gallagher, M. Hekkert, K. Jiang, L. Mytelka, L. Neij, G. Nemet, and C. Wilson. (2012). 'Policies for the Energy Technology Innovation System (ETIS)'. In Global Energy Assessment: Toward a Sustainable Future. Cambridge (UK) and Laxenburg (Austria): Cambridge University Press and Institute for Applied Systems Analysis, pp. 1665–744.
104. Naudé, W. (2011). 'Foreign Aid for Innovation: The Missing Ingredient in Private Sector Development?'. MSM Working Paper 2011/35. Maastricht: Maastricht School of Management.

105. McLean, J., J. Tan, D. Tirpak, V. Sonntag, and E. U. O'Brien (2008). 'Public Finance Mechanisms to Mobilize Investment in Climate Change Mitigation'. New York, NY: United Nations Environment Programme.
106. McLean, J., J. Tan, D. Tirpak, V. Sonntag, and E. U. O'Brien (2008). 'Public Finance Mechanisms to Mobilize Investment in Climate Change Mitigation'. New York, NY: United Nations Environment Programme.

References

Adaptation Fund (2012). *The Adaptation Fund and Direct Access: Supporting Developing Countries in Undertaking Concrete Measures to Adapt to the Adverse Effects of Climate Change*. Washington, DC: The Adaptation Fund. www.adaptation-fund.org/sites/default/files/DirectAccessMemo29_Oct_2012_0.pdf

ADB (Asia Development Bank). 2011. *Energy for All: Viet Nam's Success in Increasing Access to Energy through Rural Electrification*. ADB Working Paper. Mandaluyong City: Asian Development Bank.

Auer, M. 2005. Foreign Aid to Promote Energy Efficiency in Mexico: An Institutional Analysis. *The Journal of Energy and Development* 31 (1): 85–100.

Ballesteros, A., S. Nakhooda, J. Werksman, and K. Hurlburt. 2010. Power, Responsibility, and Accountability: Re-thinking the Legitimacy of institutions for Climate Finance. Final Report. Washington, DC: World Resources Institute.

Bauer, P. 1972. *Dissent on Development*. Cambridge, MA: Harvard University Press.

BIAC (Business and Industry Advisory Committee to the OECD). 2009. *Addressing Perception and Communication in the Energy Sector*. Paris: BIAC. www.biac.org/statements/energy/FIN09-04_Addressing_Perception_and_Communication_in_the_Energy_Sector.pdf. Accessed 9 April 2013.

Birner, S., and E. Martinot. 2003. *Market Transformation for Energy Efficient Products: Lessons from Programmes in Developing Countries*. Washington, DC: Global Environment Facility. www.keionline.org/misc-docs/Birner_Martinot_EP.pdf. Accessed 13 April 2013.

Boone, P. 1994. *The Impact of Foreign Aid on Savings and Growth*. Centre for Economic Performance Working Paper 677. London: London School of Economics.

Buchner, B., A. Falconer, M. Herve-Mignucci, and C. Trabacchi. 2012. *The Landscape of Climate Finance 2012*. CPI Report. San Francisco, CA: Climate Policy Initiative.

Burnside, C., and D. Dollar. 2000. Aid Policies and Growth. *American Economic Review* 90 (4): 847–868.

Canavan, A., J. Toonen, and R. Elovainio. 2008. *Performance Based Financing: An International Review of the Literature*. Amsterdam: KIT Development Policy & Practice.

China (The People's Republic of China). 2012. *China's Energy Policy 2012*. Beijing: Information Office of the State Council. www.news. xinhuanet.com/english/china/2012-10/24/c_131927649.htm. Accessed 13 April 2013.

Collier, P. 1999. Aid Dependency: A Critique. *Journal of African Economies* 8 (4): 528–545.

Easterly, W. 2001. *The Elusive Quest for Growth: Economists Adventures and Misadventures in the Tropics*. Cambridge, MA: Cambridge University Press.

———. 2006. *A White Man's Burden: Why the West's Efforts to Aid the Rest Have Done So Much Ill and So Little Good*. New York, NY: The Penguin Press.

Easterly, W., R. Levine, and D. Roodman. 2004. New Data, New Doubts: A Comment on Burnside and Dollar's Aid, Policies and Growth. *American Economic Review* 94 (3): 774–780.

ENAN (Encyclopaedia of the New American Nation). n.d. Foreign Aid: The Origins of Foreign Aid. www.americanforeignrelations.com/E-N/Foreign-Aid-The-origins-of-foreign-aid.html. Accessed 10 April 2012.

ESMAP (Energy Sector Management Assistance Programme). n.d. *Results-Based Approaches for Energy Sector Development*. Washington, DC: World Bank. www.esmap.org/Results_Based_Approaches. Accessed 13 April 2013.

Ethanol Producers Magazine. 2010. *International Ethanol Report, 2010*. www.ethanolproducer.com/articles/6696/international-ethanol-report-2010. Accessed on 11 April 2013.

Ferrer, N.J., and A. Behrens. 2011. *Innovative Approaches to EU Blending Mechanisms for Development Finance*. CEPI's Special Report for the German Federal Ministry for Economic Cooperation and Development. Brussels: The Centre for European Policy Studies.

Fontaine, A.L., M. Spearman, I. Househam, R. Gunning, and O. Beucher. 2012. *External Evaluation of ESMAP 2007–2011*. Final Evaluation Report. Gatineau, QC: Le Group—Conseil Baastel Itee.

Friedman, M. 1958. Foreign Economic Aid. *Yale Review* 47 (4): 501–516.

Frot, E., and J. Santiso. 2008. *Development Aid and Portfolio Funds: Trends, Volatility and Fragmentation*. OECD Working Paper 275. Paris: OECD Development Centre.

———. 2010. *Crushed Aid: Fragmentation in Sectoral Aid*. OECD Working Paper 284. Paris: OECD Development Centre.

GEA (Global Energy Assessment). 2012. *Global Energy Assessment: Toward a Sustainable Future*. Cambridge (UK) and Laxenburg (Austria): Cambridge University Press and International Institute for Applied Systems Analysis.

GEF (Global Environment Facility). 2008. *Transfer of Environmentally Sound Technologies: The GEF Experience*. Washington, DC: Global Environment Facility. http://www.thegef.org/gef/node/1549

———. n.d.-a. *Operational Program Number 5: Removal of Barriers to Energy Efficiency and Energy Conservation*. Washington, DC: Global Environment

Facility. www.thegef.org/gef/sites/thegef.org/files/documents/document/OP_5_English.pdf. Accessed 13 April 2013.

———. n.d.-b. *Operational Program Number 6: Promoting the Adoption of Renewable Energy by Removing Barriers and Reducing Implementation Costs.* Washington, DC: Global Environment Facility. www.thegef.org/gef/sites/thegef.org/files/documents/document/OP_6_Englishpdf. Accessed 13 April 2013.

———. n.d.-c. *Mitigation: Energy Efficiency. Good Practices and Lessons.* Washington, DC: Global Environment Facility. www.thegef.org/gef/node/2202. Accessed 13 April 2013.

Gomez-Echeverri, L. 2010. *National Funding Strategies: Their Role in the Transition to a New Paradigm of Global Cooperation in Climate Change.* Ebci Policy Report. www.eurocapacity.org/downloads/NFEsPolicyReport.pdf. Accessed 13 April 2013.

Goody, A. 2009. Direct Budget Support. In *International Development: The Aid Effectiveness Debate*, ed. A. Goody. Ottawa: Library of Parliament Research Publications, Parliament of Canada.

Grubler, A., F. Aguayo, K. Gallagher, M. Hekkert, K. Jiang, L. Mytelka, L. Neij, G. Nemet, and C. Wilson. 2012. Policies for the Energy Technology Innovation System (ETIS). In *Global Energy Assessment: Toward a Sustainable Future*, 1665–1744. Cambridge (UK) and Laxenburg (Austria): Cambridge University Press and Institute for Applied Systems Analysis.

IEA (International Energy Agency). 2011. *World Energy Outlook 2011.* Paris: IEA.

IMF. 2011. *Sub Saharan Africa Sustaining the Expansion.* Regional Economic Outlook, World Economic and Financial Survey. Washington, DC.

International Energy Agency (IEA) and the World Bank. 2017. *Sustainable Energy for All 2017—Progress Toward Sustainable Energy.* Washington, DC: Summary World Bank.

IRENA (International Renewable Energy Agency). n.d. About IRENA. Abu Dhabi, United Arab Emirates: International Renewable Energy Agency. www.irena.org/Menu/index.aspx?PriMenuID=13&mnu=Pri. Accessed 11 April 2013.

Johansson, T., N. Nakicenovic, A. Patwardhan, and L. Gomez-Echeverri. 2012. *Global Energy Assessment: Toward a Sustainable Future.* Cambridge (UK) and Laxenburg (Austria): Cambridge University Press and International Institute for Applied Systems Analysis.

Jolly, D. 2010. Energy Access Seen as Vital to Abolishing Worst Poverty. *The New York Times*, 21 September. www.nytimes.com/2010/09/22/business/global/22energy.html?ref=david_jolly. Accessed 10 April 2013.

Juergens, I., H. Amecke, R. Boyd, B. Buchner, A. Novikova, A. Rosenberg, K. Stelmakh, and A. Vasa. 2012. *The Landscape of Climate Finance in Germany.* CPI Report. San Francisco, CA: Climate Policy Initiative.

Juselius, K., N.F. Møller, and F. Tarp. 2011. *The Long-Run Impact of Foreign Aid in 36 African Countries: Insights from Multivariate Time Series Analysis.* WIDER Working Paper 2011/51. Helsinki: UNU-WIDER.

Knack, S., and A. Rahman. 2007. Donor Fragmentation and Bureaucratic Quality in Aid Recipients. *Journal of Development Economics* 83 (1): 176–197.

Lancaster, C. 1999. Aid Effectiveness in Africa: The Unfinished Agenda. *Journal of African Economies* 5 (8): 487–503.

Little, H.M. 2010. The Role of Private Assistance in International Development. *International Law and Politics* 42: 101–119.

McLean, J., J. Tan, D. Tirpak, V. Sonntag, and E.U. O'Brien. 2008. *Public Finance Mechanisms to Mobilize Investment in Climate Change Mitigation.* New York, NY: United Nations Environment Programme.

Modi, V., S. McDade, D. Lallement, and J. Saghir. 2005. *Energy Services for the Millennium Development Goals.* New York: Millennium Project, UNDP, World Bank and ESAP. www-lutw.org/wp-content/uploads/Energy-services-for-the-millennium-development-goals.pdf.

Morgenthau, H. 1962. A Political Theory of Foreign Aid. *The American Political Science Review* 56 (2): 301–309.

Morris, E., J. Winiecki, S. Chowdhary, and K. Cortiglia. 2007. *Using Microfinance to Expand Access to Energy Services: Summary of Findings.* SEEP Network Report. Washington, DC: The SEEP Network.

Mwega, F.M. 2009. *A Case Study of Aid Effectiveness in Kenya: Volatility and Fragmentation of Foreign Aid, with a Focus on Health.* Wolfenshon Center for Development Working Paper 8. Washington, DC: Brookings Institution.

Naudé, W. 2011. *Foreign Aid for Innovation: The Missing Ingredient in Private Sector Development?* MSM Working Paper 2011/35. Maastricht: Maastricht School of Management.

Norad (Norwegian Agency for Development Cooperation). 2002. *Direct Budget Support.* Norad Position Paper 16. Oslo: Norwegian Ministry of Foreign Affairs.

Nygaard, I. 2010. Institutional Options for Rural Energy Access: Exploring the Concept of the Multifunctional Platform in West Africa. *Energy Policy* 38 (2): 1192–1201.

OECD (The Organisation for Economic Co-operation and Development). 2010. *Better Aid: Evaluation in Development Agencies.* Paris: OECD Publishing.

———. 2012. *Development Cooperation Report: Lessons in Linking Sustainability and Development*, OECD.

———. n.d. Aid to Energy Generation and Supply Sector. Paris: OECD. www.oecd.org/dac/stats/energy.htm

OECD-DAC. 2010. *Measuring Aid for Energy.* Paris: OECD. www.oecd.org/dac/stats/45066235.pdf. Accessed 10 April 2013.

OECD-IEA (OECD-International Energy Agency). 2010. Energy Poverty: How to Make Modern Energy Access Universal. Special Early Excerpt of the *World*

Energy Outlook 2010 for the UN General Assembly on the Millennium Development Goals. Paris: International Energy Agency.

Pachauri, S., A. Brew-Hammond, D.F. Barnes, D.H. Bouille, S. Gitonga, V. Modi, et al. 2012. Energy Access for Development. In *Global Energy Assessment–Toward a Sustainable Future*, 1401–1458. Cambridge and Laxenburg: Cambridge University Press and the International Institute for Applied Systems Analysis.

Radelet, S. 2004. *Aid Effectiveness and the Millennium Development Goals*. CGD Working Paper 39. Washington, DC: Center for Global Development.

Radelet, S., M. Clemens, and R. Bhavnani. 2006. Aid and Growth. The Current Debate and Some New Evidence. In *Macroeconomic Management of Foreign Aid: Opportunities and Pitfall*, ed. P. Isard, L. Lipschitz, A. Mourmuras, and P. Heller. Washington, DC: International Monetary Fund.

Reiche, K., A. Covarrubias, and E. Martinot. 2000. Expanding Electricity Access to Remote Areas: Off-Grid Rural Electrification in Developing Countries. *World Power 2000*. London: Isherwood Production, 62–60.

Roland, G. 2008. *Privatization: Successes and Failures*. New York, NY: Columbia University Press.

Sachs, J., J. McArthur, G. Schmidt-Traub, M. Kruk, C. Bahadur, M. Faye, and G. McCord. 2004. *Ending Africa's Poverty Trap*. Brookings Papers on Economic Activity, 1: 117–240.

Sandalow, D.B. 2006. *Ethanol: Lessons from Brazil*. Brookings Paper. www.brookings.edu/research/articles/2006/05/energy-sandalow. Accessed 13 April 2013.

Schelesinger, S. 2012. Brazilian International Cooperation and Investments: The Internationalization of Ethanol and Biodiesel. FASE: Solidariedade e Educaco. http://issuu.com/ongfase/docs/sergio_schlesinger_eng

Stern, N. 2007. *The Economics of Climate Change: The Stern Review*. Cambridge: Cambridge University Press.

Stern, N. 2002. Making the Case for Aid, Savings and Growth in LDCs: Some Recent Evidence. In *A Case for Aid: Building a Consensus for Development Assistance*, ed. World Bank. Washington, DC: World Bank.

Stern, E., L. Altinger, O. Feinstein, M. Maranon, D. Ruenbergerg, N.S. Schulz, and N.S. Nielsen. 2008. *Thematic Study on the Paris Declaration, Aid Effectiveness, and Development Effectiveness*. Copenhagen: DARA, Ministry of Foreign Affairs.

Stiglitz, J. 2002. Overseas Aid Is Money Well Spent. *Financial Times*, 14 April.

Svensson, J. 2006. The Institutional Economics of Foreign Aid. *Swedish Economic Policy Review* 13: 115–137.

The New York Times. 2010. Editorial, 'Cleaner, Healthier Cookstoves'. 23 September. www.nytimes.com/2010/09/24/opinion/24fri4.html?_r=0. Accessed 10 April 2013.

Truman, H.S. 1949. Inaugural Address, 20 January. Inaugural Addresses of the Presidents of the United States. www.bartleby.com/124/pres53.html. Accessed 10 April 2013.

UK DFID (UK Department of International Development). 2004. *Poverty Reduction Budget Support*. DFID Policy Paper. London: DFID.

———. 2011. *EU Blending Facilities: Implications for Future Governance Options*. London: DFID.

UN Energy. 2010. *Delivering on Energy: An Overview of Activities of UN Energy and Its Members*. New York: UN. www.unido.org

UN Energy Knowledge Network. n.d. What Is UN-Energy? The United Nations' Inter-agency Mechanism on Energy. www.unn-energy.org Accessed 11 April 2013.

UN Office of the Special Adviser on Africa. 2010. *Africa's Cooperation with New and Emerging Development Partners: Options for Africa's Development*. New York, NY: United Nations.

UNCTAD. 2010. South South Cooperation: Africa and the New Forms of Development Partnerships. Economic Development of Africa report, New York

UNDP (United Nations Development Programme). 2011. *Going Renewable: China's Success Story in Capacity Development—Towards an 'Energy Plus' Approach for the Poor: A Review of Good Practices and Lessons Learned from Asia and the Pacific*. Case Study 6. New York, NY: UNDP.

———. 2012. *Micro-credit for Energy Services: Transforming Basic Needs of Poor Households into Effective Market Demand. Towards an "Energy Plus" Approach for the Poor: A Review of Good Practices and Lessons Learned from Asia and the Pacific*. Case Study 3. New York, NY: UNDP.

Woods, N. 2008. Whose Aid? Whose Influence? China, Emerging Donors and the Silent Revolution in Development Assistance. *International Affairs* 84 (6): 1–18.

World Bank/GEF. 2006. *World Bank GEF Energy Efficiency Projects: Synthesis Report*. Washington, DC: World Bank, Global Environment Fund.

CHAPTER 6

An Analysis of the Links between Foreign Aid and Co_2 Emissions in Cities

Sandrine Kablan

6.1 Introduction

The sorts of development that have occurred in cities over the last two centuries will be realized twice as fast over the next 50 years. These changes will take place most markedly in emerging and developing countries, where 80% of the urban population is expected to reside. Especially for Africa, the projections for the next 20 years are 3.1% urban population growth rate, versus 1.7% for the global average (World Bank 2013a). This high rate of urbanization will require huge investments in urban infrastructure. Indeed, economic development is effected through urbanization, since most of the economic activities of the secondary and tertiary sectors are concentrated in cities. According to World Bank statistics, 70% of greenhouse gas emissions come from cities. Moreover, three urban sec-

The author is grateful to the attendees of the sustainable economic development seminar at Ecole Polytechnique and especially to Eric Strobl and Patrick Lenain for their helpful comments.

S. Kablan (✉)
University Paris-Est Créteil, Paris, France

Y. Huang, U. Pascual (eds.), *Aid Effectiveness for Environmental Sustainability*, https://doi.org/10.1007/978-981-10-5379-5_6

tors are responsible for the bulk of global greenhouse gas emissions: energy used for heating and cooling, which contributes 37% of global energy-related emissions; buildings with 25%; and lastly, urban transport at 22% (World Resource Institute 2009).

This raises the issue of the potential conflict between economic development and the problem of reducing emissions. For developing and emerging countries, the new challenge facing them is that of how to achieve environmentally friendly economic development. The increasing urbanization of these countries must therefore be carried out in a well-organized, ecological and environmentally friendly way. This introduces the concept of the sustainable city. But what do we mean by this term? As defined by Ciccone (2002), Ciccone and Hall (1996), and the World Bank (2010), in economic terms, sustainable cities attempt to maximize and share the significant economic benefits that derive from increased population concentration, while trying to avoid its negative externalities—that is, congestion, loss of resources, pollution and natural disaster risks. This idea of a sustainable city is therefore an incontestable instrument of sustainable development, which the Brundtland Commission defines as "meeting the needs of the present without compromising the ability of future generations to meet their own needs" (World Commission on Environment and Development 1987). This definition refers to the impact of human consumption and production activities on the local and global environment. In a broader sense, it also encompasses social and equity issues, institutional capacity and participation, and fiscal sustainability. There are three pillars of sustainable development: economic, environmental and social. There is a relationship between these three pillars, since human activities, all of which have an economic aspect, often lead to environmental degradation. More specifically, when the poorer parts of the population are not sufficiently integrated into urban programs, their struggle for survival or a better life can be harmful to the environment. Hence the close link between these three pillars, which ultimately contribute to the wellbeing of city dwellers.

For emerging and developing countries, urban infrastructure demands are becoming increasingly important and represent a major financial and environmental challenge. It is estimated that US$1 to 1.5 trillion would be needed for developing countries to satisfy basic needs and provide infrastructure for sustained growth (European Investment Bank 2010). Hence, the importance of addressing such a topic and questioning how foreign aid is used to foster green city procedures.

To the best of our knowledge, no papers have been written on the impact of foreign aid on green cities in developing countries, apart from Kablan (2016). Therefore, we will present some initiatives of the use of foreign aid to fund sustainable cities, in order to assess the effectiveness of foreign aid practices. Subsequently, we will make an empirical analysis by trying to establish the link between foreign aid and carbon dioxide emissions in cities.

Our results show that when official development aid (ODA) is associated with ecological projects, it helps to reduce carbon dioxide emissions in cities, even though the aid variable taken alone is not significant. Indeed, this last aspect of the results is consistent with those found by Kretschmer et al. (2013). But more specifically, foreign aid has a negative impact on carbon dioxide emissions, after a certain threshold, or when, for example, sector-specific aid for energy is considered.

The rest of this chapter is organized as follows. Section 6.1 will be devoted to the literature review. In Sect. 6.2, we will use a GMM (generalized moment method) model to assess the impact of official development assistance on carbon dioxide emissions in recipient countries' cities. Lastly, the conclusion will summarize our work and will raise some policy recommendations.

6.2 Literature Review: Foreign Aid Effectiveness and Green City Procedures

6.2.1 Aid Effectiveness

Foreign aid effectiveness has been the subject of several discussions and studies. For many economists, if foreign aid increases development, it is deemed effective. Burnside and Dollar (1997, 2000) were pioneers in investigating the econometric relation between aid and growth. Their results show that aid had a positive impact on growth, in countries with good fiscal monetary and trade policy. After their pioneer articles, several economists followed in considering the impact of economic shocks, political instability and institutional quality in terms of their dampening of aid effectiveness (Collier and Dehn 2001; Collier and Hoeffler 2004; Clemens et al. 2012). Results of these studies suggest that all these situations matter for aid effectiveness, but no evidence is given on which matters the most.

Besides empirical studies made by researchers, donors have also been interested in the issue of aid effectiveness. Some principles were defined in the 2005 Paris Declaration on Aid Effectiveness: support for national ownership of the development process, promotion of donor harmonization, alignment of donor systems with national systems, management for results and mutual accountability between donor and recipient. In more detail:

- Support to national ownership refers to the extent to which developing countries exercise leadership over their development policies and strategies.
- Harmonization encourages donors to implement shared arrangements and simplified procedures, with the goal of reducing the transaction costs imposed by donors on recipient governments.
- Aid alignment with country policies and systems implies the use of the relevant country's public financial management systems, use of country procurement systems, avoiding parallel project implementation units, aid predictability, untying aid, and coordinating technical assistance with national development strategies.
- Accountability means that foreign aid should be administered in both a transparent and accountable manner.

Since then, there have been the Accra Agenda of 2008 and the 2010 Paris Declaration targets, which both showed that, according to the assessment based on the above-mentioned principles, progress on aid effectiveness has been quite mixed. Therefore, several economists made some assessments on how those four principles are integrated in foreign aid allocation. Knack et al. (2011) construct an aid quality index. This index takes into account subindices on aid selectivity, alignment, harmonization and specialization. The four dimensions of aid quality emerge from the aid effectiveness literature and from international agreements. More specifically, aid selectivity refers to the notion that aid has greater development impact where it is needed the most—that is, where there are large numbers of poor people—and where policy and institutional environments are favorable for growth and development. As for specialization, this means that a proliferation of donors and projects leads to mismanagement by recipient governments, and that this is ultimately inefficient in terms of the relationship between recipient countries and their different donors—as there may be overlap and/or conflicts between projects, Knack and Rahman (2007). Again, donors' reluctance to specialize may reduce aid's effectiveness because of the lack of expertise that this suggests. Another study, by Easterly and Williamson (2011), attempts to assess aid effectiveness among bilateral, multi-lateral and UN agencies. The authors based their analysis on aid transparency, specialization, selectivity, ineffective aid channels and overhead costs. They show that performance of all agencies with regard to transparency, fragmentation and selectivity is still very poor. However, best practices modestly improve transparency, without clear improvements in specialization, fragmentation or selectivity.

6.2.2 *Initiative of Sustainable Cities in Developing Countries*

Four aspects emerge in the process of urbanization and must be taken into account in the issue of sustainable cities. These are: (1) urban design and public policy, with an emphasis on buildings and infrastructure; (2) transport systems; (3) pollution and waste treatment; (4) water supply and sewage. As already pointed out in the introduction to this chapter, three of these urban aspects are large emitters of greenhouse gases; hence, there is a need to make these aspects of urbanization align with ecological principles.

6.2.2.1 Energy Supply

Cities are major consumers of energy, which means that they can be vectors for its efficient use. As a result, they can be seen as major players in the quest to reduce greenhouse gas emissions, through a more rational use of energy. The challenge is to satisfy energy needs in a clean, efficient, inclusive and resilient way. Clean, in terms of the greater use of renewable and non-polluting sources of energy. Efficient, insofar as the economic cost must be low and not favor the waste of resources. Inclusive, in order to provide energy to the most vulnerable populations; and to avoid the searching for energy by poor people which degrades the environment in which they live. Finally, it should be resilient by integrating issues related to climate change (World Bank 2013b). Many cities have therefore initiated policies for the efficient use of energy and the promotion of cleaner energy. Local strategies are numerous but depend on the institutional capacity of each city. They consist in improving energy conservation, increasing the use of renewable energy, improving the efficiency of fossil fuel-based power-generation facilities, transitions to less carbon-intensive fuels, and employing carbon capture and storage. Cities can also diversify the energy sources they use to reduce overall emissions, through a greater use of hydropower, natural gas and non-hydroelectric-renewable resources. As for illustrating foreign aid used to finance energy supply projects in developing countries, we can refer to the Clean Technology Fund. This is designed to finance energy project initiatives with climate change mitigation components. As a product of cooperation between the World Bank and several multi-lateral development banks, the fund is allocated through coordination with these institutions and harmonized policy support. When countries apply to the fund, they must be eligible for ODA and be in a program with a multi-lateral development bank. In addition, funded projects must be part of, and complementary to, the country's development strategy. Such projects must also be high potential, cost effective and impact oriented.

6.2.2.2 Urban Design and Public Policy

Releases of greenhouse gas emissions from buildings are estimated to be at about 8.6 million metric tons of carbon dioxide, in 2004 (Levine et al. 2007). Cities use nearly 40% of the world's energy, UNEP (2009). While the bulk of urban emissions are concentrated in North America, Europe and Central Asia, city emissions in developing countries are expected to surpass those of these three regions by 2030 (World Bank 2013). There are two ways of reducing urban emissions: adapting the technical characteristics of existing buildings so that they meet the new ecological standards; and constructing new buildings to be less energy intensive through the careful selection of building materials, design, equipment and appliances, and during buildings' operation. Several developing countries are trying to implement greenhouse gas emissions reduction policies. For instance, in South Africa, national subsidies are used to encourage public bodies to rely on renewable energy. Similarly, we can cite the Zero-Carbon Buildings initiative, where energy provided by on-site renewable sources is equal to the energy used by the building. In this framework, some projects have been piloted by Worldwide Federation for Nature, such as the Malaysia Energy Center (Pusar Tenaga Malaysia) headquarters in Kuala Lumpur; and the town of Pedra Branca, in Brazil, which is to be a zero-carbon community. Finally, in Indonesia, the World Bank has financed the establishment of a green building code. It aims to fix energy and water efficiency requirements for large commercial and residential buildings, by including climate change adaptations in building designs. The details of this code were conceived in close collaboration with the government and the private actors concerned. As for renovation of old buildings, one can cite the examples of the two historic cities of Qufu and Zoucheng in Shandong Province, China, financed by the World Bank and the Chinese government. One important part of the project is the conservation and adaptive reuse of two large historic buildings that are underutilized.

6.2.2.3 Transport

Transport contributes 13% of global emissions of greenhouse gases and 23% of all energy-related carbon dioxide emissions globally (Metz et al. 2007). The demand for transport, especially private transport, is likely to grow exponentially in developing countries, owing to the emergence of the middle class. Therefore, a way of fighting against climate change is to tax polluting fuels and replace them with less polluting energy sources, such as natural gas or electricity. Governments in developing countries may also prohibit the use of excessively polluting vehicles, as was the case in New Delhi in India in 1998. This particular government measure led to a 34.8% reduction in emissions of sulfur dioxide from 2001 to 2003. Another way to promote clean transport in urban areas is by investing in bus rapid transit solutions, as in Bogota, Parana and Sao Paulo. Again, developing countries could benefit from the Global Environment Facility fund (GEF) linked to United Nations Framework Convention on Climate Change (UNFCCC). This multi-lateral fund has been in existence since 1999 and is dedicated to urban transport projects that integrate the environmental aspect and fight against global warming. Projects submitted for GEF funding must be part of the UNFCCC and must be consistent with the relevant government's priority objectives and the national development agenda.

6.2.2.4 Water Supply and Sewage

Sustainable cities face two main issues regarding the water sector: increased water stress or scarcity and declining water quality (Miller and Yates 2006). Water extraction, treatment and distribution are energy intensive, as are treatment and distribution of wastewater. An effort can therefore be made at this level by improving the various installations and using renewable energy, such as solar or wind power, even though initial costs may be quite high. It is also important to take into account rainwater harvesting, separation of wastewater by source and water-efficient fixtures, in order to maintain the rationale of saving energy and reducing greenhouse gas emissions. Again, in fast-growing cities in Africa and Asia, governments can promote decentralized systems for piped water supply without losing the benefits of economies of scale (World Bank 2013). To give an example of foreign aid regarding sewage, the World Bank has funded the Olandes Sewage Treatment Plant in Manilla, Philippines. The project aims at cleaning up domestic waste water from Marikina and Quezon City, and help reduce pollution in Marikina River. The project will have positive impacts on Metro Manilla waterways and the Vanilla bay, as well as for public health, as it will reduce hazards caused by human exposure to sewage.

6.2.2.5 *Pollution and Waste Treatment*

Pollution and waste directly affects environmental and public health. This is particularly the case for people who live near or work with solid waste, who have far higher probability of suffering related diseases (Giusti 2009). Cities combined produce 1.3 billion tons of waste annually, and by 2025 it is estimated that this will rise to 2.2 billion tons (Hoornweg and Bhada-Tata 2012). Besides, solid waste contributes to climate change by emitting 5% of total greenhouse gases (Tchobanoglous and Kreith 2002). Dealing with such issues will have to be integrated into the municipal budgets of developing countries. Indeed, sub-Saharan Africa and South Asia will reach peak waste generation and the highest urban populations by 2100 (World Bank 2013). As stated in the World Bank report "Building sustainability in an urbanizing world", 2013, biogenic waste can be transformed into more useful products, such as nutrients, electricity or liquid fuel. Especially, lower-income countries that have a higher relative organic fraction in their waste can use technologies that transform biogenic waste, since they are easily applicable in less industrialized cities. As an example of foreign aid for pollution and waste treatment, we can look to a joint-funded project for reduction of industrial pollution undertaken by Agence française de développement with other donors. Therein, green credit lines were offered through the National bank of Egypt to industries to foster their use of clean technologies. Assessments of results revealed potential reductions in pollution comparable to the emissions from 250,000 cars.

6.2.3 *Scaling Up and Transferring Good Practices*

In this section, we discuss sustainable cities projects financed by foreign aid and which can be scaled up or are transferable. While presenting those initiatives, we show that they are good examples to be replicated.

A program that follows this approach and has already been replicated in several countries is the Scaling up Renewable Energy Program (SREP). As the name suggests, it aims to fund projects with a renewable energy component. As we have seen, cities are energy-intensive consumers, and this consumption can have negative effects on the environment if the energy source is not renewable. This program is associated with the Strategic Climate Fund (SCF), which itself is a component of the Climate Investment Fund (CIF). It has been implemented in low-income countries such as Ethiopia, Honduras, Kenya, Maldives, Mali, Nepal and Tanzania. It is implemented in these countries as a country-led project and is included in their national development strategies, which at the same time incorporate other initiatives of the same type. Therefore, SREP respects the principles of ownership and alignment. As SREP is funded by a multi-lateral source, harmonization and good management are guaranteed, as results of its assessments show (Kablan 2015).

Similarly, there are multi-lateral funds that can be extended. These are the Clean Technology Fund and the afore-mentioned GEF. These funds meet the criteria of foreign aid effectiveness. Indeed, concerning harmonization, as they are funded by several multi-lateral development banks, they are unlikely to suffer from being duplicated. In terms of alignment and national ownership, one of the conditions for a climate change project to benefit from these funds is that this project must be part of a country's national climate change strategy. Technical assistance is also provided to project promoters, which contributes in capacity building. Finally, accountability and transparency are also taken into account, as there is monitoring and control implemented when a project is accepted.

Another interesting initiative is Eco2 Cities, which was launched by the World Bank to help cities achieve greater ecological and economic sustainability. The idea is to promote urbanization in line with economic growth and poverty reduction, while integrating the environmental constraint. This involves proposing an analytical and operational urbanization framework to take into account the specific challenges faced by cities. Furthermore, each city has a specific socioeconomic, political and institutional context and faces resources and capacities constraints. The program provides technical and financial assistance to cities in developing countries, for the provision of strategic infrastructure. The effectiveness of aid is reflected in the fact that each country must develop its own Eco2 program at the national level, as well as at the city level. Countries can therefore choose pilot cities as a starting point for the national program. Moreover, the financing of investments is monitored through cost–benefit indicators.

Finally, we present a project by the World Bank concerning the issuing of green bonds to finance urban projects in relation to climate change and mitigation. The first bonds were paid out since 2008. In June 2016, the institution issued US$9.1 billion in 18 currencies. The sectors up for consideration are: renewable energy and energy efficiency with 37%, transport with 35%, agriculture, forestry and ecosystems with 13%, water and waste management with 9%, and resilient infrastructure and others with 6%. In order to be eligible, projects follow a selection process during which they ensure that they meet a country's development priorities. This process consists of (1) early screening to identify potential environmental or social impacts and designing policies and concrete actions to mitigate any such impacts; and (2) approval by the board of executive directors (World Bank 2016). The final phase of approval involves analysis by environmental specialists. They ensure that mitigation and adaptation to climate change are key elements of the project. Finally, the World Bank is accountable to its shareholders and the public through a set of feedbacks and accountability mechanisms. For this purpose, a set of indicators is used in order to track progress and the effectiveness of funded operations. Such an initiative could also be replicated by other development banks and agencies.

6.3 Empirical Analysis

This part of the study is intended to give some empirical evidence of the impact of foreign aid effectiveness on green cities procedures. We will focus on green buildings, because variables on other of green cities' features are not available.

Even if the above literature review shows that some practices of official development assistance tend to promote green cities, to the best of our knowledge, no empirical studies on the impact of foreign aid on green cities have been conducted. However, some authors have focused on the impact of foreign aid on the environment. Arvin et al. (2006) show that given a developing country-level of external debt, aid has a negative impact on pollution. This result holds especially for upper-income developing countries (newly industrialized countries), as they are in a transitionary phase of their development. Yet, for lower-income countries, the picture is different, as more aid fosters less pollution. These results are mitigated by the fact that they do not incorporate other variables which can enter the relationship between aid and environment. Moreover, they do not take into account countries' heterogeneity. Kretschmer et al. (2013) try to account for those shortcomings by adding to their regression other explanatory variables, such as the level of development, the investment ratio, the industry share, import ratio and foreign direct investment in a panel regression. They are especially interested in studying the relationship between foreign aid and carbon as well as energy intensities. Particularly, they account for the composition of aid, by distinguishing aid for industry and aid for energy. Their results show that foreign aid reduces energy use intensity, especially when the variable aid for energy is used. However, aid does not have a significant impact on carbon emissions, even when the variable of aid for industry is taken into account in the regression.

6.3.1 *The Model*

In this empirical analysis, we are particularly interested in the impact of foreign aid on green cities procedures. The idea is to link carbon dioxide emissions in cities of developing countries with its first lag, foreign aid and other control variables relevant to explain it. The ordinary least square fixed-effects model gives some biased estimators because of the inclusion of the lagged dependent variable. Nickell (1981) pinpoints the fact that the bias is serious for sample taken over only a small period. However, the lagged dependent variable bias becomes less serious when the time period increases. In our study, the span of time is about 38 years, from 1971 to 2008. However, the average number of years per country available for estimation ranges between 15 and 20, because our sample is unbalanced.

Our preferred technique to deal with this bias is the generalized moments method (GMM). This can provide solutions to simultaneity bias, reverse causality and omitted variables. Moreover, Kretschmer et al. (2013) use the GMM estimator to deal with the possible endogeneity of aid, when they assess the impact of aid on energy and emission intensities.

There are two types of GMM estimators for dynamic panel: the first difference GMM estimator and the system GMM estimator. The first estimator consists in taking, for each period, the first difference of the estimated equation in order to eliminate the country-specific effects. Then, the explanatory variables of the first difference equation will be instrumented by their lagged level values. The second estimator combines the equation in first difference with level equations in which variables are econometrically instrumented by their first differences. Blundel and Bond (1998) show that this GMM system estimator is more efficient than the one used in calculating first difference. Indeed, the first difference GMM estimator gives biased results with a small sample when the chosen instruments are weak. We therefore select the GMM system estimator for our estimations. Our model will be as follows:

$$\ln(CO_{2it}) = \propto \ln(CO_{2it-1}) + \beta \ln(\text{aid}_{it}) + \gamma X_{it} + u_i + v_t + e_{it}$$

With CO_{2it}, carbon dioxide emissions from residential buildings and commercial and public services for country *i* at year *t*.

aid is net official development assistance divided by GDP for country *i* at year *t*.

X_{it} is a vector of control variables in logarithm.

u_i are the country-specific fixed effects, v_t are the year-specific fixed effects and uit is the error term.

We chose carbon dioxide emissions from residential buildings and commercial and public services as a proxy for green buildings. This variable measures all emissions from fuel combustion in households in millions of metric tons. It thus measures efforts made toward having eco-friendly buildings in each country, because if these emissions are reduced, it means that households are using alternative, more ecologically sound sources of energy. It also means that buildings have been built using more ecological sources of energy, and in respect of green procedures. In addition, most residential buildings and commercial and public services are constructed in cities. So, this calculation is an acceptable proxy for green buildings in each country.

As explanatory variables, we use aid (defined as net official development assistance) divided by GDP. More specifically, we calculate a moving average of aid over three years to take into account aid fluctuations. The most appropriate aid variable would have been aid for green cities or building projects. However, such a variable is not available. Therefore, we also consider aid to specific sector, such as energy, to see how this aid-specific variable could impact carbon dioxide emissions in cities. Indeed, an improvement in energy efficiency could have a decreasing impact on carbon dioxide emissions. Then, we add other control variables. We believe that the size of urban population (*urbpop*) can negatively impact carbon dioxide emissions in residential buildings and commercial and public services. So, we suspect a positive relationship between the two. We also include in the regression electricity production from renewable sources (*renewable*) to control for the policy of the country in question in terms of fostering projects with renewable energy sources. Our case studies of Bangladesh and Indonesia, for example, show that those two countries really have the political will to address climate change. Therefore, we posit that such a variable can have a negative impact on carbon dioxide emissions from residential buildings and commercial and public services; the more a country uses electricity from renewable sources, the lower its carbon dioxide emissions will be. We also consider GDP per unit of energy use as the purchasing power parity GDP per kilogram of oil equivalent of energy use (*energyuse*). This variable is intended to account for productive efficiency in energy use. We believe that an improvement in this aspect of economic performance—the use of energy—has a negative impact on carbon dioxide emissions. In our regression, we cross *aid* with *renewable* and *energyuse* to apprehend the impact of aid allocated to projects with an ecological component or energy efficiency. This is a way to grasp how effective aid policy is in reducing carbon dioxide from residential buildings, when it is associated with policies and/or projects for renewable energy sources or promoting efficient energy use.

Lastly, literature on aid shows an emphasis on the role of institutions and policies. For instance, Burnside and Dollar (2000) and Boone (1996) introduce in their regression, respectively, variables concerning the political-economic environment and an index of political participation and civil liberties. In particular, Burnside and Dollar use aid (as a share of GNP) as well as aid interacted with their policy variable in a standard growth regression. Hansen and Tarp (2001) introduce in the regression a quadratic term of aid, showing that there is a threshold beyond which aid has no more positive impact on growth. Here, we introduce to the regression a quadratic term of *aid* to account for such an effect of official development aid. We also take into account as institutional variable the quality of government index (*qog*). It is a composite index, which covers corruption, bureaucracy quality, law and order, and is often used to measure quality of governance. We also cross this variable with *aid*, to take into account the quality of aid management by the government and what impact well-managed aid has on carbon dioxide emissions from residential buildings.

6.3.2 The Data

Data on carbon dioxide emissions from residential buildings and commercial and public services, electricity from energy use, GDP per unit of energy use, GDP overall, net official development assistance and urban population are taken from world development indicators (WDI). Data related to sector-specific aid for energy come from the Organisation for Economic Co-operation and Development.[1] Studies on aid effectiveness preferably use aid disbursements rather than aid commitments. In fact, commitments do not lead to actual resource flows to the recipient country, and sometimes the actual flow may be considerably delayed (Michaelowa and Weber 2007; Dreher et al. 2008). However, sector-specific aid data are available on a commitment basis. Lastly, the index "quality of government" is provided from the Quality of Government dataset. The period of analysis is from 1971 to 2008 for the variables extracted from WDI and from 1984 to 2008 for the institutional variable "quality of government". Statistics are presented in Table 6.2 and the correlation matrix is displayed in Table 6.3. Our panel covers 144 countries, for which the relevant data are available and which are classified by the World Bank as low-income, lower middle-income and upper middle-income countries. Even though small low-income countries and least developed countries contribute little to global emissions, we examined a full sample of developing countries in order to study the general impact of aid on developing countries' urban carbon dioxide emissions.

6.3.3 Results

Our method of analysis consists of using a basic model represented by regression (1) and to add new variables or crossed variables in order to take into account the impact of foreign aid on carbon dioxide emissions through some variables.

GMM results in Table 6.1 show that carbon dioxide emissions from residential buildings and commercial and public services are impacted by

Table 6.1 Generalized method of moment regressions for the impact of foreign aid on carbon dioxide emissions in cities of developing countries

Dependent variable:	Carbon dioxide emissions from residential buildings and commercial and public services							
	(1)	(2)	(3)	(4)	(5)	(6)	(7)	(8)
CO_2^{t-1}	0.843	0.858	0.898	0.857	0.889	0.892	0.694	0.670
	(10.74)***	(15.41)***	(24.44)***	(19.54)***	(24.78)***	(27.78)***	(5.74)***	(5.47)***
aid	0.021	−0.196	−0.105	−0.145	−0.118	−0.126		
	(0.55)	(1.06)	(0.91)	(0.92)	(1.01)	(1.05)		
aid2				0.007	−0.002	−0.002		
				(1.65)*	(0.47)	(0.54)		
Aene							−0.470	7.230
							(0.10)	(1.46)
Aene2								−137.262
								(2.31)**
renewable	−0.022						−0.051	−0.058
	(2.09)**						(2.39)**	(2.58)**
energyuse	0.144						0.279	0.314
	(1.68)*						(2.50)**	(2.76)***
*aid*renewable*		−0.007	−0.007	−0.011	−0.008	−0.007		
		(1.48)	(2.09)**	(2.18)**	(2.25)**	(2.16)**		
*aid*energyuse*		0.056	0.04	0.055	0.043	0.043		
		(2.19)**	(2.33)**	(2.63)**	(2.63)**	(2.84)***		
urbpop1	0.201	0.176	0.132	0.183	0.144	0.141	0.427	0.467
	(2.24)**	(2.64)**	(2.76)***	(3.26)***	(3.08)***	(3.31)***	(2.61)**	(2.80)***

(*continued*)

Table 6.1 (continued)

qog			0.049		0.051		0.129	0.143
			(0.52)		(0.50)		(0.93)	(1.00)
*qog*aid*						0.021		
						(0.53)		
Constant	−3.633	−2.744	−2.046	−2.846	−0.571	−2.166	−7.432	−8.168
	(1.86)*	(2.61)**	(2.72)**	(3.26)***	(2.77)***	(3.32)***	(2.50)**	(2.71)***
N	2090	2090	1499	2090	1499	1499	843	843
AR (2)	0.61	0.64	0.97	0.65	0.97	0.95	0.083	0.11
Hansen test	1	1	1	1	1	1	0.99	0.99

t statistics in parentheses; *, **, ***, significant respectively to 10, 5 and 1%

Aene: foreign aid share specific to energetic projects

the lagged dependent variable, electricity production from renewable sources, and urban population. Carbon dioxide emissions in developing countries' cities are strongly path dependent. Indeed, the co-efficient is high relatively to other variables' co-efficients and strongly significant in regression 1 as well as all other regressions. Otherwise, in our basic model represented by regression 1, urban population and electricity production from renewable sources are significant with the expected signs. In particular, in regression 1, the co-efficient of *urbpop* is tenfold the one of *renewable*. This means that an increase of urban population will necessitate much more effort in electricity production from renewable sources to dampen the induced increase of carbon dioxide emissions in cities. Unfortunately, our variable of interest *aid* is not significant. This result is consistent with those of Kretschmer et al. (2013). They find that aid does not reduce carbon dioxide emissions but rather energy intensity. However, aid is significant when it is crossed with the variable renewable, even though the co-efficient is far smaller than the co-efficient of renewable taken alone. This means that when aid is allocated to projects intended to favor electricity production from renewable sources, it can effectively reduce carbon dioxide emissions in cities. Besides, when aid is introduced under a quadratic form in the regression, the co-efficient associated with it becomes significant. This confirms the idea that there is a threshold beyond which aid has no positive impact on carbon dioxide emissions from residential buildings and commercial and public services. When aid is crossed with GDP per unit of energy use, it has a significant positive impact. This is unexpected. An explanation could be that the production effect is greater than the energy efficiency effect. Therefore, development assistance allocated to a more efficient use of energy in the production process increases carbon dioxide emissions in cities. This can be due to a change in households' behavior in fuel consumption, derived from a wealth effect from a more productive country. Finally, the variable quality of governance is not significant, even if it is crossed with aid.

In order to check more precisely the impact of aid on cities' carbon dioxide emissions, we run additional regressions with some specific aid variables: regressions 7 and 8. In this context, net official aid collected from WDI refers to total aid flows to beneficiary countries per year. The share of aid devoted to ecological projects is embedded in this variable. We were not able to include variables related to foreign aid for green cities (because they don't exist in The OECD databse that we used) however, we could have variables related to sector-specific aid for energy. To get a better idea of how energy-specific aid projects can impact cities' carbon dioxide emissions, we use sector-specific aid for energy—in other words, the share of aid devoted to energy projects. This variable is expected to have a negative impact on carbon dioxide emissions. Replacing net official aid by sector-specific for energy in our basic model, we found that sector-specific aid for energy has a negative impact on carbon dioxide emissions, after a certain threshold. Indeed, the co-efficient of sector-specific aid under a quadratic form is negative and significant. All other co-efficients' variables remain consistent with previous estimates.

6.3.4 *Robustness Tests*

We carried out some robustness tests, as shown in Table 6.4, by replacing the variable aid, which is the ratio of a three-year moving average of net official development assistance to GDP, with its absolute term (aid:abs). The amount of foreign aid may be related to the underlying technology transfer, such as energy efficiency, and not to the size of the country. Again, we use an alternative institutional variable ("government effectiveness index" available from 2002 to 2008) in the "quality of government" dataset. It measures the ability of the government to deliver public goods and implement good policies.

The co-efficients of the regression with aid in absolute terms as explanatory variable do not change that much. Net development assistance is not significant when it enters the regression alone or as a quadratic term. However, it is significant with a negative sign when it is crossed with *renewable*. This confirms our idea, according to which foreign aid is effective in promoting green cities when it is allocated in a country that has greater ecological consciousness. However, the magnitude of our co-efficients is smaller compared to the core regressions. Urban population always has the positive expected sign, and GDP per unit of energy use is not significant. Lastly, our institutional variable is not significant, even when we change for the index of government effectiveness.

We conducted a second set of robustness tests to check whether our results are robust with respect to two subsamples: a sample of middle-income countries and another of least developed countries. It appears that middle-income countries are experiencing a particular phase of their development in which they are changing their production technology to make it cleaner. In addition, their level of industrialization is higher than that of the least developed countries. In Tables 6.5 and 6.6, regressions with the *aid* variable show that it is not significant, even under its quadratic form. Of all variables included in the regressions, only urban population is significant. Therefore, the conclusions we drew for the whole sample are not necessarily robust when we consider these subsamples. However, regressions with sector-specific aid variables performed on subsamples of middle-income countries and least developed countries are consistent with those that have been made on the whole sample. They thus confirm the robustness of our results. Moreover, for the two subsamples, signs and coefficients' significance are roughly the same. Sector-specific aid for energy is significant for the least developed countries in its quadratic form, clearly confirming our previous results. Finally, it is worth noting that *renewable* is significant for middle-income countries, confirming the role of the use of cleaner energy at this stage of development.

6.4 Conclusion

This chapter tries to analyze how foreign aid contributes to sustainable cities. To this end, a literature review provides an understanding of the importance of development assistance for promoting sustainable cities in developing countries. Indeed, cities produce more than 60% of all carbon dioxide and significant amounts of other greenhouse gas emissions. They are, at the same time, powerful engines for development. Thus, the idea of foreign aid effectiveness regarding sustainable cities procedures is to ensure that ODA for urbanization takes place in an ecological direction. In this regard, it is important to know whether official development assistance for green cities is effective. To understand this idea of aid effectiveness, we rely on the criteria presented in the literature. Those criteria are used in several articles related to aid effectiveness but also are subject to an international consensus. We are also interested in the criteria of climate finance effectiveness. Finally, we use as the main criteria for our assessment of foreign aid regarding sustainable cities, those which are consistent both at the level of aid effectiveness but also in terms of climate finance. These are: national ownership, harmonization, alignment and mutual accountability, and results management.

We then rely on reports of development agencies, and articles related to foreign aid effectiveness, to draw our analysis and answer questions raised in the introduction. Several foreign aid initiatives are effective according to our chosen criteria. This helps us to define how best practices could be replicated in the context of green cities. In particular, common funds managed by multi-lateral agencies together with a government agency could answer all the requirements for aid effectiveness.

The second part of our study aims to establish an empirical relationship between foreign aid and green cities procedures. In order to grasp sustainable (or green) cities procedures, we chose as variables carbon dioxide emissions from residential buildings and commercial and public services. Our results show that there is a threshold after which ODA will not influence carbon dioxide emissions in cities. Moreover, ODA is only significant when it is associated with a political will to promote urban green practices. This means that development aid will be effective if it is used to promote projects of electricity generation by renewable sources, for example. Indeed, such projects reduce carbon dioxide emissions in cities. By leveraging such projects, ODA will foster the spread of green cities procedures in developing countries. We do not have data related to green cities projects; however, we use sector-specific aid for energy, to see how this variable impacts on carbon dioxide emissions. What emerges is that it has a negative impact, after a certain threshold. Furthermore, our results are globally robust when they are drawn on subsamples of middle-income countries and least developed countries.

To sum up, this chapter shows clearly that efforts made by the international community to promote green cities procedures are effective and could be even more so. Indeed, many programs designed for this purpose meet the criteria of aid effectiveness. Moreover, our empirical analysis reveals that the variables that most affect carbon dioxide emissions in cities are the lagged dependent variable, urban population and electricity production from renewable sources. Among those variables, the latter one has a negative impact on (reduces) carbon dioxide emissions in cities. Again, sector-specific aid for energy also shows a negative impact on carbon dioxide emissions. This means that more efforts should be made to encourage energy production from renewable sources. Therefore, we recommend that development aid be allocated to ecological projects that are known to strongly contribute to the greening of cities.

Appendices

Table 6.2 Descriptive statistics

Variable	*Variable name*	*Number of observations*	*Mean*	*Standard deviation*	*Minimum*	*Maximum*
CO_2 emissions in residential buildings and commercial and public services	CO_2	3008	9.400309	37.26702	0	404.35
Net official development assistance/GDP	*aid*	3939	55.81705	268.1432	−8.836013	5649.187
Electricity production from renewable sources	*renewable*	3077	1.21e+10	3.96e+10	0	6.03e+11
GDP per unit of energy use	energyuse	3150	919.7615	816.5754	9.021294	5928.793
Urban population	*urbpop*	5537	1.09e+07	3.80e+07	1611.03	6.17e+08
Quality of government	*qog*	2185	0.4366688	0.152311	0.0416667	0.8981481
Government effectiveness	gee	1381	−0.4799525	0.6261736	−2.510756	1.392398
Sector-specific aid for energy	aene	1755	0.0126834	0.0597667	−0.0020467	2.20407

Table 6.3 Correlation matrix

	CO_2	*Aid*	*Renewable*	*Energyuse*	*Urbpop*	*Gee*	*Qog*	*Aene*
CO_2	1.0000							
aid	0.0175	1.0000						
renewable	0.6222*	−0.0207	1.0000					
energyuse	0.1973*	−0.0937*	0.1530*	1.0000				
Urbpop	0.8231*	0.0249	0.7752*	0.0525*	1.0000			
Gee	0.0876*	−0.0729*	0.1331*	0.1927*	0.1095*	1.0000		
qog	0.1243*	−0.0054	0.1122*	0.2532*	0.1420*	0.7249*	1.0000	
Aene	−0.0913*	−0.0239	−0.1092*	−0.1453*	−0.0599*	−0.1744*	−0.0789*	1.0000

*Significant at 10%

Table 6.4 Some robustness checks on the impact of foreign aid on carbon dioxide emissions in cities of developing countries, whole sample

Dependent variable	*Carbon dioxide emissions from residential buildings and commercial and public services*							
	(1)	*(2)*	*(3)*	*(4)*	*(5)*	*(6)*	*(7)*	*(8)*
CO_2t−1	0.944	0.959	0.969	0.959	0.810	0.803	0.735	0.739
	(77.13)***	(97.54)***	(76.54)***	(98.75)***	(10.47)***	(9.20)**	(5.12)**	(5.90)**
Aidabs	−0.001	0.010	0.005	0.062	0.010	−0.005	0.590	0.515
	(0.10)	(1.61)	(0.86)	(1.42)	(0.05)	(0.02)	(1.15)	(1.19)
Aidabs2				−0.001	−0.001	−0.002		−0.010
				(1.22)	(0.18)	(0.35)		(0.80)
renewable	−0.023							
	(2.22)**							
energyuse	0.012							
	(0.53)							
Aidabs**renewable*		−0.001	−0.000	−0.001	0.000	0.002	−0.008	−0.007
		(3.23)***	(3.36)***	(3.28)***	(0.01)	(0.31)	(0.41)	(0.43)
Aidabs**energyuse*		0.001	0.001	0.001	0.001	−0.001	−0.068	−0.046
		(1.61)	(1.04)	(1.49)	(0.09)	(0.07)	(0.84)	(0.87)
Urbpop1	0.092	0.064	0.044	0.065	0.232	0.241	0.344	0.333
	(3.51)***	(3.99)***	(2.52)**	(4.19)***	(2.44)**	(2.17)**	(1.86)*	(2.13)**
qog			0.023		0.295			
			(0.52)		(1.28)			
*qog**aidabs						0.006		
						(0.72)		

(*continued*)

Table 6.4 (continued)

Dependent variable	*Carbon dioxide emissions from residential buildings and commercial and public services*							
	(1)	*(2)*	*(3)*	*(4)*	*(5)*	*(6)*	*(7)*	*(8)*
Gee							0.042 (0.48)	
Gee*aidabs								0.040 (0.61)
Constant	−0.990 (2.75)***	−0.997 (3.86)***	−0.652 (2.70)***	−1.512 (4.24)***	−3.764 (2.38)*	−3.836 (2.18)**	−5.272 (1.87)*	−5.138 (2.12)**
N	2531	2531	1657	2531	1499	1499	738	738
Ar(2)	0.52	0.52	0.92	0.51	0.93	0.95	0.49	0.49
Hansen Test	1	1	1	1	1	1	1	1

*, **, *** significant at 10, 5 and 1%

aid:abs: three-year moving average of net official development assistance in absolute terms

Table 6.5 Some robustness checks on the impact of foreign aid on carbon dioxide emissions in middle-income countries

Dependent variable:	*Carbon dioxide emissions from residential buildings and commercial and public services*							
	(1)	*(2)*	*(3)*	*(4)*	*(5)*	*(6)*	*(7)*	*(8)*
L.lco22	0.671	0.694	0.635	0.689	0.579	0.605	0.842	0.844
	(4.86)***	(6.18)***	(4.92)***	(6.85)***	(4.70)***	(5.73)***	(12.56)***	(13.36)***
Laid	−0.006	−0.043	−0.009	−0.086	0.000	0.061		
	(0.19)	(0.16)	(0.04)	(0.31)	(0.00)	(0.23)		
Laid2				0.006	−0.005	−0.006		
				(1.34)	(0.53)	(0.63)		
Aene							2.586	−0.795
							(0.87)	(0.05)
Aene2								241.409
								(0.20)
renewable	−0.059						−0.044	−0.044
	(1.99)**						(2.50)**	(2.57)**
energyuse	0.140						0.107	0.106
	(1.52)						(2.19)**	(2.22)**
Aidrenew		−0.002	0.003	−0.001	0.006	0.000		
		(0.26)	(0.50)	(0.16)	(0.61)	(0.03)		
Aidenergy		0.018	−0.002	0.018	−0.007	−0.007		
		(0.65)	(0.08)	(0.64)	(0.20)	(0.23)		
Lurbpop	0.461	0.376	0.430	0.375	0.495	0.475	0.236	0.233
	(2.34)**	(2.76)***	(2.80)***	(3.14)***	(3.21)***	(3.51)***	(2.53)**	(2.66)**
qog			−0.039		−0.032		−0.032	−0.032
			(0.18)		(0.12)		(0.33)	(0.33)
*qog*aid*						0.091		
						(1.23)		
Constant	−6.662	−5.729	−6.498	−5.724	−7.489	−7.204	−3.364	−3.323
	(2.32)**	(2.77)***	(2.73)**	(3.15)***	(3.11)***	(3.49)***	(2.49)*	(2.62)*
N	923	923	682	923	682	682	313	313
Ar(2)	0.13	0.16	0.83	0.16	0.80	0.77	0.25	0.28
Hansen	1	1	1	1	1	1	1	1

*, **, *** significant at 10, 5 and 1%

Table 6.6 Some robustness checks on the impact of foreign aid on carbon dioxide emissions in least developed countries

Dependent variable	*Carbon dioxide emissions from residential buildings and commercial and public services*							
	(1)	*(2)*	*(3)*	*(4)*	*(5)*	*(6)*	*(7)*	*(8)*
CO_2t–1	0.880	0.827	0.875	0.828	0.881	0.883	0.715	0.692
	(13.58)**	(13.65)**	(15.96)**	(17.40)**	(17.46)**	(16.55)**	(6.71)**	(6.19)**
aid	−0.015	0.430	−0.088	0.290	−0.163	−0.242		
	(0.24)	(0.99)	(0.39)	(0.71)	(0.74)	(1.20)		
aid2				−0.000	0.003	0.003		
				(0.01)	(0.66)	(0.68)		
Aene							−0.269	7.234
							(0.06)	(1.53)
Aene2								−132.576
								(1.88)*
renewable	−0.029						−0.017	−0.026
	(1.51)						(0.62)	(0.88)
energyuse	0.099						0.224	0.267
	(1.23)						(2.14)**	(2.33)**
*aid*renewable*		−0.010	−0.004	−0.006	0.003	0.005		
		(0.77)	(0.63)	(0.58)	(0.48)	(0.73)		
*aid*energyuse*		−0.042	0.026	−0.031	0.012	0.018		
		(0.86)	(0.95)	(0.58)	(0.48)	(0.92)		
urbpop	0.166	0.232	0.149	0.222	0.134	0.128	0.378	0.419
	(2.55)**	(2.47)**	(2.06)**	(3.09)***	(2.04)**	(1.77)*	(2.82)***	(2.96)***

(*continued*)

Table 6.6 (continued)

Dependent variable	*Carbon dioxide emissions from residential buildings and commercial and public services*							
	(1)	*(2)*	*(3)*	*(4)*	*(5)*	*(6)*	*(7)*	*(8)*
qog			0.022		0.025		0.165	0.179
			(0.14)		(0.17)		(1.10)	(1.13)
*qog*aid*						−0.002		
						(0.06)		
Constant	−2.593	−3.613	−2.355	−3.459	−2.117	−2.010	−7.066	−7.851
	(1.54)	(2.44)**	(1.95)*	(3.12)***	(1.94)*	(1.72)*	(2.58)**	(2.70)**
N	1167	1167	817	1167	817	817	530	530
AR(2)	0.43	0.43	0.96	0.43	0.96	0.95	0.14	0.2
Hansen	1	1	1	1	1	1	1	1

*, **, *** significant at 10, 5 and 1%

Notes

1. Those data where directly taken from Kretschmer et al. (2013). The period span is 1973–2005.

References

Agence française de développement. 2010. AFD and Environmental Finance: Partnering with Banks to Finance Green Growth.

Arvin, B.M., P. Dabir-Alai, and B. Lew. 2006. Does Foreign Aid Affect the Environment in Developing Economies? *Journal of Economic Development* 31 (1): 63–87.

Auer, M.R. 2006. Foreign Aid to Promote Energy Efficiency in Mexico: An Institutional Analysis. *Journal of Energy and Development* 31 (1): 85–100.

Binsted, A., D. Bongardt, H. Dalkmann, and K. Sakamoto. 2012. Accessing Climate Finance for Sustainable Transport: A Practical Overview. Sustainable Urban Transport Technical Document 5.

Bird, N., and J. Glennie. 2011. Going Beyond Aid Effectiveness to Guide the Delivery of Climate Finance. Background Note. Overseas Development Institute.

Bird, N.M. 2011. *The Future for Climate Finance in Nepal*. ODI Report. London: ODI.

Blundell, R., and S. Bond. 1998. Initial Conditions and Moment Restrictions in Dynamic Panel Data Models. *Journal of Econometrics* 87 (1): 115–143.

Boone, P. 1996. Politics and the Effectiveness of Foreign Aid. *European Economic Review* 40: 289–329.

Brown, J., and L. Peskett. 2011. *Climate Finance in Indonesia: Lessons for the Future of Public Finance for Climate Change Mitigation*. EDC 2020 Working Paper 11. Bonn: EADI.

Burnside, C., and D. Dollar. 1997. *Aid Policies and Growth, World Bank Policy Research Paper.* Working Paper No. 1777, World Bank, Washington, DC.

———. 2000. Aid, Policies, and Growth. *American Economic Review* 90: 847–868.

Ciccone, A., and R. Hall. 1996. Productivity and the Density of Economic Activity. *American Economic Review* 86: 54–70.

Cicoone, A. 2002. Agglomeration Effects in Europe. *European Economic Review* 46: 213–227.

Clemens, M., S. Radelet, R. Bhavanani, and S. Bazzi. 2012. Counting Chickens When They Hatch: Timing and Effects of Aid on Growth. *Economic Journal* 122 (561): 590–617.

Climate Investment Funds. 2011. Scaling Up Renewable Energy Program: Investment Plan for Kenya.

Cohen, N., and P. Robbins. 2011. *Green Cities: An A to Z Guide*. Thousand Oaks, CA: Sage Publications.

Collier, P. and J. Dehn. 2001. *Aid Shocks and Growth*, World Bank Policy Research Working Paper No. 2688, World Bank, Washington, DC.

Collier, P., and A. Hoeffler. 2004. Aid, Policy and Growth in Post-Conflict Societies. *European Economic Review* 48 (5): 1125–1145.

Dreher, A., P. Nunnenkamp, and R. Thiele. 2008. Does Aid for Education Educate Children? Evidence from Panel Data. *World Bank Economic Review* 22 (2): 291–314.

EIB (European Investment Bank). 2010. *Public and Private Financing of Infrastructure: Policy Challenges in Mobilizing Finance*. Luxembourg: EIB.

Easterly, W., and C. Williamson. 2011. Rhetoric versus Reality: The Best and Worst of Aid Agencies Practices. *World Development* 35 (11): 1930–1949.

Giusti, L. 2009. A Review of Waste Management Practices and Their Impact on Human Health. *Waste Management* 29 (8): 2227–2239.

Hansen, H., and F. Tarp. 2001. Aid and Growth Regressions. *Journal of Development Economics* 64: 547–570.

Hedger, M. 2011. *Climate Finance in Bangladesh: Lessons for Development Cooperation and Climate Finance at National Level.* EDC 2020 Working Paper 12. Bonn: EADI.

Hicks, R.L., B.C. Parks, J.T. Roberts, and M.J. Tierney. 2008. *Greening Aid? Understanding the Environmental Impact of Development Assistance*. Oxford: Oxford University Press.

Hoornweg, D., and P. Bhada-Tata. 2012. *What a Waste: Waste Management Around the World*. Washington, DC: World Bank. http://www.unepfi.org

Kablan, S. 2016. Foreign Aid to Foster Green Cities Procedures, What Do We Know? In *Handbook on the Economics of Foreign Aid*. Cheltenham: Edward Elgar Publishing.

Knack, S., and A. Rahman. 2007. Donor Fragmentation and Bureaucratic Quality in Aid Recipients. *Journal of Development Economics* 83 (1): 176–197.

Knack, S., H. Rogers, and N. Eubank. 2011. Aid Quality and Donor Rankings. *World Development* 39 (11): 1907–1917.

Kretschmer, B., M. Hubler, and P. Nunnenkamp. 2013. Does Foreign Aid Reduces Energy and Carbon Intensities of Developing Economies? *Journal of International Development* 25 (1): 67–91.

Levine, M., D. Urge-Vorsatz, K. Blok, L. Geng, D. Harvey, S. Land, et al. 2007. Residential and Commercial Buildings. In *Climate Change 2007: Mitigation, Contribution of Working Group III to the Fourth Assessment Report of the Intergovernmental Panel on Climate Change*, ed. B. Metz, O.R. Davidson, P.R. Bosch, R. Dave, and L.A. Meyer. Cambridge and New York, NY: Cambridge University Press.

Metz, B., O. Davidson, P. Bosch, R. Dave, and L. Meyer. 2007. *Climate Change 2007: Mitigation of Climate Change, Contribution of Working Group III, to the Fourth Assessment Report, of the Intergouvernemental Panel of Climate Change.* Cambridge: University press.

Michaelowa, K., and A. Weber. 2007. Aid Effectiveness in the Education Sector: A Dynamic Panel Analysis. In *Theory and Practice of Foreign Aid*, ed. S. Lahiri, 357–385. Amsterdam: Elsevier.

Nickell, S. 1981. Biases in Dynamic Models with Fixed Effects. *Econometrica* 49 (6): 1417–1426.

Sylwester, K. 2008. Foreign Aid and Urbanization in Developing Countries. *International Journal for Business and Economics* 7 (2): 153–166.

Tchobanoglous, G., and F. Kreith. 2002. *Handbook of Solid Waste Management.* 2nd ed. New York: McGraw-Hill.

Thornton, N. 2011. Realizing Development Effectiveness: Making the Most of Climate Change Finance in Asia and the Pacific. CDDE Facility.

UNEP (United Nations Environment Programme). 2009. *Buildings and Climate Change: Summary for Decision-Makers.* Paris: UNEP, Sustainable Consumption and Production Branch.

———. 2012. Financing Renewable Energy in Developing Countries: Drivers and Barriers for Private Finance in Sub-Saharan Africa.

World Bank. 2010. Climate Finance in the Urban Context. Development, Climate and Finance.

———. 2013a. *Building Sustainability in an Urbanizing World.* Urban Development Series Knowledge Paper. Washington, DC.

———. 2013b. Inclusive Green Growth in Latin America and the Caribbean, Washington, DC.

———. 2016. Green Bond Impact Report. Washington, DC.

World Commission on Environment and Development. 1987. *Our Common Future.* Oxford: Oxford University Press.

WRI (World Resources Institute). 2009. *World Greenhouse Gas Emissions in 2005.* WRI Working Paper. World Resources Institute.

CHAPTER 7

Foreign Aid, Urbanization and Green Cities

Jun Li

7.1 Introduction

A 'green city' or a sustainable city is one that has been designed with due consideration of its environmental impact, and is inhabited by people dedicated to minimizing necessary inputs and waste outputs. More specifically, it means creating the smallest possible ecological footprint, producing the lowest quantity of pollution, using land and materials efficiently, encouraging biodiversity by preserving natural habitats and so on. Green cities function in the spirit of self-reliance and self-sufficiency, and they offer their residents an excellent quality of life.

It is estimated that more than 50% of the world's population now live in cities and urban areas, and this is set to rise to 60% within the next few decades. These large communities provide both challenges and opportunities for achieving the goals of green city development. This is particularly the case in developing countries and may have dramatic economic and environmental effects on these cities.[1]

Foreign aid plays a vital role in improving the lives of people across the developing world. In the early 21st century, as concern for the environment has become prominent, more foreign aid has been diverted to the

J. Li (✉)
School of Mathematics and Computer Science, University of Wolverhampton, Wolverhampton, UK

Y. Huang, U. Pascual (eds.), *Aid Effectiveness for Environmental Sustainability*, https://doi.org/10.1007/978-981-10-5379-5_7

sustainable development of certain localities (e.g. cities) in the developing world. For example, with the help of foreign technologies, management and capital, cities in China and Brazil have been working towards building sustainable city models for the benefit of the whole country. However, while the world encourages sustainable development and supports it through financial means or aid, there are suggestions that this assistance has not made a regular and predictable contribution to overall development. Implementation strategies and their effectiveness are the subject of increasing debate. Furthermore, there are no comprehensive studies at present in the literature on foreign aid and green cities, which could report oin the types of schemes and implementation methods, or the amounts of foreign aid, that contribute directly to green city development.

7.2 Literature Review

7.2.1 Introduction of Foreign Aid

Foreign aid (also known as international aid, overseas aid or just aid) is the international transfer of capital, goods or services from a country or an international organization for the benefit of a recipient country or its population. Aid can be economic, military or emergency humanitarian (e.g. aid given in the aftermath of a natural disaster) (Encyclopædia Britannica 2012). Aid may be given by individuals, private organizations or governments. In general, it can be classified into two major types: humanitarian aid and development assistance. Humanitarian or emergency aid is speedy assistance given to people in immediate distress, while development aid is given to support development in general, which can be either economic or social development in developing countries (Wikipedia 2012a).

Foreign aid in its modern form dates back from the early 1940s, when it intensified after the Second World War as Europe faced a critical shortage of capital and the need for reconstruction. After success in Europe, aid in the 1950s and 1960s focused on the developing countries because it was widely believed that economic growth was its key objective. It was assumed that, based on capital investment and increasing savings through a 'big push', long-run economic performance would launch countries into self-sustaining growth or 'take off', resulting in the elimination of poverty and inequality. In the 1970s there was increasing interest in employment, income distribution and poverty alleviation through multilateral channels such as the United Nations (UN) and the World Bank. In the 1980s, as developing countries faced setbacks from high oil prices, internal macroeconomic imbalances and slower external global growth, financial programme aid and adjustment loans became fashionable instruments for achieving external and internal macroeconomic balance. International financial institution policymakers advocated the necessity to develop countries of a better economic policy environment for capital accumulation and technological progress. In parallel, poverty alleviation slipped out of view in the mainstream agenda, although continuing to remain at the centre of unorthodox thinking. In the 1990s, aid was directed more towards development assistance. Public finance analyses suggested that funds should go to activities that generated positive externalities, advocated sustainability and promoted projects that required only start-up funding after which they would survive without external support. This, however, was not always true for projects such as disease control (Kremer and Miguel 2008). In the new millennium, development related to environmental issues has become the most prominent purpose of aid in many countries. In practice, there was an agreement to sideline growth as the basic measure of aid effectiveness, to be replaced by a stronger focus on the principles of recipient ownership, alignment, harmonization, managing for results, and mutual accountability. Multiple goals were laid out. For example, United Nations' Millennium Development focuses on poverty alleviation and environmental quality as well as literacy, health and women's right.

Aid can be classified as official aid and private aid or non-governmental aid, which reaches recipients through bilateral or multilateral delivery systems. Bilateral aid refers to government-to-government transfers through agencies, such as the US Agency for International Development, while multilateral institutions, such as the World Bank and the United Nations Childrens Fund, pool aid from many sources and disperse it to numerous recipients. Private aid includes help from charities, philanthropic organizations and businesses to recipient countries or programmes within recipient counties (Wikipedia 2012a). There are dozens of bilateral and multilateral agencies and hundreds of non-governmental organizations (NGOs). Between 2000 and 2009, members of Development Assistance Committee (DAC) of the Organisation for Economic Co-operation and Development (OECD) have continued to dominate governmental aid. The governments include, notably, the USA, European institutions and the UK (GHA 2012). Non-DAC countries such as China, India and Saudi Arabia are increasingly engaged in strategic 'south–south' aid programmes. Private foundations and NGOs are another significant addition to ODA. Multilateral aid is mainly used to support development programmes of the UN agencies, the European Union, the World Bank and regional development banks. Bilateral aid is disbursed largely through grants to government ministries or through national and international development NGOs (OneWorld 2012).

Today, as many as 180 countries or territories receive foreign aid. Between 2000 and 2009, the main recipients have included Africa(a large proportion has gone to sub-Saharan countries) and Asia, accounting for 46% and 24% of total assistance, respectively.[2] The next two major recipients are the Middle East and Latin America (GHA 2012). In Asia and the Middle East, Afghanistan, Pakistan, Indonesia, Palestine/Occupied Palestinian Territories and Iraq dominate the aid scene, largely owing to conflict and natural disasters.

Foreign aid has been justified in public policy pronouncements in widely differing ways, ranging from pure altruism to the shared benefits of economic development in poor countries, political ideology, or support of the foreign policy and commercial interests of donor countries (Tarp 2010). It is also true that aid is rarely given for purely altruistic reasons. For example, 'only about one-fifth of US aid goes to countries classified by the OECD as "least developed"' (Singer 2009). The motives are multifaceted assistance, ranging from the selfish to the generous. In principle, aid programmes are inspired by four broad motives (OneWorld 2012), although these vary between donors and are influenced by the global political and economic climate. The first is poverty alleviation. Historically, humanitarian aid has attracted a relatively high level of assistance allocated during various stages from emergency relief to recovery. Now, food and emergency relief still remain an important form of aid. The second motive is the donor country's pursuit of strategic political and economic interests, such as strategic self-interest and development of markets. The third motive stems from tackling terrorism within so-call fragile states, whose vulnerability to poverty is believed to be a causality of cyclical violence and terrorism. The fourth motive is rooted in the recognition of a country's interdependence in the context of economic, environmental and security issues. Economic development and growth, poverty and healthcare are major concerns in aid-receiving countries.

Foreign aid may involve transfers of financial resources (e.g. loans or cash), commodities (e.g. food), and technology and training. Resources can take the form of grants or concessional credits (e.g. export credits). Grants and loans with at least a 25% grant element are defined as official development assistance (ODA), which is the most common type of foreign aid (Encyclopædia 2012). A proportion of aid goes to international research, such as experimental development conducted within the green revolution or many vaccines.

7.2.2 Foreign Aid, Environment, and Development

Many people in developing countries are living in ecologically fragile environments, and the concern for the protection or improvement of the environment came increasingly to the fore during the second part of the 1980s and the early 1990s. Its origins can be traced back to the publication of *Limits to Growth* (Meadows et al. 1972) in the early 1970s and the United Nations Conference on Human Environment in Stockholm. But the issue was not a priority theme, nor was it linked to development assistance until the publication of the report *Our Common Future* by the World Commission on Environment and Development (WCED 1987). The WCED in 1992 further developed the link through the mechanism of a Global Environment Facility (GEF). Following the mandate of its 2011 ministerial council meeting, the OECD forged an ambitious, institution-wide development strategy to help countries achieve sustainable growth by helping them to mobilize their own resources, put in place sustainable solutions and become the authors of their own development. The OECD Report 2011 called for new targets, highlighting concerns and global challenges such as transport, energy and climate change, and redefining the Millennium Development Goals (MDGs) to include elements of global public goods. Even though ecology has a broader scope, aid agencies perceive poverty to constitute a major threat to the environment (Stokke 1996).

Environmental sustainability is essential for strong economic development. Many environmentalists deplore the sacrifice of the environment to promote development, and some critics call for aid to be invested in sustainable projects that include built-in measures to protect the environment (Purvis 2003). Emerging countries such as China, Brazil and India are undergoing vast industrialization and urbanization processes similar to what happened early in the last century in the developed world, and the transformation process needs external support. But protection of the environment is often opposed as the developing countries fear they cannot compete in world markets. Several foreign aid projects in environment and development are emerging to promote the integration of environment and climate change into all aspects of development cooperation. For example, the Green Growth Strategy by the OECD endeavours to provide policy advice to developing countries with respect to economic growth, job creation, environmental improvement and social equality. In addition, a series of 'good practice guidance notes' on the role of donor agencies in assisting developing counties in green growth transition has been generated by the DAC team. These notes highlight innovation and green technology adoption, governance, policy coherence for development, and welfare generation from natural capital (OECD n.d.). The programme suggested that policy and measurement focus worked.

Natural disasters caused by climate change affect poor people and poor countries particularly badly. A key challenge for the development community is to ensure that climate change mitigation and adaptation are integrated at all levels of development decision-making. The Cancun Agreement promises to mobilize financial resources for developing countries. The Green Climate Fund was established later as a channel dedicated to climate finance. However, as no reporting standards exist to determine whether aid projects include climate mitigation or adaptation components, there is potential for double-counting conventional assistance as aid for climate change (OneWorld 2012). Thus critics argue that the primary problem in climate change adaption is not foreign aid but rather institutional reform (Raymond/IHC 2009).

7.2.3 The Pros and Cons of Foreign Aid, and the Lessons Learned

Foreign aid and its effectiveness in tackling poverty and health issues, or promoting growth and development in the emerging countries, have triggered intense controversy. Proponents pressing for enhanced foreign aid commitments are convinced that aid does work. Examples from recent studies point to success in many poor countries in public health (Demombynes and Trommlerova 2012; Murray et al. 2012), emergency food supplies and security, access to safe drinking water and so forth. According to Arndt, Jones and Tarp (2010), aid has a positive and statistically significant causal effect on growth in the long run, and it remains a key tool for enhancing the development prospects of poor countries. Reviews by the United Nations Development Programme (2007) and the UN Millennium Project (2005) observe that poverty in the poorest countries can be dramatically reduced through international aid. However, some analyses (Boone 2006; Burnside and Dollar 2000) find both positive and negative outcomes for aid. For instance, countries that received large aid flows between 1970 and 1993 fared no better in terms of growth or measures of extreme poverty than countries with small aid flows. Moreover, critics argue that the public successes in these countries are unsustainable without effective governance and that the right economic and social policies need to be in place. Some critics have argued that long-term development aid does not work, when in fact they prove the opposite (Sachs 2005; UN Millennium Project 2005). Further, some critics such as Anderson (2007) and Sachs (2005), have pointed out that the donor community has failed to meet the established ODA international target of 0.7% of national income.

The evidence from the failures—as well as successes—suggests that improvement is needed. On the donor side, this includes coherent strategies and objectives, accountability, alignment (i.e. linking donor programmes with the goals, objectives and strategies of the beneficiary country, and avoiding conflicts of interest), and sustainability of aid. On the recipient side, improvements in the effective implementation of aid objectives, good governance, accountability, respect for human rights and other norms, and reduction of aid dependency are major concerns. Furthermore, capacity building of recipients, harmonization within donors and coordination among donors and recipients are essential if aid is to succeed.

Boone (2006) and his research team suggest that to make aid work it needs to be focused on the specific domains that work. They believe that one of the failures of aid projects is the lack of post-project assessment by donors, but targeting aid to the right projects is also important. Easterly (2008) suggests that historically the 'search' approach of exploring for solutions through 'trial and error' performed better than centrally planned methods, which work according to predetermined goals and large-scale choices. On the other hand, Easterly notes that all human activity, including foreign aid projects based on the 'search' approach, involve some degree of planning. Banerjee and He (2008) lament the lack of evaluation as one of the key weaknesses of aid agencies. Recently the World Bank (n.d.) and its Impact Evaluation Group endorsed the idea of randomized controlled trials on a limited scale. Sachs (2005) considers aid to be ineffective, and, according to him, coherent strategies and objectives from donors, good governance within recipient countries, and alignment between donors and recipients through top-down board-based methodology are the key factors for a successful aid project. Collier and Dollar (2002) show that good policies and efficient aid allocation play a primary role in aid effectiveness. Aid is often tied to conditions with respect to its allocation, and, as academic research has shown, aid is often tied because of political motives (e.g. human rights) rather than concerns about proper policy implementation (Wikipedia 2012a). Some critics (Oxfam 2002; Santiso 2001) argue that conditionality is detrimental to developing countries, or has no chance of success (Moss et al. 2008). Graham and O'Hanlon (1997) also note that studies in the 1990s found no positive relationship between tied financial aid and economic growth. Santiso believes that a more appropriate approach to strengthening good governance and democracy is to cede control to the recipient country but only within a framework of agreed objectives. Hoffman (2008) considers defects of the aid system to originate from the inability to assume an entrepreneurial outlook. Consequently, he suggests that donor accountability to recipients should resemble a business to include risk assessment, awareness of targets and the knowledge of how to meet these at the lowest cost possible. A recent pilot experiment (ibid.) tested the theory and the objective is now to scale it up substantially.

Buss and Gardner (2008) suggest that assistance, other than for short-term crisis programmes, should be undertaken only if it is sustainable. However, as Graham and O'Hanlon (1997) point out, the benefit of certain sustainable projects, such as those supporting sustained economic growth, may become visible only in the long term, which could be an unpopular option politically for recipients. Subedi (2005) also indicates that this principle could lead in the long term to an unfair distribution of benefits. Graham and O'Hanlon call for greater selectivity in aid allocation and a sustainable framework with well-designed strategies and objectives to help country compliance. Subedi, on the other hand, advocates capacity building and the engagement of local people. Moreover, research also suggests that the lack of a framework to harmonize aid programmes (CIDA 2004) and coordinate donor–recipient efforts (ECOSOC 1999) may cause failures. According to Burnell (1997), both donors and recipients are to blame because donors shun coordination efforts which could significantly reduce their freedom to pursue their own policies and objectives while recipients feel constrained. Dialogue and communication between donors and recipients to produce guidelines for donors and to share donor experiences have been suggested, but the lack of leadership is challenging.

The criticism of aid has been a source of longstanding anxiety among the international donor community. Against this background, principles for aid effectiveness were created in the 2005 Paris Declaration and Accra Forum. In particular, the declaration required action on donor–recipient alignment, harmonization among donors, the establishment of reporting standards and an improvement in accountability (OneWorld 2012; Atlantic-Community 2012; OECD 2005/2008). All the lessons learnt from various practices need to be given greater consideration so that successful experiences can be transferred, incorporated and scaled up in future aid strategies and approaches.

7.2.4 Foreign Aid, Urbanization and Green Cities

Urbanization is a process that shifts society from a rural environment to an urban one, and it involves increasing numbers of people and the physical growth of urban settlements. The process is largely driven by market forces and government policies, which result in changes in land use, health and natural resources management, including water, soil and forests. Sustainable development—embracing social, environmental and economic dimensions—is defined as development that meets the needs of the present generation without compromising future generations (WCED 1987). A green city, or sustainable city, is one designed with attention to its environmental impact, and its inhabitants are dedicated to minimizing the needed inputs and waste outputs. It is an ecologically healthy human settlement modelled on a self-sustaining resilient structure and functioning of natural ecosystems and living organisms (Ecocity Builders 2012). Rapid urbanization—and in particular the associated problems of urban poverty, unsustainable development and environmental degradation—poses a formidable challenge to many developing countries. For example, issues related to farmlands and watershed conversion, pollution and rubbish disposal have a major impact on the quality of urban life. Green cities could offer a solution to rapid urbanization and global climate change. The development of green cities, as a new target of foreign aid, has strong implications for poverty reduction, tackling climate change and pollution, economic growth and sustainable development, and social equality, and promoting green cities as a sustainable development process integrates political, social, economic and environmental domains. This requires a better understanding of whether, where and what works with foreign aid—or could work—in the developing countries, what types of foreign aid practice have delivered successful outcomes, and what could be scaled up further and transferred across regions and countries.

7.2.4.1 How does Foreign Aid Work for Green Cities?

According to Stren (2012), experiences of the international assistance regime in supporting city development in sub-Saharan Africa and elsewhere are positive albeit modest. From his study of the development of sub-Saharan African cities since the 1970s, Stren suggests that the delivery of essential services, capacity-building at the local level with the help of aid agencies, harmonization between donors, and coordination of donors and recipients with a demand-driven approach seem to have had a positive effect on urbanization. Similarly, Adelman and Eberstadt (2008) recommend the application of a demand-driven business model based on the recipients' own contribution commitment to aid projects. The Urban Management Programme (UMP) introduced the so-called city development strategies (CDS) in certain cities, Kisumu in Kenya being one. The overall purpose was to enable the municipal authorities to achieve sustainable urbanization with a long-term citywide strategy. The ongoing experience suggests that poverty reduction, good governance, locally driven participation of stakeholders in decision-making, a well-designed action plan with reliance mainly on local resources, and early feedback of results are key to a successful urban city project. The UMP city consultation programme (UN-HABITAT n.d.-a) seems to have shared these principles.

In a specific scenario, the CDS for Lake Victoria City (UN-HABITAT 2008) reviewed strategies for improving the urban environment, preserving biodiversity and reducing urban poverty in the cities around Lake Victoria. The study highlights the success of in situ conservation measures, including public awareness and environmental education, and the requirement of further planning and links between the local and global levels. However, the scope of CDS is limited in terms of supporting green city development because it views the enabling conditions for sustainable urban development to constitute only good urban governance and fiscal balances. Moreover, the CDS focuses mainly on policies targeted at sustainable city development. Although the CDS encompasses an economic aspect to urban development, there is no clear guidance to link foreign aid to sustainable urban finances to support a city's urbanization and development process. Foreign aid could be the vehicle to help developing countries achieve sustainable city development. However, as an ongoing programme, the UMP achievements and lessons need to be evaluated further. This experience shows that the efforts to build both capacity and a forum between donors and aid-related institutions have worked, but that performance evaluation as well as the sustainability of finances have been weak throughout the programme (UN-HABITAT 2005).

Environmental concerns have been integrated into the aid programmes of some donors, in which targets need to pass environmental standards. Furthermore, protection of the environment has been established as a goal in its own right (Stokke 1996). However, as Arvin, Dabir-Alai and Lew (2006) point out, there is a noticeable gap in research in explaining how aid flows are linked to the environment in developing economies. Their later work notes that aid projects and pollution are linked, although only in certain countries. Based on this, these authors suggest that there is a need to promote policies which could facilitate income and employment generation through environmental/natural resource management. As aid at present focuses on poverty and related development and the environment, not much money has been allocated to the full spectrum of green city building in developing countries. Many recent research reports (Raymond/IHC 2009) urge for greater attention and resources to be targeted at urban issues. The Cities Alliance (2012), which concentrates primarily on urban dwelling improvement and urban development policies, was set up as a global partnership between various stakeholders (e.g. governments, NGOs and slum dwellers) to promote the vision of 'sustainable cities without slums'. Its country programme highlights the importance of a framework to enhance cooperation between urban stakeholders, and public and private investments in urban communities. This concerns, in particular, secondary cities where early, decisive action provides the best opportunity to manage rapid urbanization and to ensure a better urban future for all. The Cities Alliance keeps abreast of new practices through project repositories and a review of past experiences. Although urban poverty and economic growth are still priority issues, urban environment and environmental health are also a major concern, as, for example, in Sudan where litter and polluted waterways plague most of the country's urban centres. The United Nations Environment Programme (UNEP) has investigated issues such as rapid urbanization, urban planning, water and sewage, waste management, air pollution and transport, urban energy, sustainable construction and governance. It concludes that the main obstacle to improvement in these areas is the lack of investment. However, other problems, such as a widespread lack of adequate urban planning, also play a role (UNEP 2007). The lessons suggest that planning, governance and the often ignored social sustainability are essential for the success of a green city.

Climate change adds a new dimension to the concept of foreign aid and urbanization. Climate change is rooted in the cities; they are responsible for 70% of carbon dioxide emissions. Surveys suggest that because of recently intensifying natural disasters and rapid urbanization, aid agencies will invest more in disaster prevention for urban areas (Nguyen and Rowling 2012). However, despite the fact that international negotiators have repeatedly pointed to the transfer of technical and financial aid as a possible remedy for global and local environmental problems, little systematic research exists on whether donor commitments have been honoured (Timmons Roberts et al. 2009). For example, the Green Climate Fund was to deliver financial and technical assistance for climate change activities in the developing world through bilateral and multilateral programmes. Its goal has not been met. However, earlier research suggests that a policy framework, accountability, assessment and sustainability of projects, as well as stakeholder involvement constitute key measures for programme implementation (Pittock and Hartmann 2011; Kerkhoff et al. 2011; Pittock 2010; Hallegatte 2009). A two-country model developed by Hatzipanayotou, Lahiri and Michael (2002) reveals that with performance-driven aid transferred from a developed to a developing country, the impact on cross-border pollution from production activities in the recipient country can lead to reductions in total emissions over the medium and longer term. Critics, however, argue that such transfers could lead to unsustainable development, and eventually to environmental and ecological degradation. A contrary view by Asafu-Adjaye (1999) suggests that these transfers may not only reduce poverty but also encourage greater care of natural resources by poorer nations.

In view of the problems faced by developing countries with regard to environmental and resource protection, research suggests that priority in city planning should be given to measures which support ecologically sustainable development in specific fields such as drinking water supply, waste disposal (solid and liquid) and transportation. Timmons Roberts et al.'s (2009) research based on the PLAID database emphasizes accountability for both donors and recipients. Microlevel measures should be linked with those at the macrolevel, for example, by elaborating and implementing a sustainable environmental policy (Kausch 1996). Since some countries lack sufficient technical and financial means, there is an urgent need for the provision of this support. Aid through central governments is constrained by bureaucracy, corruption and so forth, thus direct funding transfers to local community organizations to improve their access to basic social and economic infrastructure and income-generating activities is an alternative solution. Researchers and practitioners generally agree (Collier and Dollar 2004) that aid allocated along political lines has less chance of leading to positive development outcomes than aid allocated according to need or with government commitment to good policy.

Timmons Roberts et al. (2009) find the overall influence of most eco-functional variables on environmental aid to be small compared with the more traditional determinants of foreign aid allocation, such as political, commercial and various historical factors. A UNEP (2007) assessment of the environmental impact of various aid programmes (including humanitarian, recovery, development and environmental) on post-conflict communities in Sudan was considered to have been fairly negative. Two core problems were identified. First, the impact of good individual projects and efforts were greatly weakened by the lack of integration with the core government and international aid programmes. Second, management of the environment and natural resources sector was burdened by the paucity of funds and continuity of funding. This shows again that the common factors—donor commitment, harmonization, cooperation and alignment and sustainability of foreign aid—are essential components for the success of an aid project.

What Could Work (in Foreign Aid) on Urbanization and Green Cities?

Harris (1989) notes that the main problems of urbanization fall into two interrelated areas: various constraints on urban productivity growth, and poverty. He argues that aid should become technical assistance-led within a policy framework or development plan rather than capital-led. Hoffman (2008) suggests adapting a business-like approach which perceives donor accountability to recipients similarly to a business relationship that includes risk assessment, acknowledged targets and the ability to meet these at the lowest possible cost. Based on experiences from China, Qiu (2011) suggests that development of a sustainable city should focus on forming an alliance of nature and traditional cultures, developing sustainable-city qualifications, urban planning, an international cooperation system, and establishing an evaluative and inspection system. He concludes that fundamental prerequisites should include compact and mixed land-use regulations with certain population densities, a high proportion of green buildings, biodiversity, green transportation and environmental thresholds for industries. Green city researchers believe that the development of sustainable cities is currently driven by climate-change challenges and rapid urbanization in tandem with growing business interests and advances in technological innovation. Based on some scenario studies, Joss (2011) considers the presence of governance to be a central and defining feature. He adds that development should be based on technical innovation within each city's particular context (geographical, political and economic) and interaction among multiple stakeholders. As the ecocity concept is still new, the implementation process of the green city is problematic (ECMM 2011).[3] In several instances (as in Dandong, China), projects have had to be scaled back or postponed as a result of financial and political problems. Questions also arise over the issue of transferability: it is yet to be seen how green city principles and approaches can be applied to other urban centres.

Whitfield (2009) offers a different angle for simplifying and reorienting aid and aid practices. In order to reorient donor staff and expertise, and to introduce specialization within their areas of assistance, Whitfield advocates reducing the intensity of donor engagement, areas of donor intervention, the number of donors and projects in a country, and the size of donor organizations. Some other new ideas for effective aid also exist in this area: some researchers have advocated creating an aid 'market', where donors pay only for outcomes that have actually been achieved, while other aid agencies have echoed the need for a performance-based approach. However, these ideas have not been implemented or tested in detail so their feasibility is not known.

What Is Scalable (in Foreign Aid) on Urbanization and Green Cities?
'Scalability' is defined as the capability of being easily upgraded or expanded, as needed (MW 2012). Based on a study of aid in urban hazard mitigation, Vincent (2011) believes that codes and standards are an essential component of scalability, and in order for an effort to be scalable, a flexible but systematic process must be in place. He suggests the following approach with regard, for example, to city construction: analyse the current state first; next, identify and initiate steps to correct deficiencies in regulation (e.g. replacing or modifying unsuitable regulations); and, third, integrate efforts across related sectors. Victor concludes that in order for urban improvement to be scalable, it must take into account economic, social and political issues.

The 'Millennium Villages' project in Ghana (UN-MV n.d.) targets poverty in parallel with a focus on health, education, agriculture, rural infrastructure and economic development. The aim is to create one coordinated aid programme instead of several different aid agencies tackling these issues separately. The project implements an evidence-based approach and emphasizes sustainability and scalability through major policies to secure the involvement of local government, implementation by the communities themselves, guiding communities towards self-sustainment and rigorous evaluations with an exit strategy. It recommends action plans based on 'human needs' (EICU n.d.), and it follows the Millennium Project framework. Although the outcome is mixed, the Millennium Village project and aid in Ghana as a whole have made huge progress in meeting the MDGs on poverty and hunger, and boosting the country's economic growth (Mulholland 2012).

In designing sustainable solutions, Isaac (2012) suggests working with local entrepreneurs or local governmental institutes because these continually consider such factors as scalability and expense. A recent pilot experiment tested the concept of aid adopting a business-like approach that included risk assessment and targets (Hoffman 2008), and the objective is now to scale it up substantially.

Research (Easterly 2008; Sachs 2005; CIDA 2004; Graham and O'Hanlon 1997) suggests that a scalable framework or top-level action plan that includes capacity building, assessment, and sustainable funding and technology support is important for any aid agency working in a new field, in our case urbanization and green city development. For developing countries, poverty alleviation, environment protection and economic growth are intertwined problems. The essential services generated by urbanization and green city development (related to farmlands and watershed conversion, water supply, rubbish disposal, transportation and pollution reduction) while concurrently being aware of the need for poverty reduction and social equality have to upscale on the basis of experiments.

But there is no straightforward design for upscaling a project. The green city concept is new and implementation can face difficulties (ECMM 2011) as a result of financial and political problems, as mentioned previously. Rogerson (2011) describes the upscaling process as helping 'countries pilot changes carefully for themselves, evaluate pilots rigorously, debate and learn lessons inclusively and finally scale up a few proven and popular experiments'. After examining aid cases related to education in Africa, Gillies (2010) proposes that instead of looking for a formula and rushing to scale up, policymakers should acknowledge the human process of developing ownership or strengthening new behaviour, or that changing to a system is a case-by-case endeavour. To scale a successful aid project up or down, one needs to be able to foresee potential changes that could originate from pilot project documentation or through negative effects such as the destruction of local initiative (Samoff et al. 2001, 2011).

What Is Transferable (in Foreign Aid) on Urbanization and Green Cities? The 'transferability' of aid programmes is closely linked to the scalability of pilot projects. The previous discussion has indicated that a transferable framework with a focus on planning, accountability, alignment, aid sustainability, governance, capacity building, donor harmonization and donor-recipient coordination is essential if foreign aid through various international agencies is to succeed. Some characteristics are shared by successful aid projects, as Adelman and Eberstadt (2008) note. These include the focus on local ownership and initiative, partnerships, flexibility and anticipation of funding allocations, peer-to-peer approaches, technology adaptation and adoption, and continuous information feedback. Harris (1989) suggests that aid should become technical assistance-led rather than capital-led. An urban strategy in the form of a policy framework or a development plan is one instrument of assistance that is easily transferred.

Based on aid's educational experiences, Riddell (2012) believes that the contextualization of a recipient country, its capacity development, local ownership and leadership, and stakeholders' involvement are keys features in considering transferability. On the one hand, transferability is derived from what has worked in aid, particularly aid targeted at urbanization and green cities. On the other hand, transferability is closely linked to a local political, social, environmental and economic context. In the implementation of a framework, aid projects should look for a demand-led approach based on local needs and self-evaluation of development. But it must be realized that in supporting advanced green city development, efforts to promote technological transfers and staff capacity building of both the aid agency and the recipient country are as important as funding transfers, if not more so. The lessons learnt from past experiences of what has worked and what has not are transferable. As urbanization and green city development are a new domain for foreign aid, pilot projects on transferability should encompass both traditional aid targets (e.g. poverty eradication and health issues) and newly acquired domains (e.g. transportation and pollution reduction). Again, this means collaboration among various aid agencies, and between aid agencies, local government and entrepreneurs. As green city development embodies multiple fields of expertise, identifying the key areas that can be quickly transferred and scaled up to other developing cities is vital if foreign aid is to become more effective.

As the green city is still a novel idea in foreign aid, questions regarding its implementation affect the issue of transferability, and it is yet to be seen how these principles and approaches can be applied to other cities (ECMM 2011). Moreover, the changing nature and capabilities of the developing as well as the developed worlds, as well as the emergence of new aid sources and approaches, point to the dynamism of transferable strategies and experience. Whitfield (2009) is a proponent of change and emphasizes that not all donor agencies have to follow the same aid practices. Experiences of transferability are evidence-based, and the success of one will be copied by others.

In conclusion, aid allocated to urbanization and ecocity development is an unexplored domain, although the pros and cons of individual experiments have been examined through either research or practical application. But the link between foreign aid, urbanization and the green city is not well established. Thus successful foreign aid outcomes in other areas have to be acknowledged, duplicated and applied. It has been demonstrated that positive experiences can be duplicated and transferred to plan green cities and urbanization in the developing countries but a comprehensive scenario analysis on foreign aid, urbanization and green cities with more specific policy and implementation focuses is essential as a starting point.

7.3 Analysis of Three Case Studies

China is experiencing an urbanization process that is rapid, large scale and long term. Its central government recognized that sustainable development was vital for creating a harmonious society and has worked towards that goal. In response, local authorities have promoted an ecocity development strategy. Two examples are given here: the Sino-Singapore Tianjin Eco-City, built in a location comprising mainly saltpans, barren land and polluted waterways, and the Wenchuan and Beichuan areas, built in a region devastated by a massive earthquake. In addition, Brazil, another major developing country, shares some experiences with China. Curitiba in Brazil is heralded as one of the first ecocities in the world, and it was awarded the Globe Sustainable City Award in 2010. The case studies introduce interesting contrasts: Tianjin, Wenchuan and Beichuan areas in China were all built largely on barren land, while Curitiba rose from an existing city. The restoration of Tianjin and Curitiba started with well-designed plans, while recovery efforts for Wenchuan and Beichuan had to be rushed after a devastating disaster.

7.3.1 Tianjin City, Northern China

Urbanization in China is projected to rise to about 64% by 2025, which translates to more than 350 million additional people living in urban areas. China has recognized that economic progress has to be based on 'green growth', and thus national- and local-level programmes have been established to define climate change targets and promote the construction of ecocities, although with a variety of experimental approaches and standards.

China is currently constructing an ecocity in the coastal district of Tianjin to serve as a practical, replicable and scalable model for sustainable development. Originally, two criteria guided the selection of the ecocity site: the location was to be on non-arable land, and northern China was faced with a water shortage. The Tianjin site fitted the bill with the extra advantage of having sound infrastructure, easy accessibility and commercial viability. The project started in 2007, when China's rapid urbanization and increased global awareness of the importance of sustainable development gained momentum. With investment and technology supplied by the government of Singapore, the 'thriving city, which is socially harmonious, environmentally-friendly and resource-efficient' (Wikipedia 2012b), was to be constructed in several stages for completion around 2020.

Tianjin City is located in one of the fastest-developing districts in China in the Bohai Bay region which, after the Pearl River delta and Yangtze River delta, is the country's largest growth engine. As the site was comprised mainly of saltpans, barren land and polluted waterways, the necessary technology and expertise were provided by Singapore, based on previous experiences with respect to the Suzhou Industrial Park. According to the masterplan, the district is initially to derive energy from a waste incinerator plant as well as several other options for clean fuel, renewable and geothermal energy. A light-rail transit system, supplemented by a secondary network of trams and buses, is to be the main mode of transportation, covering 90% of public transport needs (Joss et al. 2011). All buildings are to conform to stringent energy-efficiency standards that include advanced water-saving and waste-management systems with particular emphasis on the reduction, reuse and recycling of waste. As Tianjin City is located in a low rainfall area, the ecocity is to draw a significant part of its water supply from non-traditional sources. The existing wetlands around the city are to be protected to enhance biodiversity. The city layout is based on an integrated mixed land-usage system to create variety in the landscaped 'eco-neighbourhoods' with green 'eco-valley' corridors that will serve as the main public open spaces. Private sector agents are also involved, and 125 companies in the ecocity were registered by 2010. The Sino-Singapore Tianjin Eco-City signed agreements with Hitachi, Philips, Siemens, ST Engineering and two leading property developers in Asia to develop a green central business district.

It is expected that by 2020 the ecocity will create 80,000 to 100,000 jobs, contributing a total of CNY40–50 billion to the gross domestic product (GDP).

Social harmony is a key consideration for Tianjin City, covering such areas as education, healthcare and culture. Important instruments include subsidized public housing to help meet the housing needs of low-income people and to enable different social strata to live together, catering to the needs of the elderly and the disabled, and providing public facilities and respecting local heritage.

In 2010 the city received a grant of USD6 million from the World Bank's

GEF to support the development of policy, monitoring and regulatory mechanisms in which the recipient promised to invest an additional USD57.9 million. As the World Bank has extensive experience in urban development and capacity building, it could provide Tianjin City with strategic support in developing an energy- and resource-efficient city with low greenhouse gas emissions, including:

- technical assistance, software and equipment for the implementation framework of the masterplan and dissemination activities;
- technical assistance for the public transport system;
- technical assistance and pilot investment for green building (World Bank 2012a).

Project design was founded on the experiences of past projects implemented by the GEF-World Bank in China in relation to urban transport and building energy efficiency. The main focus was on energy, transportation, public housing, infrastructure and climate change with corresponding indicators (e.g. quality of tapwater and carbon emissions per unit GDP) and timelines. However, the adoption of energy-efficient technologies in the pilot public buildings was delayed as a result of scheduling conflicts, design changes and inconsistences with the masterplan. This underscores the fact that project plans need to be practical, and cost- and market-based, while implementation must conform to standards and be well controlled, regularly inspected and evaluated. Government is the key in policy enforcement.

World Bank aid was helpful in project planning and management. By collaborating with other international donor partners such as the Australian Agency for International Development on ecological urban development and ESMAP on building energy efficiency, the World Bank gained knowledge and generated expertise which it lacks itself. IE Singapore also offered a range of financial tools, grants and tax incentives to help Singapore enterprises in the city gain access to capital, develop their financial management capabilities and defray developmental costs.

The Tianjin Eco-City project did not strive for technical advancement but was instead based on existing practices, utilizing the best and most affordable technology currently available to minimize the harmful impact of development—in other words, technology that could be scaled up and transferred to other cities in China (BBC 2012). It was designed partly as an experiment to determine how urban problems such as gridlock, water pollution and energy consumption could be resolved. Other than using public transport or recycling rubbish, the main contribution of Tianjin residents was adopting the role of a guinea pig, to participate in the experiments of their city.[4] In 2012, people started to move in when the smart grid service, renewable energy, ecoindustrial park, electric vehicle charging stations and green parks were operational. An international school will be opened later, with environmental protection as an important part of its curriculum.

However, despite progress, according to a recent study, it may not be possible to reach some of the city's key environmental indicators, such as CO and SO_2 levels, as a result of heavy pollution from the surrounding areas. The frequency of diseases caused by poor air quality and water contamination is already high, and some critics argue that the project is more a propaganda instrument than a comprehensive ecological project (Wikipedia 2012b).

7.3.2 *Wenchuan and Beichuan Areas After the Earthquake*

On 12 May 2008, a devastating earthquake hit southern China that left 87,000 people dead or missing, destroyed 5 million homes and produced direct economic losses valued at USD120 billion.[5] The earthquake, which was centred on Wenchuan county, destroyed all public service facilities such as hospitals and schools, 70% of its housing and industries, and 80% of its farmland. Economic losses totalled CNY63.4 billion (USD9.29 billion). In the post-disaster period, aid poured in both from China and from the international community for rescue work and later rebuilding.

After initial disaster relief efforts, aid was directed to rebuilding and improving public infrastructure, roads, water, transportation and electricity, and to helping economic recovery and structure transition. The Wenchuan Earthquake Recovery Project, based on a framework approach, was launched in early 2009 with the aim of facilitating the emergency response, support the government's recovery and reconstruction strategy, and restore and improve essential infrastructure and health and education facilities, while attempting to provide the basis for longer-term sustainable development. A rapid assessment project financed by the GEF to assess and mitigate the risks and potential environmental impacts was initiated. Next, a knowledge-management workshop was organized to share global expertise on post-disaster reconstruction and risk control, and to ensure that the knowledge gained from the project could be incorporated into future plans on disaster control. The project can be considered a success because the mitigation activities proposed by the GEF are now included in Sichuan's future action plan for post-disaster recovery and reconstruction, with further support forthcoming from the World Bank's emergency recovery loan (World Bank 2009a, b). In the remote, impoverished areas that received less attention, Oxfam Hong Kong has been working in the fields of reconstruction, sustainable livelihoods, ecological condition assessment for disaster prevention, social participation and capacity building (Oxfam 2011). Other NGOs such as 512 Centre, Handicap International and the World Health Organization have directed their expertise to sectors where it was needed the most, such as rebuilding, education, and health and rehabilitation.

International research communities have also set their sights on the transition to an ecocity. Feffer and Pastreich (2008) suggest that China could turn its disaster into an advantage and utilize the assistance from Japan or South Korea, for instance, to leapfrog current technology and create a new kind of city in Wenchuan. However, they also warn that in order to deal effectively with pollution, climate change and energy inefficiency, the problems currently facing China, the Wenchuan project needs to be more than simply a showcase: it must be sustainable and replicable. Baozing, China's minister for housing and building, has called for innovative city structures, which would be equipped not only to reduce pollution and save energy but also to deal with a disaster. He notes that post-disaster reconstruction standards for an ecocity need to emphasize anti-disaster capacity, environment protection, technology adaption and the creation of economic opportunities (Sina Finance and Economics 2008).

Immediately after the disaster, the State Council of China drew up a general plan to rebuild the 51 destroyed county-level regions, Wenchuan included. The immediate priorities included the reconstruction of homes, schools, hospitals, roads and temples. Tragic as the earthquake was, it also provided an opportunity for economic and environmental improvement in these regions. With help from Japan, other countries and international organizations, the central government assigned the financial responsibility for redevelopment and rebuilding of a particular area to each province. Thus in the case of Wenchuan county, Guangdong Province became its aid partner. As Wenchuan is one of the four main ethnic centres of the Qiang people, efforts were made to ensure that the town's reconstruction reflected this cultural heritage.[6] By 2011, three years after the earthquake, 95% of the post-quake reconstruction projects had been completed, although a few earthquake ruins were preserved as monuments to support tourism as one of the key pillars of economic development in the post-disaster period. Wenchuan county also plans to explore natural resources such as aluminium and hydropower.

The schemes for the Wolong area redevelopment and the Giant Panda Nature Reserve restoration were funded by Hong Kong, with a masterplan that ascribes value to ecological remediation, and landscape as a placeholder. Redevelopment of the urban and natural landscape was divided into three zones: (1) a core area; (2) buffer zones to protect core habitats and the ecological system; and (3) ecotourism and resettlement areas to house the human population and protect the local culture. Damaged ecologies will be restored through natural processes. Likewise, the scheme endeavours to re-establish the interdependent relationship between humans and nature implicit in traditional Chinese planning practices (Bassett 2009).

Remarkably, the reconstruction scheme not only envisions new towns being erected on the earthquake ruins but also hopes to see changes in the attitudes and lives of Wenchuan residents, with people being more attentive to maintaining a good community environment. But there has been criticism of the rampant pollution, misuse of funds and poor quality of newly built homes (*South China Moring Post* 2012). It has been said that reconstruction efforts in Wenchuan have destroyed the area's distinct cultural identity. The reconstruction project launched by Hong Kong Red Cross in 2009, which favoured modern building materials, has been blamed for altering the area's unique appearance. In response, the authorities attempted to restore some of the cultural elements, an effort which upset local residents even further because of cultural misunderstandings.

Reconstruction projects in these damaged zones must include input from local residents as well as experts, and cannot simply be set up for rebuilding houses and exploiting the area as a tourist attraction without respect for local wishes. On the other hand, it needs to be acknowledged that there are no national guidelines for this kind of 'starting-from-scratch' concept for ecocities, although the relevant housing ministry recommended that cities and towns hit by the earthquake should prioritize reconstruction that was aligned with ecocity principles.

Beichuan City, in the epicentre of the earthquake, was completely erased and had to be relocated and rebuilt for 50,000 inhabitants. Aiming for a new urban form, to be supported by tourism, Beichuan City was designated by the authorities for development as an ecocity even before reconstruction had started and would serve as a distinct model city of sustainability for the country. Reconstruction has followed a top-down approach in accordance with a central government masterplan, and the government at the implementation level respected the unique position of the planners with regard to power and control (Ward 2011). This is seen to broadly reflect the shift towards ecocities in China's national urbanization policy in order to support the environment and harmony of society. Beichuan's design is organized around the principles of environmentally friendly urbanism and local landscapes across a spectrum of elements, including culture preservation, environment valuation and protection, industrial economy transition, green transportation, renewable energy, green building, and water and waste systems management. Economically the development of the city is founded on green industry and tourism to take advantage of its ecological resources. Existing heavy industry will be deindustrialized and the environment assessed in economic terms.

Aid has been forthcoming from a number of agencies. The International Finance Corporation, the private sector arm of the World Bank Group, provided equity support to local small and medium enterprises. It also provided the local township bank with funds and technical support to strengthen its operations and to develop into a competitive and commercially sustainable microfinance institution to serve the lower segments of the market (World Bank 2011, 2009a, b). Forestry Innovation Investment Ltd and Canada Wood (FII/CW), on behalf of the Canada-BC initiative, constructed a comprehensive care facility for senior citizens, which was designed to incorporate elements of the traditional housing style of the Qiang with Canadian wood-framed structures. The Prince Claus Fund provided funding for the Beichuan Library, which houses a collection of works of great importance to the Qiang minority.

7.3.3 Curitiba in Brazil

The city of Curitiba has been heralded as one of the first ecocities. It began proactively to address the challenges of sustainable urban development in 1966. Designed through a commercial competition, the masterplan outlined integration between urban development, transportation and public health, and the city created an administrative agency to implement it.

Political will and skill were important factors in the success of the city's urban development. The adopted philosophy centred on the integration of functions and urban services, predominately in a tripod-type fashion: transportation, roads network and use of land for residential, commercial, industrial and service purposes. These were linked to expansion out of the city centre along arterial growth corridors to distribute settlement densities more evenly. The results have been substantial: a 22% drop in private car usage, dynamic economic growth for local shops, the development of community space for pedestrians, the lowest air pollution rates in Brazil, natural flooding reduction and a high rate of resident participation in recycling (Suzuki et al. 2010).

The plan created a modern Curitiba with green spaces, low-density residence, an effective bus system, environmental education and a waste processing system. Since then the plan has regulated the physical, economic and cultural transformation of the city. The harmonious coexistence of people, the environment and its urban ambience has made the city a tourist attraction. Curitiba's masterplan also introduced economic changes, such as the creation of the Industrial City of Curitiba, but this conforms to city criteria with regard to topography and integration with its surroundings. The city's cultural transformation gained momentum with a celebrated event when Rua XV de Novembro was converted into a popular pedestrian walkway, and people began to realize that they were part of the city and wanted to be involved. The Fundação Cultural de Curitiba was set up as a facilitating agent for the cultural production of the city and preservation of local heritage. Curitiba adheres to a practical and repetitive planning process; proposed concepts and ideas are tested and tried before being put into practice, and the generated feedback has resulted in further improvements and applications.

Curitiba's strategy focused on integrated planning (e.g. transportation, land use and efficient resource management) that prioritized people and commitment to local values such as accessibility, transparency, social justice and poverty reduction. Strong leadership guaranteed the successful, long-term implementation of the strategy (ICLEI 2002). Curitiba has faced similar urbanization problems as the rest of the world (overcrowding, poverty, pollution and limited public funding), but it addressed these in a cheaper and more integrated manner. The city's extensive bus system, costing less than a tenth of a subway network, at the same time addressed concurrent issues of pollution and poverty, industries and green space, heritage and tourism.[7] Curitiba did not become successful overnight; its masterplan has persisted and evolved over generations and through experimentation with the help of local urban planning expertise. People still worry about the plan's continuity, and related expertise and leadership (Gnatek 2003).

Curitiba's many initiatives—environmental cleanup, city restoration, job creation, education improvement, disease intervention and hunger prevention—were tackled without too great a reliance on federal government, international organizations or charity. However, the city has received financial and technical support from various sources such as the United Nations, the Inter-American Development Bank (IDB) and developed countries. The CIFAL Curitiba project, founded by the United Nations Institute for Training and Research, is helping to develop the capacities of local stakeholders in urban transport, green cities, municipal finance and infrastructure project management. The IDB loan is supporting Curitiba in developing an integrated programme to improve the quality of life for local residents, particularly in the slums. Specific efforts include improvement in housing and environmental sanitation, mobility, social and public services in poor areas, and institutional capacity. France's multisectorial aid is helping Curitiba with the urban environment and transport to promote the city's development and transformation through land-use planning (e.g. housing, urban equipment, construction of transport infrastructure and environmental preservation).

This review and these case studies underscore the need for a foreign aid framework geared towards urbanization and ecocity development to ensure project success. The framework should include both objectives and reports outlining progress towards earlier goals for a socially, environmentally and economically sustainable city. Because of the complexity of aid for urbanization and green city development, in order to simplify issues and clarify ambiguities, this chapter presents diagrams to illustrate the various roles involved, their inter-relationships, and the processes of policies, practices and implementation.

In Fig. 7.1, donors and recipients are recognized as the major actors in aid earmarked for ecocity development. However, the surrounding environment and external players such as central government (if aid is allocated directly to locals) may exert considerable influence, while external elements could come from geographical, environmental, political and economic contexts. Aid policies that are designed according to aid type coupled with specific requirements and implementation practices are founded on successful aid experiences and can thus lead to fulfilled goals. The

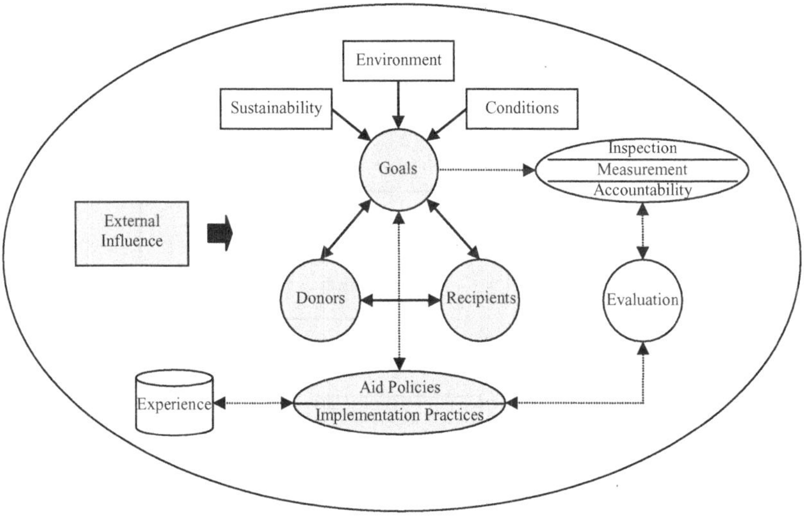

Fig. 7.1 Factors, conditions, interactions and the process of aid in ecocity development. Source: Author's illustration

experiences generated by aid in the development of ecocities, on the other hand, may differ case by case, while others may share common features, making them scalable and transferable. In practice, aid objectives are always tied to conditions regarding donors' pursuit of ideologies, political returns and so on. In the context of this chapter, the sustainability of aid, including replicability and scalability, and its environmental valuation for the development of ecocities are essential concerns, and are thus highlighted here. Figure 7.1 also shows the importance of evaluation and monitoring during the implementation of an aid project. The agreed targets may be reached through various methods, including 'trial and error', pilot projects and performance-based approaches. As implementation is a key driver for the success of an aid project, it is further illustrated by Fig. 7.2.

Here the chapter suggests a recursive aid process that prioritizes project sustainability, including transferability and scalability. As Fig. 7.2 shows, a prospective aid project designs the plan based on past experiences—that is, the pool of knowledge accumulated through earlier work in the same field. The project starts on a small scale, with either a pilot plan or one or a few subfunctional tasks (here called 'stories'), which enable early feedback of

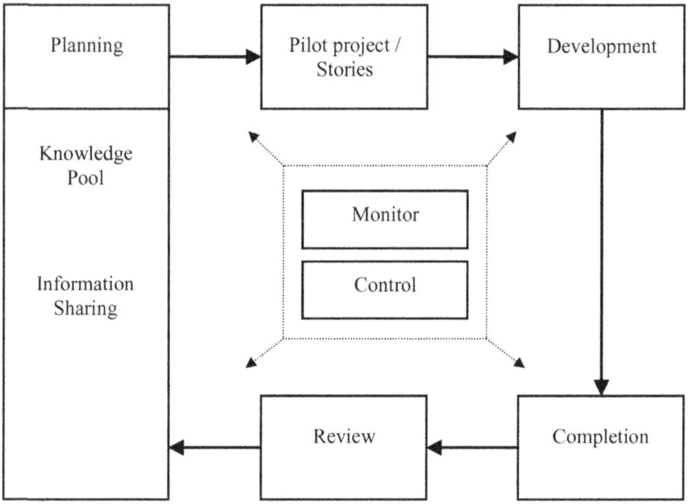

Fig. 7.2 A recursive implementation process of an aid project. Source: Author's illustration

possible results. The pilot plan or 'story' is carried out through four stages—development, completion, review and plan revision—during which the knowledge and experience collected during the ongoing analysis are identified and documented. The whole process is controlled by stakeholders through constant mutual interaction. It is also monitored and evaluated by responsible authorities or third parties. Thus in a recursive process, all changes can be properly incorporated and risks avoided or reduced to a minimum. Based on review and feedback from donors, recipients and other stakeholders, implementation results can be properly evaluated and errors eliminated. Harmonization between various schemes is achieved through information-sharing and active communication, ensuring that different programmes complement one another and avoid duplication. An exit strategy is included in an evidence-based approach: if, after proper evaluation, the trial is unsuccessful, the aid project can be aborted with minimal loss.

The case studies of Tianjin, Beichuan and Curitiba all suggest that large-scale green city development requires considerable government power and involvement. The work is generally supported by central government but implemented by local authorities. Figure 7.3 describes the two levels of involvement and support with practices and goals in ecocity development.

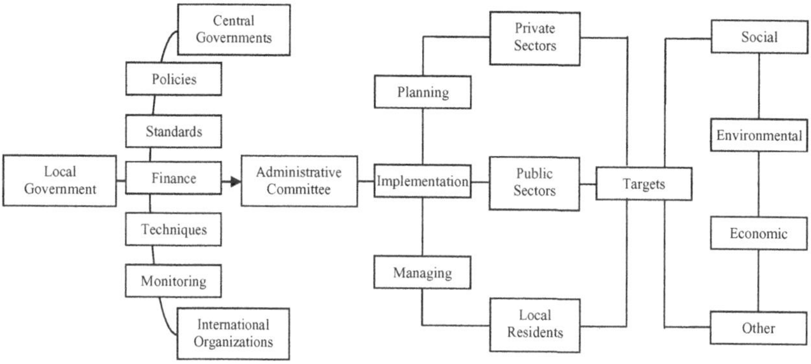

Fig. 7.3 Stakeholders, practices and goals in ecocity development. Source: Author's illustration

With respect to the timeline of ecocity development, the main body responsible for drawing up the city plan and implementing policies is the administrative or planning committee, as the experiences of Curitiba and Beichuan show. Central and local governments are responsible for policies and standards, provision of financial and technological support and monitoring. It is important to involve private and public sectors and local residents in a systematic process, although their input may be required at different stages. For example, Beichuan concentrated on the reconstruction of the city and tourism during the initial stages, while Curitiba mainly focused on transportation, roads and land use. Usually the objectives of ecocity development are defined along three target pillars (i.e. social, environmental and economic benefits). But a government may also pursue other goals, such as self-promotion and technological advancement. Figure 7.3 assigns a role for international organizations involved with aid agencies, which is similar to that of government, but it does recommend that these interact with the locals or become directly involved in the locals' development, as demonstrated in Fig. 7.1.

7.4 Conclusion

In summary, an aid framework in policy dimension that provides integrated solutions to combine poverty, environment and governance concerns is the road for foreign aid in supporting urbanization and green city building in developing countries. More specifically, the framework should focus on aid earmarked for green city planning, accountability, alignment, aid sustainability, governance, capacity building, harmonization between donors and donor–recipient coordination. On the other hand, the framework should also support aid efforts on specific issues within recipient countries (e.g. poverty) without overlooking other concerns. The framework should be transferable and able to work in most projects in this particular field. It should be flexible so that it can be tailored to any particular context (e.g. geographical, environmental, political and economic). The framework also needs to function as a bridge to link foreign aid to cities. Centrally planned or politically tied aid has not worked in the past. However, as political stability and effectiveness are important factors in any nation's development process, an aid-driven solution for gradual institutional reform should be investigated. During implementation, the search approach through trial and error with an outcome-oriented goal is recommended. If a pilot experiment succeeds, then the project can be scaled up with subsequent aid allocations. Furthermore, ongoing analyses and documentation of experiences can provide an important knowledge-management component that contributes to scaling up global cooperation at the local level.

Foreign aid for urbanization and green city development can focus on urban poverty reduction, climate change and other environmental issues, such as water and sewage, waste management, air pollution, transport, urban energy and sustainable construction. Lax funding commitments and paucity of aid are still major obstacles to urbanization and green city development. A share of the funding should also be earmarked for technical innovation for green city development in developing countries. Assistance motivated solely by technical or capital considerations is not the right solution, although a combination of the two may well be the right answer. Selectivity and sustainability can be fostered by foreign aid agencies in their allocations in support of green city development. A locally driven initiative with main reliance on local resources and participation of all stakeholders in decision-making is one way to ensure the success of green city development. There are also some new ideas worth testing, such as a foreign aid 'market' or a performance-based approach. A project always implies some risk and it should be flexible to tolerate such risks. However, as Adelman and Eberstadt (2008) suggest, this should not become an obstacle to experimenting with new approaches.

International aid could play an important role in promoting ecocities with the development of policy or major plans, or it could pursue a smaller role in filling the gap left by development authorities with regard to vulnerable groups. For instance, research shows that the points detailed below could secure the success of an aid project facing the challenge of urbanization and green city development. These could therefore be viable for further replication and scaling up:

Aid should be focused at both policy and implementation levels. At the policy level an aid framework to provide integrated solutions that include measures on poverty, environment and governance is ideal, but it should also flexible enough to allow tailoring to specific issues. Aid could help with policymaking and developing ecocity standards. For example, as no common definition exists for ecocity, it has been necessary to develop standards for individual ecocity projects by combining local realities with key sustainability features. Aid could be used to help establish quantitative and qualitative key performance indicators and so forth. At the implementation level, proper procedure is vital and should include project planning, harmonization, coordination, risk control and capacity building. A well-designed plan is equally important. For example, in connection with the environmental assessment project on the Wenchuan earthquake, the World Bank included a follow-on workshop to ensure that the newly acquired knowledge would be incorporated into the city's future risk-control plans.

Aid should emphasize that the sustainability of an aid project extends to its transferability and scalability. For example, the World Bank supports the sustainability of aid programmes through specific policies, regulatory, incentive and institutional frameworks, management systems and financing mechanisms. It tests the replicability of evaluation standards for green building through pilot projects. On the other hand, as ecocity development is a long-lasting process, stable and continuous support from international aid agencies is important.

Aid has multiple goals with multiple measures to be achieved through multiple means. Aid may involve the transfer of financial resources, commodities, technology, training and research. An aid project is usually based on multiple goals rather than pure altruism, and these can differ between donor and recipient. Thus it is important for both parties to maintain good dialogue and to be fully informed. On the other hand, it is important to have multiple measures of aid effectiveness at both macro- and microlevels that relate to the three target pillars rather than just economic growth. Moreover, the effectiveness of a policy framework with respect to transferability and scalability, and implementation practices such as alignment, harmonization and mutual accountability should also be included in the measures.

Aid is a tool for filling the gaps. International aid investments may not contribute much to ecocity development projects because of the scale, timeline, complexity and challenges of these initiatives. For example, compared with the overall estimated investments of CNY150 billion (USD22 billion) for Tianjin Eco-City, GEF's USD6 million input is small. However, aid could be engaged in various ways to assert influence. For example, international aid could be the channel for fundraising and the introduction of new technology by organizing international forums or expos through which development authorities, financial organizations, business partners and research institutes that are interested in green, environmentally friendly construction can establish contacts. Aid could adopt a demand-driven model based on contributions and promises of commitment by the recipients themselves. Aid could also work to fill whatever gaps environmental development may have overlooked, such as raising public awareness for environment protection, alignment development, or eliminating risks that may emerge during the development process.

For large aid investments, such as ecocity development, however, *a systematic and integral approach is necessary.* In the Tianjin case, the World Bank adopted a city-based approach rather than a sector approach by being attuned to more than one sector—that is, transport and building sectors. Furthermore, aid should be applied where organizational expertise is already present. For example, in Tianjian the World Bank focused on (1) knowledge transfers (i.e. incorporating best international practices into the design, planning and management of ecocities); and (2) best-practice demonstrations (which involves not only technology but also establishing benchmarks and standards for follow-on investment) (World Bank 2012b).

Aid can work to promote information transparency. In developing countries, information transparency is a technical and political issue, while information collection, transmission, and provision for ecocity development, including feedback from implementing officials and local residents, are essential for any ecocity project. This is also important for generating international aid and raising awareness of the issues encountered. For example, although China deserves praise for its handling of the Wenchuan disaster, the lack of transparency regarding aid, requirements of the disaster area and status of reconstruction have concealed some of the successes, and blurred areas where improvement was still needed. On the other hand, the literature shows that for ecocity development and related aid investments in particular, the current mechanisms for knowledge sharing, collaboration between aid organizations, guidance on where to target aid and project progress are not well established.

Urbanization and ecocity development is a three-pillar advancement process that includes social, economic and environmental aspects, which coexist and interact with each other. The development of the three pillars should be compatible with and integrated into urbanization plans. World poverty, for example, is considered to be one of the major threats to the environment, while economic growth is believed to offer the solution to social and environmental issues. However, the prerequisites of ecocity development mean alleviating poverty, respecting local identities, generating economic growth and directing modern technology towards energy efficiency, while working at the same time to protect the environment. This harmonious integration could bring about extra benefits, as in the case of Curitiba, where active, fruitful interaction among the residents, environment and the urban atmosphere made the city a tourist attraction.

Comprehensive investigation, integrated policy guidance and standards for an ecocity at the national level are needed to support the developing countries' urbanization and ecocity development process. Currently, many governments are developing their own standards and protocols, which may be reinventing the wheel, but which may not be transferable, scalable or work well for other cities.

Urbanization and ecocity development make up a systematic, long-term, practical and repetitive process. These need collaboration at all levels, with all types of organizations (governments, private agencies, research institutes and aid agencies) and on a range of issues (urban planning, economic promotion, public infrastructure, technological advance and environment improvement). For example, the Tianjin Eco-City project was a government-to-government cooperation effort that included six public-sector working committees and a consortium of private sector agents. Ecocity development calls for strong, effective governance, and implementation has to be practical and repetitive as ecocity plans evolve. New issues or ideas, as they emerge, need to be tested and then, based on feedback, put into practice. But development of an ecocity essentially relies on guidance from the city itself in terms of the scale needed. Although strong governance is a prerequisite, local residents are also an important factor. It is important to recognize that foreign aid can usually play only a complementary role. A direct appeal for foreign aid by a local community may be a good way to sidestep bureaucracy and corrupt governments.

Ecocity development must complement a country's own urbanization process and national development plans as well as local practices of the urban centre. One of the reasons for the selection of the Tianjin site was because it offered the opportunity to find solutions to problems common in northern China—that is, shortage of arable land and drinking water. It is also important to make sure that foreign aid is allocated within the context of, and integrated with, the government's core plans and policies. Ecocity development is one of the solutions for the pervasive urbanization trend in developing countries. Strong government support, involvement of firms and international organizations, financial incentives, ample opportunities and creation of markets, and advanced technologies for ecosolutions should be prioritized by planners.

The approach in *ecocity development planning can be both top down and bottom up*, and aid plays a significant role at both levels. Policies and development plans can originate with top government or be driven by the requirements of the local public and private sectors or residents. An issue can integrated into a common plan that, for example, considers pollution and poverty together, or heritage and tourism. On the other hand, the initial phases of ecocity development may concentrate on a few key sectors considered significant by local people, as in the cases of Tianjin and Curitiba, where emphasis was given to transportation and land use. The initial success of an experiment may attract more interest nationally and internationally, as happened in Curitiba.

Ecocity development authorities need to *respect local culture and identities*, and consider the needs of the end users. The focus of ecocity planning is predominantly on the reconstruction of the area and of homes. China was perceived to have failed with the development of Huangbaiyu as an ecocity because the concept was not integrated into the daily lives of the inhabitants. Furthermore, the ecological structures and urban form introduced in the Huangbaiyu area overlooked existing local economies (Bassett 2009).

While aid agencies are bound to well-defined practices, it is also *important that recipients create an effective and reliable environment* as well as define the procedures to be followed. An enabling atmosphere also includes good governance, respect for human rights and gradual reduction of aid dependency. Cooperation between aid agencies and recipients is needed. For example, during Curitiba's developmental stages, local authorities set up an administrative entity to coordinate public and private sectors as well as possible international aid organizations.

Coordination and collaboration among the evolving ecocities is currently rare but necessary, and recognition of successful or dysfunctional experiments is an important input. Curitiba's approach to controlling population density through measures in land use and transportation planning is a very good example for the cities of China that are confronted with fast development through vast immigration. International aid agencies may play an important leading role in this area.

Mitigating climate change is one of the primary goals of ecocities. Similar to the development of ecocities, climate change mitigation needs extensive collaboration between governments and international organizations, and more importantly the involvement of each individual. It must be recognized that climate change involves a series of global public good elements. As countries strive towards an international agreement, it is important that newly established policies are already embedded in the planning and implementation process of an ecocity.

As demonstrated, urbanization and green city development is a systematic process. However, the concept of a green city and the nexus of aid–urbanization–green cities are not well established yet. Developing countries need stronger financial and technological support, but data suggest that the international donor community has failed to meet the agreed targets. Meanwhile, social reform or improvement in information transparency, governance, human rights and reduction in aid dependency on the part of the recipients will encourage the international donors and build their confidence. At present, the majority of green cities are at the planning stage, with the duration from plans to full implementation lasting more than a decade (ECMM 2011). Therefore the success of aid in other domains has to be applied to green city development and urbanization in developing countries. The case studies and the analysis here have also shown that policies and implementation practices can be scaled and transferred to other projects. Comprehensive analyses of the link aid–urbanization–green cities, with a more specific focus on policy and implementation, are required.

Even though we emphasize various stakeholders in ecocity development, it needs to be remembered that residents in the relevant cities are the main participants. Therefore it is important to raise their awareness of the ecocity notion, and to secure their involvement and feedback, while at the same time respecting and protecting their identity and culture. Ancient Chinese philosophy proposes the harmony between man and nature (天人合一)—that is, the human being is an integral part of nature and this should be construed as the best target for the development of an ecocity and the involvement of aid.

Notes

1. For example, it is estimated that 93% of urban growth will occur in the developing nations in regard to future trends.
2. In Africa, Sudan, Ethiopia and the Democratic Republic of Congo have been among the largest recipients of humanitarian aid.
3. The large majority of ecocity proposals are still at the planning stage, with no commonly defined standards.
4. General Motors, for example, is experimenting in the city to determine whether electric driverless cars could provide the solution to China's pollution and traffic problems.
5. This region also includes the Wolong National Nature Reserve, which houses more than 150 endangered giant pandas.
6. In addition, the government organized various activities to ease the psychological trauma of the people by arranging, for example, guozhuang competitions (a traditional Qiang dance) or psychological lectures by professionals.
7. Tackling issues of pollution and poverty included such measures in low-income neighbourhoods as bus tokens awarded for recycling initiatives, or food and school supplies for trash collection.

References

Adelman, C.C., and N. Eberstadt. 2008. *Foreign Aid: What Works and What Doesn't*. Vol. 3. Washington, DC: American Enterprise Institute for Public Policy Research, October.

Anderson, E. 2007. *Aid Allocation and the MDGs*. ODI Briefing Paper 19. London: Overseas Development Institute.

Arndt, C., S. Jones, and F. Tarp. 2010. Aid, Growth, and Development: Have We Come Full Circle? *Journal of Globalization and Development* 1 (2): 5.

Arvin, B.M., P. Dabir-Alai, and B. Lew. 2006. Does Foreign Aid Affect the Environment in Developing Economies? *Journal of Economic Development* 31 (1): 63–87.

Asafu-Adjaye, J. 1999. The Environment and Development: Theory and Empirical Evidence. *International Journal of Development Planning Literature* 14: 117–134.

Atlantic-Community. 2012. Better Aids: Spotlight on Transparency. www.atlantic-community.org/index/articles/view/Spotlight_on_Transparency

Banerjee, A., and R. He. 2008. Making Aid Work. In *Reinventing Foreign Aid*, ed. W. Easterly. Cambridge, MA: MIT Press.

Bassett, S. 2009. *New Economies of Landscape and Its New Urban Form: Eco-Tourism and Ecological Urbanism as a New Approach to Redevelopment in Post-Disaster Reconstruction in Sichuan Province, China*. Paper Presented at the 4th International Conference of the International Forum on Urbanism (IFoU), 26–28 November, Amsterdam.

BBC World News. 2012. EPISODE 10: Exploring the Future of Sustainable Cities in the Context of the Increasing Global Population. www.horizonsbusiness.com/episode/eco-cities/

Boone, P. 2006. Effective Intervention, Making Aid Work. In *Centre for Economic Performance Centre Piece*. London: LSE.

Burnell, P. 1997. *Foreign Aid in a Changing World*. Maidenhead: Open University Press.

Burnside, C., and D. Dollar. 2000. Aid, Policies, and Growth. *American Economic Review* 90 (4): 847–868.

Buss, T.F., and A. Gardner. 2008. *Haiti in the Balance: Why Foreign Aid Has Failed and What We Can Do about It*. Washington, DC: Brookings Institution Press.

CIDA (Canadian International Development Agency). 2004. *Canadian Cooperation with Haiti: Reflecting on a Decade of "Difficult Partnership"*. Ottawa: CIDA.

Cities Alliance. 2012. 2011 Annual Report: A Strengthened Partnership for Inclusive Cities. www.citiesalliance.org/sites/citiesalliance.org/files/Annual_Reports/WB%20cities%20alliance%20AR%202012%202-27-12.pdf

Collier, P., and D. Dollar. 2002. Aid Allocation and Poverty Reduction. *European Economic Review* 46 (8): 1475–1500.

———. 2004. Development Effectiveness: What Have We Learnt? *Economic Journal* 114 (6): 244–271.

Demombynes, G., and S.K. Trommlerova. 2012. *What Has Driven the Decline of Infant Mortality in Kenya?*. WB Policy Research Working Paper 6057. Washington, DC: World Bank.

Easterly, W. 2008. Introduction: Can't Take It Anymore. In *Reinventing Foreign Aid*, ed. W. Easterly. Cambridge, MA: MIT Press.

ECMM (Eco-City Modus Magazine). 2011. www.westminster.ac.uk/?a=119912

Ecocity Builders. 2012. *Eco-city Definition*. www.ecocitybuilders.org/why-ecocities/the-solution/ecocity-definition/

ECOSOC (United Nations Economic and Social Council). 1999. *Report of the Ad Hoc Advisory Group on Haiti*. Document E/1999/103/July 2. New York: UN Economic and Social Council.

EICU (The Earth Institute Columbia University). n.d. *Extreme Poverty: A Global Emergency*. http://www.earth.columbia.edu/articles/view/1780

Encyclopædia Britannica. 2012. Foreign Aid: Types and Purposes. www.britannica.com/EBchecked/topic/213344/foreign-aid/256254/Types-and-purposes

Feffer, J., and E. Pastreich. 2008. Wenchuan as Eco-City. *Foreign Policy in Focus*. www.fpif.org/articles/wenchuan_as_eco-city

GHA (Global Humanitarian Assistance). 2012. *GHA Report 2011*. www.globalhumanitarianassistance.org/report/gha-report-2011

Gillies, J. 2010. *The Power of Persistence: Education System Reform and Aid Effectiveness, Case Studies in Long-Term Education Reform*. Washington, DC: USAID.

Gnatek, T. 2003. Curitiba's Urban Experiment. www.pbs.org/frontlineworld/fellows/brazil1203/index.html

Graham, C., and M. O'Hanlon. 1997. Making Foreign Aid Work. *Foreign Affairs* 76 (4): 96–104.

Hallegatte, S. 2009. Strategies to Adapt to an Uncertain Climate Change. *Global Environmental Change* 19 (2): 240–247.

Harris, N. 1989. Aid and Urbanization: An Overview. *Cities* 6 (3): 174–185.

Hatzipanayotou, P., S. Lahiri, and M.S. Michael. 2002. Can Cross-Border Pollution Reduce Pollution? *Canadian Journal of Economics* 35: 805–818.

Hoffman, K. 2008. Placing Enterprise and Business Thinking at the Heart of the War on Poverty. In *Reinventing Foreign Aid*, ed. W. Easterly. Cambridge, MA: MIT Press.

ICLEI (International Council for Local Environmental Initiatives). 2002. Curitiba: Orienting Urban Planning to Sustainability. *ICLEI Case Study 77*. Toronto: ICLEI. www.iclei.org.br/polics/CD/P2_4_Estudos%20de%20Caso/1_Planejamento%20Urbano/PDF106_EC77_Curitiba_ing.PDF

Isaac, C. 2012. Social Entrepreneurship and Foreign Aid: Three Ways This Model Could Work in Developing Economies. *Forbes*, 20 July. www.forbes.com/sites/worldviews/2012/07/20/entrepreneur-aid-3-ways-such-a-model-could-help-foreign-aid-work-in-developing-economies/

Joss, M. 2011. Eco-City Governance: A Case Study of Treasure Island and Sonoma Mountain Village. *Journal of Environmental Policy & Planning* 13 (4): 331–348.

Joss, S., D. Tomozeiu, and R. Cowley. 2011. Eco-Cities: A Global Survey 2011. University of Westminster. www.westminster.ac.uk/ecocities

Kausch, I. 1996. Development Cooperation with Countries in Transformation: The Case of the Central Asian CIS Republics. In *Foreign Aid Towards the Year 2000: Experiences and Challenges*, ed. O. Stokke. Geneva: EADI.

Kerkhoff, V.L., I.H. Ahmad, J. Pittock, and W. Steffen. 2011. Designing the Green Climate Fund: How to Spend $100 Billion Sensibly. *Environment: Science and Policy for Sustainable Development* 53 (3): 18–31.

Kremer, M., and E. Miguel. 2008. The Illusion of Sustainability. In *Reinventing Foreign Aid*, ed. W. Easterly. Cambridge, MA: MIT Press.

Meadows, D.H., D.L. Meadows, J. Randers, and W.W. Behrens. 1972. *The Limits to Growth, a Report for the Club of Rome's Project on the Predicament of Mankind.* London and New York: Earth Island Press and Universe Press.

Moss, T., G. Pettersson, and N. Van de Walle. 2008. An Aid-Institutions Paradox? A Review Essay on Aid Dependency and State Building in Sub-Saharan Africa. In *Reinventing Foreign Aid*, ed. W. Easterly. Cambridge, MA: MIT Press.

Mulholland, J. 2012. In 10 Years' Time Ghana May Not Require Any Aid at All. *The Guardian/The Observer*, 15 January. www.guardian.co.uk/world/2012/jan/15/ghana-aid-10-years

Murray, C., L. Rosenfeld, and S. Lim. 2012. Global Malaria Mortality between 1980 and 2010: A Systematic Analysis. *The Lancet* 379 (9814): 413–431.

MW (Merrian-Webster). 2012. Merrian-Webster Dictionary. www.merriam-webster.com/dictionary/scalability

Nguyen, K., and M. Rowling. 2012. How Climate Change, Urbanization Are Changing Disaster Aid. *Reuters*, 26 January. www.reuters.com/article/2012/01/26/us-disasters-poll-idUSTRE80P0LB20120126

OECD. 2005/2008. *The Paris Declaration on Aid Effectiveness' (2005) and the 'Accra Agenda for Action' (2008)*. Paris: OECD. http://www.oecd.org/dataoecd/11/41/34428351.pdf

———. 2011. *Development Co-operation Report 2011, 50th Anniversary Edit.* Paris: OECD.

———. n.d. *Environment and Development: Green Growth and Development.* Paris: OECD. www.oecd.org/document/43/0,3746,en_2649_34421_44309739_1_1_1_1,00

OneWorld. 2012. Foreign Aid Guide; Aid Statistics. www.uk.oneworld.net/guides/aid/#Statistics

Oxfam. 2002. Oxfam Annual Report and Accounts 2001–2002, April. London: Oxfam. www.oxfam.org.uk

———. 2011. *Wenchuan Earthquake Three Years On*. London: Oxfam. www.oxfam.org.hk/filemgr/890/512report3e.pdf

Pittock, J. 2010. A Pale Reflection of Political Reality: Integration of Global Climate, Wetland, and Biodiversity Agreements. *Climate Law* 1 (3): 343–373.

Pittock, J., and J. Hartmann. 2011. Taking a Second Look: Climate Change, Periodic Relicensing and Improved Management of Dams. *Marine and Freshwater Research* 62 (3): 312–320.

Purvis, N. 2003. *Greening US Foreign Aid Through the Millennium Challenge Account.* Policy Brief 119. Washington, DC: Brookings Institution.

Qiu, B. 2011. Speech at the 2nd China Eco-City Conference, 24 September. www.westminster.ac.uk/?a=151853

Raymond, J. S./International House Coalition. 2009. The Challenge of an Urban World: An Opportunity for US Foreign Assistance. www.intlhc.org/docs/The_Challenge_of_an_Urban_World.pdf

Riddell, A. 2012. *The Effectiveness of Foreign Aid to Education: What Can Be Learned?*. WIDER Working Paper 2012/75. Helsinki: UNU-WIDER.

Rogerson, A. 2011. What If Development Aid Were Truly "Catalytic"?. *ODI Background Notes*. London: ODI. www.odi.org.uk/resources/docs/7373

Sachs, J.D. 2005. *The End of Poverty: Economic Possibilities for Our Time*. New York: Penguin Books.

Samoff, J., M. Dembélé, and E.M. Sebatane. 2001. *Scaling Up by Focusing Down: Creating Space to Expand Education Reform*. Paper Presented at ADEA Biennial Meeting: Reaching Out, Reaching All-Sustaining Effective Policy and Practice for Education in Africa, 7–11 October, Arusha.

———. 2011. *Going to Scale': Nurturing the Local Roots of Education Innovation in Africa*. EdQual Working Paper 28. Bristol: EdQual.

Santiso, C. 2001. Good Governance and Aid Effectiveness: The World Bank and Conditionality. *The Georgetown Public Policy Review* 7 (1): 1–22.

Sina Finance and Economics. 2008. Cities in Wenchuan Earthquake Will Be Rebuilt into Eco-Cities [in Chinese]. www.finance.sina.com.cn/roll/20080620/02402286572.shtml

Singer, P. 2009. *The Life You Can Save*. New York: Random House.

South China Morning Post. 2012. Earthquake Anniversary Leaves Bitter Taste, 27 May. www.scmp.com/article/1000768/earthquake-anniversary-leaves-bitter-taste

Stokke, O. 1996. Foreign Aid: What Now? In *Foreign Aid towards the Year 2000: Experiences and Challenges*, ed. O. Stokke. Geneva: EADI.

Stren, R. 2012. *Donor Assistance and Urban Service Delivery in Africa*. WIDER Working Paper 2012/49. Helsinki: UNU-WIDER.

Subedi, M.S. 2005. *Foreign Aid, Sustainable Development and Rapti IRDP*. Occasional Papers in Sociology and Anthropology, 9.

Suzuki, H., A. Dastur, S. Moffatt, N. Yabuki, and H. Maruyama. 2010. *Eco²Cities: Ecological Cities as Economic Cities*. Washington, DC: World Bank Publications.

Tarp, F. 2010. Aid, Growth and Development. In *Foreign Aid for Development: Issues, Challenges and the New Agenda*, ed. G. Mavrotas. Oxford: Oxford University Press.

Timmons Roberts, J., B.C. Parks, M.J. Tierney, and R.L. Hicks. 2009. Has Foreign Aid Been Greened. *Environment Magazine*, January–February. www.environmentmagazine.org/Archives/Back%20Issues/January-February%202009/RobertsParks Tierney Hicks-full.html

UN Millennium Project. 2005. *Investing in Development: A Practical Plan to Achieve the Millennium Development Goals*. New York: United Nations.

UNDP. 2007. Urban Environment and Environmental Health. *Sudan Post-Conflict Environmental Assessment*. New York: UNDP. postconflict.unep.ch/publications.php?prog=sudan

UN-HABITAT. 2005. *Habitat Debate—The Legacy of the Urban Management Programme*. United Nations Human Settlement Programme, December 2005, Vol. 11, No. 4.

———. 2008. Promoting Biodiversity in and Around the Lake Victoria Basin. *United Nations Human Settlements Programme*. www.unhabitat.org

———. n.d.-a City Consultation Methodology. *Urban Management Programme*. www.unhabitat.org/content.asp?typeid=19&catid=374&cid=186

United Nations Environment Programme. 2007. Chapter 6: Urban Environment and Environmental Health [Online]. In *Sudan Post-Conflict Environmental Assessment*, June 2007. http://postconflict.unep.ch/publications.php?prog=sudan. Accessed 31 August 2012.

UN-MV (UN Millennium Villages). n.d. Millennium Villages. www.millennium-villages.org/

Vincent, A. 2011. Beyond Band-Aids, the Contribution of Codes and Standards to the Long Term Viability of Foreign Assistance. *Journal of Engineering and Public Policy* 15.

Ward, E. 2011. Sustainable Landscapes. www.planning.gatech.edu/news/sustainable-landscapes

WCED (World Commission on Environment and Development). 1987. *Our Common Future*. London: Oxford University Press.

Whitfield, L. 2009. *Reframing the Aid Debate: Why Aid Isn't Working and How It Should Be Changed*. DIIS Working Paper 2009:34. Copenhagen: Danish Institute for International Studies.

Wikipedia. 2012a. Aid. www.en.wikipedia.org/wiki/Aid

———. 2012b. Sino-Singapore Tianjin Eco-City. www.en.wikipedia.org/wiki/Sino-Singapore_Tianjin_Eco-city

World Bank. 2009a. *From Rubble to Knowledge: Lessons Learned from the Wenchuan Earthquake*. Washington, DC: World Bank. www.go.worldbank.org/VRED8Z4DB0

———. 2009b. *Wenchuan Earthquake Recovery*. Washington, DC: World Bank. www.worldbank.org/projects/P114107/wenchuan-earthquake-recovery?lang=en

———. 2011. *Technical Assistance on Microfinance Operations to Beichuan Fumin Village and Township Bank*. Washington, DC: World Bank. www.afci.de/content/technical-asstiance-microfinance-operations-beichuan-fumin-village-and-township-bank

———. 2012a. *GEF Sino-Singapore Tianjin Eco-City*. Washington, DC: World Bank.www.worldbank.org/projects/P098915/gef-sino-singapore-tianjin-eco-city?lang=en

———. 2012b. *Sino-Singapore Tianjin Eco-City Project*. Project Appraisal Document Report 55016-CN. Washington, DC: World Bank.
———. n.d. *What Are Governance and Institutional Reforms?* Poverty – Growth of Employment: Evaluation of Governance and Institutional Reforms. www.worldbank.org/WBSITE/EXTERNAL/TOPICS/EXTPOVERTY/EXTISPMA/0,,menuPK:384336~pagePK:149018~piPK:149093~theSitePK:384329,00.html

CHAPTER 8

Opportunities and Conditions for Successful Foreign Aid to the Forestry Sector

Unai Pascual, Eneko Garmendia, Jacob Phelps, and Elena Ojea

8.1 Introduction

8.1.1 Global Forest Loss and Degradation

Forests provide a range of key ecosystem services, including those essential to climate change mitigation (Canadell and Raupach 2008; IPCC 2007) and adaptation (MA 2005; Naidoo et al. 2008; Turner et al. 2009). Forests also provide widely recognized ecosystem services related to biodiversity conservation (FAO 2000), provision of water resources (Fischlin

U. Pascual (✉) • E. Garmendia
Basque Centre for Climate Change (BC3), IKERBASQUE, Basque Foundation for Science, Leioa, Spain

J. Phelps
Lancaster Environment Centre, Library Avenue, Lancaster University, Lancaster, UK

E. Ojea
Universidade de Vigo, Campus Lagoas Marcosende, Vigo, Spain

Y. Huang, U. Pascual (eds.), *Aid Effectiveness for Environmental Sustainability*, https://doi.org/10.1007/978-981-10-5379-5_8

and Midgley 2007) and soil protection (Sidle et al. 2006; Garzia-Ruiz et al. 2008; Stickler et al. 2009). Moreover, forests directly contribute to the livelihoods of more than half a billion people (Chhatre and Agrawal 2009),[1] many of whom are the resource poor in tropical developing countries (Sunderlin et al. 2005; Campbell 2009).

Despite its global importance, the forestry sector continues to be under intense pressure. The United Nations Food and Agricultural Organization (FAO) reports that between 2000 and 2010, approximately 13 million hectares of forest were converted to other uses annually, or were lost through natural causes (FAO 2010). Furthermore, many remaining forests are subject to relatively weak governance and mismanagement that allow for unsustainable timber harvesting and encroachment into forests. The International Tropical Timber Organization (ITTO) reports that only 5% of tropical forests are sustainably managed (Nasi et al. 2011), and recent estimates suggest that up to 15% of internationally traded roundwood might originate from illegal sources (Contreras-Hermosilla et al. 2008).

The general lack of global progress on improving tropical resource management can be attributed to a number of proximate and indirect factors, including widespread agricultural expansion and increasing global agricultural consumption; global demands for tropical hardwoods and forest resources; a general lack of on-the-ground monitoring and enforcement; and cumulative anthropogenic environmental stressors that amplify natural hazards (Geist and Lambin 2002; Dellasala et al. 2012). Moreover, tackling deforestation and degradation poses enormous challenges, especially given the context-specific drivers (e.g. socioeconomic, demographic or political), the multitude of norms, rules and policies that shape resource use and regulation across different countries and levels of governance, and the difficulty for developing countries to forego conventional, extraction-based economic growth strategies (Corbera and Schroeder 2011).

These various drivers are responsible for extensive deforestation and degradation, habitat fragmentation, soil degradation, depletion of biomass and associated carbon stocks, transformation of stand age and species compositions, species loss, species introductions and a range of cascading effects such as increased fire risk (Thomson et al. 2009). Cumulatively, these types of land use changes in the forestry sector also represent the second-largest source of global greenhouse gas (GHG) emissions, accounting for 10–20% of total anthropogenic carbon emissions (Houghton 2008; Harris et al. 2012). Additional accounting of below-ground biomass, which represents more than half of the carbon storage in tropical forests, could dramatically increase our understanding of the importance of forests to climate change mitigation (Verchot and Petkova 2009; Ziegler et al. 2012).

Given the importance of forest resources to the global environment, economic development and to the wellbeing of human communities at multiple scales, forestry sector sustainability has been an area of focus for various forms of foreign aid and official development assistance (ODA).

8.1.2 Scope

In this chapter, we refer to a wide range of approaches and interventions for promoting tropical forest conservation and sustainable management. However, we focus principally on sustainable forest management (SFM) as a strategy for protecting forest resources for their associated ecosystem services (e.g. water, carbon stocks, biodiversity) while still allowing for some forms of multiple use (e.g. restricted resource harvest). SFM in tropical developing countries has received considerable foreign assistance and attention within global multi-lateral institutions. While multiple use is not appropriate in all conservation contexts, SFM is a strategy through which to balance various environmental and human objectives, and it also accommodates forestry sector climate change mitigation efforts (see Sect. 8.4).

This chapter examines the role of climate finance as foreign aid used for the purposes of climate change mitigation and adaptation going from developed to developing countries, which is 'new and additional to existing development finance' (Arndt and Bach 2011). More specifically, we focus on the role of foreign aid in helping to achieve the SFM multiple objectives, especially given the costs associated with improving sector management and reducing deforestation. As such, we specifically consider the potential for additional donor support for the forestry sector associated with new climate change finance. Indeed, the scale of foreign aid directed towards forests has increased dramatically in recent years, in recognition that forests have a significant potential to help mitigate climate change through avoided emissions and through enhanced carbon sequestration. This includes traditional ODA to improve forestry sector management, as well as a number of new and proposed financing strategies. In this context, it is noteworthy that policies for reducing emissions from deforestation and forest degradation, and through the conservation, enhancement and sustainable management of forest carbon stocks (REDD+) have emerged as the principal foreign aid mechanism for promoting forest sustainability.

REDD+ is a prominent climate change mitigation strategy and mechanism that aims to reduce carbon emissions in the forestry sector and to enhance carbon stocks (e.g. through reforestation). The mechanism's function and financing are under negotiation through the United Nations Framework Convention on Climate Change (UNFCCC), as well as through a number of non-governmental, national and bilateral and multi-lateral donor initiatives (see Sect. 8.3.3). Generally, REDD+ policies propose to channel funds from industrialized nations to developing tropical nations in order to support forest conservation interventions that would (1) decrease GHG emissions by reducing deforestation, reducing forest degradation and conserving existing forests and their biodiversity, and (2) increase carbon sequestration by promoting sustainable forest management (e.g. by selective logging) and enhancing carbon stocks (e.g. by tree planting). The proposals are novel in their broad geographic scope, significant financing and use of performance-based payments—payments that would be delivered only if emissions reductions were measured, reported and verified.

The chapter is structured following the four main questions: Sect. 8.2 considers 'what works (and what does not)?', assessing the potential for SFM and multiple forest use to help balance human and environmental objectives within the forestry sector, and the historical role of foreign aid in supporting related strategies. The section draws lessons from these experiences to help inform future interventions. Section 8.3 considers 'what could work?' by assessing recent policy developments in the forestry sector related to REDD+ policies. It considers how REDD+ policies both integrate lessons from previous instruments and differ from them, creating the potential to overcome the challenges that have limited previous interventions. Section 8.4 addresses 'What is transferable?', and introduces a conceptual framework to explore the conditions for promoting SFM and forest conservation across the developing tropics. The framework identifies the financial conditions, governance requirements and policy trade-offs that need to be considered to increase the effectiveness of foreign aid in the forestry sector. Section 8.5 considers 'What is scalable?' and how REDD+ policies represent a dramatic financial and geographic scaling up of SFM and forest conservation. Importantly, it explores this scaling up in the context of complex, multiple stakeholder interests and negotiations, including those of local resource users and international donors.

8.2 What Works (and What Does Not)?

8.2.1 Forest Management for Multiple Objectives

In practice, forest conservation and management include a broad range of strategies. Traditional exclusionary conservation strategies, such as protected areas in which resource access is heavily restricted or prohibited, may continue to be appropriate in many contexts. However, there is also broadening recognition that many other management arrangements can also be successful at achieving conservation objectives, and that long-term SFM often has to balance multiple human development and environmental conservation objectives (Hutton et al. 2006). This is represented in the diversity of conservation strategies that have developed over the past 20–30 years, including voluntary 'payment for ecosystem services' (PES) programmes that have emerged to incentivize landholders to engage in environmentally friendly practices. Forest management strategies also encompass diverse state, private and community-based arrangements that include multiple uses, such as selective logging, harvest of non-timber forest products or regulated hunting.

SFM has presented one of the most promising frameworks through which to balance multiple, often competing objectives within the forestry sector. As with definitions of 'sustainability' in other sectors (WCED 1987), defining 'forest sustainability' and 'sustainable forest management' is challenging, and meanings tend to vary given conservation objectives, development goals and stakeholder interests. SFM is specifically understood as promoting forest management for multiple objectives, and implies some degree of sustainable resource extraction in a way that also maintains multiple ecosystem functions (Wiersum 1995). Principles for SFM were first established during the 1992 United Nations Conference on Environment and Development (UNCED) in Rio de Janeiro. Since then, SFM has remained a somewhat flexible concept, defined by the 2008 UN Resolution 62/98 (UN 2008):

> as a dynamic and evolving concept that aims to maintain and enhance the economic, social and environmental value of all types of forests, for the benefit of present and future generations. It is characterized by seven elements, including: extent of forest resources, forest biological diversity, forest health and vitality, productive functions of forest resources, protective functions of forest resources, socioeconomic functions of forests and legal, policy and institutional framework.

The United Nation Forum on Forests (UNFF) has also acknowledged these elements and recommended them to national governments as a framework for the development of SFM policies. McDermott et al. (2007) provide a detailed review of how international forest policies include these themes.

The results of SFM initiatives have been mixed, and have generally lagged behind expectations, particularly in the tropics (Garcia-Fernandez et al. 2008). For example, forest certification schemes to promote sustainably sourced wood products have failed to become firmly established in tropical areas, which have also often lacked robust audits, compliance and enforcement (Rametsteiner and Simula 2003). Indeed, the viability of SFM seems to depend heavily on a strong governance context, including tenure clarity, stakeholder buy-in, and capacity, which have achieved limited progress in many tropical developing regions (Nasi et al. 2011). Moreover, the high opportunity costs associated with tropical deforestation in many regions have limited the viability of SFM to compete on economic grounds (Nasi et al. 2011, see Box 8.3 in Sect. 8.3.3). Nevertheless, there is broad evidence that when SFM is operationalized, the resulting logged forests can retain significant biodiversity, carbon and timber stocks (Putz et al. 2012). As we discuss in Sect. 8.3, properly implemented SFM principles have the potential to greatly improve tropical forest management.

8.2.2 *Foreign Aid for SFM*

Traditionally, financial support for SFM has come from a number of sources, including central and local government budget allocations of revenues from the sale of forest-related goods and services and private sector investments. However, foreign aid has also been a leading catalyst of the SFM agenda, principally channelled through financial contributions to bilateral and multi-lateral partners, as well as through non-governmental organizations (NGOs).

Foreign aid flows into forestry-sector management and conservation emerged strongly following the 1992 Rio Earth Summit, through a range of international conventions, instruments and financing facilities, such as the United Nations Environment Programme, the Framework Convention on Climate Change, the Convention on Biological Diversity, Forum on Forests and the International Tropical Timber Organization. One of the most important international financing mechanisms for forests has been the Global Environmental Facility (GEF), which provides recipient countries with unconditional grants to cover the incremental costs of actions to protect the environment (Parker et al. 2009). The GEF programme has disbursed US$1.6 billion since 1991, primarily to initiatives that address biodiversity conservation and land degradation, although direct SFM investments have become an increasingly important part of their portfolio (GEF 2010). The GEF-4 SFM programme was started in 2007 and has allocated more than US$300 million (GEF 2010). More recently, the GEF-5 programme has explicitly incorporated climate change mitigation and has provided a separate funding envelope of specific donor support for SFM/REDD+ actions that target climate change mitigation and adaptation.

Donor aid specifically for biodiversity conservation has also supported SFM initiatives. Parker et al. (2012) estimate that the global scale of funding for biodiversity and ecosystem services in 2010 was US$51.5–53.4 billion. Much of this funding originated from government support to agricultural subsidies and greening commodities to benefit biodiversity conservation, sustainable forest management and, most recently, REDD+ (Stephenson 2011; Streck 2012). However, the ODA contribution is smaller and in that same year it was estimated that US$6.3 billion arose from bilateral ODA to developing countries (Parker et al. 2012).

8.2.3 *Lessons from Forest Conservation and Management*

Decades of experience with forest resource management merit careful consideration in the context of designing new policies and maximizing the effectiveness of foreign aid, especially given limited resources and shortfalls in the performance of historical investments (Winterbottom 1990; Kanowski et al. 2011). SFM principles have been embodied in a wide range of conservation instruments in various forms, including integrated conservation and development projects (Blom et al. 2010), community-based natural resource management (Agrawal and Angelsen 2009) and forest certification schemes (Upton and Bass 1995). Increasingly, international efforts have drawn on novel, voluntary programmes such as payment for ecosystem services (PES) initiatives (Engel et al. 2008; Blom et al. 2010, see Box 8.2 in Sect. 8.3.2). While many previous initiatives have been relatively local and project based, there is equally the experiences of how broader dynamics, including national governance, politics and global resource demand, shape the forestry sector (McDermott et al. 2007; Corbera and Schroeder 2011).

We present a synthesis of key lessons derived from this broad array of initiatives to develop a rough list of considerations for donors investing in the forestry sector through REDD+. These include insights related to long-term financial sustainability of mechanisms; the acknowledgement of policy trade-offs; and the importance of landscape-level approaches to conservation; monitoring and enforcement; forest governance and conditionality of payments.

Forest conservation and management require long-term, stable funding (Phelps et al. 2011). However, it is generally agreed that historical financing for SFM and forest conservation efforts have been inadequate (Tomaselli 2006), and have been critiqued as 'long on rhetoric and short on achievements' (Holopainen and Wit 2008). For example, while the estimated global financing cost of implementing SFM ranges between US$70–160 billion per year (Tomaselli 2006; Global Witness 2012), actual funding has been estimated at around US$18 billion per year, of which ODA represented only US$1.1 billion in 2004 (Tomaselli 2006).

Crucially, forest management generally involves high opportunity costs (associated with foregoing conventional land use practices) and so often requires up-front investments (Holopainen and Wit 2008; Nasi et al. 2011). This is especially true in the context of growing commodity prices, which have increased the opportunity costs associated with conservation (Butler et al. 2009; Venter et al. 2009). Where forest management is profitable—through community management, multiple resource use, PES programmes or donor subsidies—deforestation has generally been reduced/halted. Investing in and fostering the profitability of conservation and sustainable resource management are fundamental to reducing deforestation.

It is also important to acknowledge the existence of policy trade-offs. Win-win solutions are commonly proposed in the forestry sector to accommodate, for example, forest conservation with resource harvest, poverty alleviation and agricultural intensification (Garcia-Fernandez et al. 2008; Phelps et al. 2012a). Such approaches are also common in SFM planning. However, win-win solutions are often unrealistic (McShane et al. 2011; Hirsch et al. 2010; Lindenmayer et al. 2012). In the context of REDD+ there are trade-offs between carbon sequestration and biodiversity conservation (Phelps et al. 2012a; Grainger et al. 2009; Harvey et al. 2010), climate change mitigation and adaptation targets (Locatelli et al. 2011) as well as conservation and development goals (Adams et al. 2004) (c.f. Sect. 8.4). This implies that foreign aid instruments for sustainable forestry should allow for negotiating related trade-offs early during planning processes (Blom et al. 2010; McShane et al. 2011).

There is growing recognition that while site-specific interventions can have positive conservation outcomes, broader-scale management is also necessary to maximize biodiversity conservation and maintain diverse ecosystem function (UNESCO 2007; Brown et al. 2008; CBD 2010). Thus, forest management should also take into account a landscape-level approach. For instance, in the specific context of climate change mitigation, donors should also consider that larger-scale management potentially reduces cases of 'carbon leakage' (Oliveira et al. 2007), in which interventions to reduce deforestation at one site simply displace pressures and increase emissions elsewhere (Wunder 2008).

Prior experiences also highlight the importance of regular monitoring and rule enforcement to project success (Blom et al. 2010; Persha et al. 2011), especially given that natural resources management is susceptible to rule-breaking and corruption in institutionally weak environments (Smith and Walpole 2005; Bennett 2011). Broader forestry-sector governance is also key factor to improving forest management (Ostrom 2010; Agrawal et al. 2008; Kanowski et al. 2011). However, many forest-rich tropical developing countries have low governance scores that fundamentally limit their ability to improve management (Karsenty and Ongolo 2011). Specifically, good forest-sector governance translates not only into enforcing regulations and adjudicating infractions but to decisions on land use planning, resource access and land tenure—factors that are also crucial to determining forest management regimes and conservation outcomes (Nasi et al. 2011; Grieg-Gran et al. 2005; McElwee 2012).

There is evidence that good forest governance also specifically includes local engagement, which can involve a wide range of types and levels of participation (Arnstein 1969). In particular, evidence highlights the importance of addressing local-level distributional issues and ensuring that benefits to local communities are both visible and equitable (Chan et al. 2007). Yet, local engagement also extends beyond tangible benefits, and includes procedural issues such as active engagement in resource management decisions. Evidence also shows that local support, accountability and legitimacy are crucial to enhancing forest regrowth, protecting carbon stocks and conserving biodiversity (Chhatre and Agrawal 2009; Ostrom 2010; Persha et al. 2011).

8.3 What Could Work?

8.3.1 Emergence of REDD+

Policies for 'reducing emissions from deforestation and forest degradation' and through the conservation, enhancement and sustainable management of forest carbon stocks (REDD+) are now at the forefront of climate change mitigation policy (Box 8.1; e.g. UNFCCC 2013). The policies have also quickly come to dominate broader forest-sector policy, including links to biodiversity conservation and forest management, largely because of the unprecedented scale of donor finance that is supporting REDD+ initiatives (see Sect. 8.3.3). REDD+ policies have been widely hailed as a potential 'game changer' for tropical forest management conservation (e.g. Venter and Koh 2012). Moreover, REDD+ policies are highly synergistic with SFM policies, as multiple use (e.g. selective and reduced impact logging) can be compatible with efforts to protect and enhance forest carbon stocks (e.g. Nasi et al. 2011).

Policies associated with REDD+ are preliminary, evolving and heterogeneous across the dozens of participating countries. However, there is evidence that many new REDD+ policies are taking on lessons from previous forest management and conservation initiatives, offering potential success where previous initiatives have failed or come up short (Sect. 8.4; see Blom et al. 2010). Furthermore, REDD+ policies seek to exploit the potential of PES conservation instruments (Sect. 8.3.2) and have the potential to overcome many of the financial limitations that have hindered previous efforts (Sect. 8.3.3).

Box 8.1: REDD+ Forest Carbon Policies

In 2007, parties to the UNFCCC agreed that improved forest management and conservation would play a major role in future efforts to reduce greenhouse gas emissions. While, to date, negotiators have failed to reach a binding agreement on reducing climate change emissions, there has been broad consensus on efforts to reduce emissions from the forest sector, and to enhance forest carbon stocks. REDD+ policies continue under negotiation in the UNFCCC and are progressing in parallel through a number of national, multi- and bilateral agreements that are supporting a wide range of 'readiness' activities in preparation for a future, more formal mechanism through the UNFCCC.

A future REDD+ mechanism proposes to channel funds from industrialized nations that have historically been responsible for the majority of GHG emissions (UNFCCC Annex nations; see Sect. 8.5.2) to tropical developing countries, in an effort to incentivize improved forest management. It proposes to support five types of REDD+ activities in tropical developing countries.

Five REDD+ activities:

- reduce emissions from deforestation
- reduce emissions from forest degradation
- conserve existing forest carbon stocks
- carbon stock enhancement
- sustainable management of forests.

REDD+ carbon stock verification:

The UNFCCC has established guidelines for monitoring and reporting on forest carbon stocks, and working groups continue to generate recommendations for new standards. A number of third party organizations have also established independent carbon verification standards and services.

REDD+ safeguards:

The UNFCCC and a number of third party agreements and organizations have also established social and environmental safeguards to ensure that REDD+ does not result in unintended negative consequences (see Sect. 8.5).

REDD+ co-benefits:

Recognizing that improved forest management has the potential to yield additional benefits for biodiversity and forest-dependent communities, many REDD+ policies have sought to jointly address human and environmental priorities within forest management (see Sects. 8.4.4). The distinctions between safeguards and co-benefits, however, are not yet clearly established within UNFCCC negotiations.

Source: Compiled by the authors.

8.3.2 Leveraging PES Instruments for SFM

The UNFCCC has specifically included SFM within the list of REDD+ activities (UNFCCC 2010; Box 8.1), recognizing the multiple uses of forests and that carbon gains can arise not only from traditional conservation but from dynamic management that includes human use (e.g. Berry et al. 2010; Nasi et al. 2011). In this respect, REDD+ and SFM objectives are highly compatible. Moreover, REDD+ policies are based heavily on the logic of international PES schemes, which leverage incentives to promote voluntary conservation actions (Box 8.2). Recent evidence suggests that voluntary but conditional payments can potentially overcome some of the limitations of other conservation policy instruments (Engel et al. 2008) to catalyse effective and cost-efficient conservation outcomes (Wunder 2008). Such PES schemes are a prospective strategy through which to promote SFM and conservation, as many of the benefits of SFM (e.g. biodiversity and carbon conservation) are external to forest owners and managers (Chipeta and Joshi 2001). Moreover, unlike many traditional conservation instruments, PES involves an incentive system that is conditional on the actual provision of the environmental service, which promises to increase instrument effectiveness (Wunder 2008). In comparison with traditional command-and-control regulation (e.g. protected areas), PES schemes also have the potential to offer alternative livelihoods for local communities, are more flexible, and allow for better targeting (focusing on areas/ecosystems with higher value in terms of service provisioning) and thus potentially increase efficiency (Wunder 2008).

A meta-analysis of incipient government-led PES schemes to reduce deforestation revealed that these have yielded uncertain conservation outcomes (Pattanayak et al. 2010). However, REDD+ policies have the potential to excel where many existing PES schemes have led to few or uncertain results (Pattanayak et al. 2010; Venter and Koh 2012).

Box 8.2 Payment for Ecosystem Services (PES)

Payment for ecosystem services schemes can be defined as voluntary transactions where a well-defined environmental service (or a land use likely to secure that service) is being 'bought' by a service buyer(s) from a service provider(s), if and only if the service provider secures service provision (Wunder 2008). This contingent payment is known as conditionality.

However, PES can also refer to a much wider set of conservation instruments that leverage incentives to promote conservation, not necessarily through a traditional buyer–seller relationship, and can include donor-supported conservation efforts that are based on incentives.

Types of mechanisms:

- regulation and penalty by limiting access
- cap and trade, like the carbon markets
- direct payments, where providers receive payment for supplying services
- voluntary agreements (Kinzig et al. 2011).

Implementation examples:

- carbon sequestration in China
- watershed protection in South Africa and Mexico
- biodiversity conservation in Costa Rica and Nicaragua (Kinzig et al. 2011).

Challenges of PES:

- conditionality on service provision
- additionality
- enrolment
- monitoring, reporting and verification (MRV)
- social justice (Pattanayak et al. 2010).

Source: Compiled by the authors.

REDD+ involves national-level land use planning, and is being deployed across the developing tropics, which represents a more integrative approach to conservation and a significant scaling up over previous conservation efforts (see Sect. 8.5 on scalability; Pattanayak et al. 2010). Notably, REDD+ policy in many contexts has also adopted a strong focus on conditionality, by which funds are fully disbursed only after conservation outcomes are demonstrated. This represents a sharp departure from previous conservation and ODA initiatives (ibid.). For example, only a small fraction of Norway's US$1 billion commitment to Indonesia has been disbursed due to lack of demonstrable progress in reducing emissions. Similarly, the UNFCCC 18th Conference of Parties in Doha established a strong focus on 'results based finance', with donors requiring clear outcomes for their investments. This focus on results extends to the REDD+ focus on additionality. In principle, REDD+ financial aid and investments for REDD+ should demonstrate conservation outcomes that surpass business-as-usual or do-nothing scenarios. This is a significantly more stringent requirement than has traditionally been placed on foreign aid or conservation efforts.

Because of this strong performance-based nature, monitoring and enforcement are at the core of most REDD+ plans. There are considerable resources being dedicated to country-specific carbon accounting, baselines and monitoring against which to measure success. To this end, donor support in the context of REDD+ is very heavily focused on strengthening forestry-sector governance and building the local capacity requisite for successful forest management (Cerbu et al. 2011).

Forest PES is also an attractive instrument because of its potential to balance human and environmental considerations with the forestry sector. Unlike traditional exclusionary approaches to conservation, in principle it seeks voluntary cooperation (although see Beymer-Farris and Bassett 2012). Many schemes seek to provide local land managers with sustainable development and livelihood options and/or fair compensation for their conservation actions, which is compatible with multiple resource use within forests. While outcomes are variable, many PES schemes in Latin America have increased household incomes and tenure security and helped to strengthen local resource management institutions (Grieg-Gran et al. 2005; Pagiola et al. 2005), although there are exceptions (see Sect. 8.5.2).

Forest sustainability efforts based on PES face considerable challenges (see Sect. 8.4). However, contemporary REDD+ policies are potentially integrating lessons from previous instruments and are leveraging incentives in ways that *could* serve to overcome many of the challenges that have limited previous initiatives and *could* serve to widely mainstream SFM objectives.

8.3.3 Increasing Foreign Aid for the Forestry Sector

The potential for REDD+ to succeed in improving forestry sector management and sustainability across the tropics relies on the scale of financial investment that the mechanism has received. To date, the vast majority of these resources represent new, additional and increased foreign aid, and encompass a major financial scaling up for the forestry sector (Phelps et al. 2011). In this section, we briefly present an overview of the actual architecture of foreign aid regarding REDD+, highlighting (1) the direction, type and scale of foreign aid flows, (2) the primary actors in terms of donors and recipients, and (3) the recent financing trends.

It should be noted that ODA disbursements into the forestry sector increased by an average of 125% between the periods 2002–4 and 2008–10, mostly attributed to financing related to REDD+ (CPF 2012a). In 2009, the Copenhagen Accord committed developed countries to contributing US$3.5 billion in fast-start climate finance in the 2010–12 period, especially for REDD+. Since then, commitments have increased to more than US$7 billion, surpassing all previous ODA into the forestry sector (Creed and Nakhooda 2011; Simula 2010). However, by the end of 2011 (when the pledges had reached US$4.17 billion), only US$446 million had been actually allocated and approved (Nakhooda et al. 2011). In many cases, there is also little clarity over whether conservation financing has been additional to estimates of original ODA pledges, leading to some double-counting of funding towards ODA and REDD+ (Global Witness 2012), and creating debate over future financing scale for the forestry sector (Stephenson 2011).

To date, financial support for REDD+ has mainly been channelled through new bilateral agreements, which have amounted to US$4.8 billion since 2008 (Simula 2010). Recent ODA support has been overwhelmingly Norway's Climate Change Initiative (Table 8.1), which has pledged over US$2.3 billion, including US$1 billion to Indonesia, US$250 million to Guyana and US$72 million to Tanzania (Government of Norway 2011).

Many other donors, particularly smaller ones, have chosen to combine climate and forestry-sector financing with traditional ODA (Streck and Parker 2012), and support has represented natural extensions of existing donor–recipient relationships developed over decades of partnership on sustainable forest and development programmes. However, the use of ODA budgets to deliver climate finance is a matter of much political debate, given the concerns that support for the forestry sector will substitute development aid. Stephenson (2011) differentiates three possible financing options in this respect. First, he proposes that a strict mitigation approach in which REDD+ finance would be separated from development activities so as to avoid compromising ODA. Second, a co-benefits approach would use REDD+ finance for climate mitigation together with poverty alleviation and development, seeking to jointly catalyse improved forest management, sustainable development and biodiversity conservation (Creed and Nakhooda 2011; Sect. 8.4). The last scenario involves use of ODA for forestry-sector climate change mitigation but only if these funds represent additional ODA.

Multi-lateral agreements represent about US$2.6 billion of ODA to REDD+ since 2008 (climatefundsupdate.org 2012; see Annex). Funding agreements include the REDD+ Partnership (a group of developed and developing countries with a commitment for international cooperation on REDD+ mitigation); the UN–REDD programme partnership among UNDP, FAO and UNEP (US$119.9 million pledged/$118.9 million deposited); the World Bank's Forest Carbon Partnership Facility (FCPF), with funds divided into the REDD+ Readiness Fund (US$229.6 million pledged and deposited) and the Carbon Fund (US$204.5 million pledged/$179.3 million deposited); the Forest Investment Programme (FIP) (US$644 million pledged/$459 million deposited); the Congo Basin Forest Fund (US$165 million pledged and deposited); and the Amazon Fund (with US$1.32 billion pledged/$57 million deposited). The GEF has established a funding envelope that specifically draws on ODA to support SFM and REDD+ actions for climate change mitigation and adaptation in the forestry sector. This includes US$250 million in funding to incentivize developing countries to invest up to US$750 million of their allocations for biodiversity, climate change and addressing land degradation projections into SFM/REDD+ projects and programmes (GEF 2010). The upcoming GEF6 replenishment (2014–18) is likely to further expand foreign aid focus on SFM/REDD+ (CPF 2012b), and the UNFCCC Green Climate Fund, currently under negotiation with plans to provide US$100 billion in annual climate change mitigation and adaptation financing by 2020, is likely to increase support to the forestry sector (GCF 2013).

This multi- and bilateral ODA is currently supporting REDD+ forest management efforts in over 40 developing countries, although ODA has been targeted to a subset of forest-rich countries with forests under immediate threat of deforestation (Table 8.1). Funding has principally supported capacity building within the forestry sector, REDD+ demonstration and early implementation activities, as well as support for increasing the scale of REDD+ efforts (Creed and Nakhooda 2011). Even given the prospect of increased private-sector interest, ODA finance is likely to remain a strong contributor to the forestry sector in the near term (Phelps et al. 2011; Streck 2012).

Table 8.1 Major donor and recipient countries in REDD+ funds (percentage of total funds)

REDD+ major donors		*REDD+ major recipient countries*	
Norway	71.56%	Guyana	23.50%
Australia	6.93%	Brazil	13.18%
UK	6.76%	Indonesia	11.04%
US	4.48%	Democratic Republic of Congo	5.52%

Source: Climate funds update. URL: http://www.climatefundsupdate.org/data

Even though REDD+ represents a major increase in foreign aid for the forestry sector, for which it has been hailed as transformative, it is uncertain whether REDD+ financing will be enough to tip the balance in favour of widespread forest conservation and sustainable management (Boccucci et al. 2008). Although more than US$7 billion was pledged for REDD+ between 2008 and 2012, the required resources are substantially greater (Box 8.3).

There are two main proposed financing approaches through which policymakers anticipate recruiting adequate financing for REDD+: a fund-based and a market-based system. Some participating countries, including Brazil, support the establishment of an international REDD+ fund into which ODA could continue to be channelled. However, it remains very uncertain whether ODA could ever provide long-term, stable funding of the scale needed to operationalize REDD (Phelps et al. 2011).

Box 8.3 Cost of Implementing REDD+

The costs of improving forest conservation and management across the developing tropics are cumulatively high. For example, offsetting deforestation in the Brazilian Amazon would cost between US$7–18 billion per year (Nepstad et al. 2009). Similarly, halting deforestation across all developing countries is estimated in the range of US$25–185 billion per year (Parker et al. 2009, 2012; UNFCCC 2007). The Eliasch Review estimated the global costs of REDD+ to be between US$17–33 billion per year, assuming a 50% abatement of forest-related emissions by 2020 (Eliasch 2008). Kindermann et al. (2008) estimated the costs to halve deforestation by 2030 to be between US$17.2–28 billion per year, while the European Commission established an annual price tag until 2020 of US$19.7–32.8 billion (EC 2008). These studies estimate the total economic abatement potential from REDD+ activities, assuming a certain price per ton of carbon dioxide and a certain cost associated with land use conversion. The figure for actual abatement potential, however, is likely to be smaller than this, due to the various constraints on generating emission reductions through REDD+. As such, global cost estimates illustrate the maximum potential of forests and other land use activities to remove or retain GHG at a certain price point, rather than a realistic potential for emission reductions in the short to medium term (Streck and Parker 2012).

Source: Compiled by authors.

In lieu, countries such as Costa Rica, Indonesia, USA and Australia have favoured a market-based REDD+ mechanism, and envision an increase in global trading of emissions credits derived from emissions reductions and carbon stock enhancement in the forestry sector. Some projections suggest that as a result, REDD+ financing could increase to US$6.2–39 billion per year by 2020 (CPF 2012b; Streck 2012). However, the projections anticipate viable carbon markets and increased private-sector participation, which to date has only contributed US$0.15 billion to REDD+ via voluntary carbon markets (Simula 2010; Hamilton et al. 2010). The small scale of private-sector financing can be attributed to the slow on-the-ground development of REDD+, the gross unreliability of carbon markets (e.g. Clark et al. 2013) and overall uncertainty over carbon investments (Streck and Parker 2012; Phelps et al. 2011; Sect. 8.4.1).

The European Union has generally supported a combined market-fund financing pathway (Rosendal and Andresen 2011), and it is likely that REDD+ will be funded through both public and private finance, including ODA, results-based payments from public sector overseas (e.g. Norway Partnership), and private investments in future carbon markets (Streck 2012). Indeed, independent of the potential for a market-based approach, ODA is likely to remain an important source of funding for the forestry sector. For example, SFM/REDD+ require up-front financial support for a preparation and development phase for REDD+ (Stephenson 2011).

8.4 What Is Transferable?

8.4.1 Integrating Lessons Learned into REDD+

We argue that in the context of unprecedented financial flows into the forestry sector, REDD+ funding should be leveraged not only to reduce carbon emissions but also to ensure that REDD+ achieves multiple objectives—in recognition of the social ecological complexity of environmental management, in line with SFM objectives, and acknowledging that win-win outcomes are hard to come by. In this section, we discuss the necessary conditions for leveraging ODA to achieve SFM objectives.

We use a conceptual framework based on lessons learned from previous conservation and SFM interventions (Sect. 8.2.3) that identifies the necessary conditions for delivering forest sustainability via REDD+ as catalysed by donor finance. This framework analyses REDD+, taking into account (1) a range of financial considerations, (2) institutional arrangements and governance processes and (3) trade-offs among climate change mitigation, adaptation and other ecosystem services, extractive industries and livelihoods. All considerations are instrumental to maximizing donor finance and ensuring long-term sustainable outcomes. Figure 8.1 illustrates the interconnection between the financial mechanisms of REDD+, the institutional and governance environment and the trade-offs that need to be acknowledged for the effective leverage of REDD+ finance flows.

While REDD+ has been popularized as a cost-effective approach to mitigating the impacts of climate change (Stern 2006), there remains significant uncertainty about how REDD+ will be financed in the future (Sect. 8.3.3). The unprecedented scale of foreign aid required for REDD+ implementation aggravates the risks associated with financial uncertainties and instability. Financing also interconnects with how (and whether) the mechanism rewards diverse ecosystem services (carbon storage, biodiversity conservation, development goals etc.), the multiple levels of resource

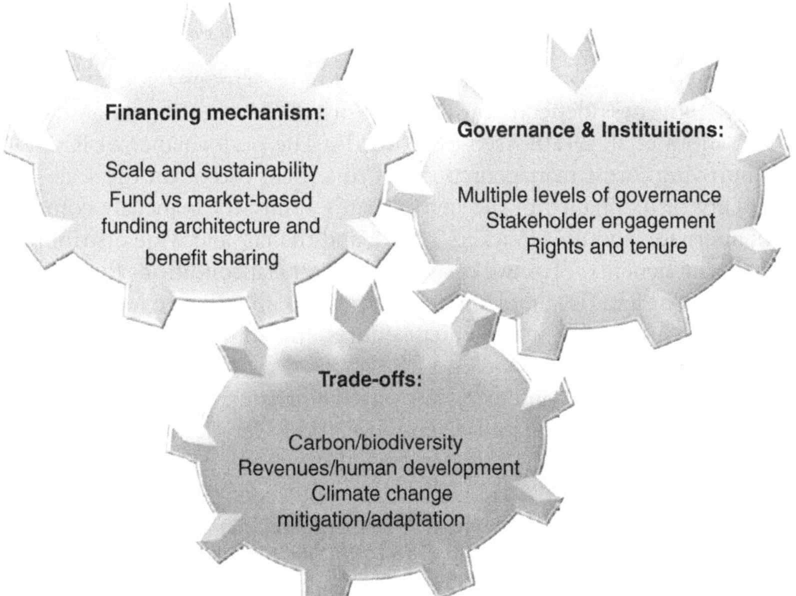

Fig. 8.1 Multiple dimensions shaping forest management in the context of multiple objectives. Source: Authors' elaboration

management (local, regional, national and international), and the diversity of forest actors (donors, governments, rural and indigenous communities, investors and forestry sector, conservationist and diverse NGOs, intergovernmental bodies). These factors, mediated through forest governance and institutional structures, are shaping key decisions about REDD+, including site selection, monitoring requirements, payments levels and the requirements for participation.

Effective mechanism governance and financial architecture must recognize the associated social and ecological trade-offs in order to identify clear objectives, fairly evaluate outcomes and maximize donor financial resources (Stickler et al. 2009; Ghazoul et al. 2010a; Harvey et al. 2010; Hirsch et al. 2010; Phelps et al. 2012a). Given the central role of international donors in both financing REDD+ and policy development through bi- and multi-lateral engagement with forested countries, donors are uniquely positioned to shape how these conditions are assessed, and to consider competing stakeholder interests in mechanism design (see Sect. 8.5).

8.4.2 *Financing Mechanism: Building on Foreign Aid*

Previous conservation and SFM initiatives have demonstrated that long-term effectiveness depends on financial sustainability (Chipeta and Joshi 2001; Blom et al. 2010; Asen et al. 2011). The performance of REDD+ in improving forest management relies to a great extent on the scale and reliability of its financing, the mechanism's ability to financially compete with alternate land uses (Vickers 2008), and the fair and wide distribution of financial benefits. Here we consider the financial conditions for successful long-term REDD+ implementation to scale up effective forest conservation and sustainable management efforts, including the role of benefit sharing.

There is growing consensus that public financing is needed to operationalize REDD+ (Creed and Nakhooda 2011). Yet, despite considerable donor pledges, long-term, large-scale voluntary public finance cannot yet be assured (Phelps et al. 2011). The closest precedent for this type of voluntary giving is ODA, on which countries are increasingly relying for REDD+ funds (Streck and Parker 2012) but which has historically been highly unreliable and erratic (see Phelps et al. 2011). As such, the current architecture of REDD+ financing poses risks to financial sustainability. Sustainable financing relies on diversifying finance beyond voluntary donor-based support, including increased investment from the private sector, addressing dysfunction within international carbon markets, increasing emissions reduction commitments through the UNFCCC, adopting more resilient financial management structures and potentially restraining REDD+ projects to scales that can be actively funded.

It is therefore widely expected that the private sector will contribute a significant share of future REDD+ finance (Streck and Parker 2012). Most proposals involve the establishment of carbon markets for ensuring demand for carbon credits generated by REDD+. As in the clean development mechanism (CDM), industrialized countries, private individuals and industries would purchase credits in compliance with mandated emissions mitigation efforts (UNFCCC 2010; Corbera and Schroeder 2011). However, to date, participation of the private sector has remained limited due to the associated high risks regarding carbon price fluctuation and significant uncertainties over the long-term demand for carbon emission reductions (Creed and Nakhooda 2011; Phelps et al. 2011). This is largely a result of global disagreements over UNFCCC emissions reduction targets and the lack of clear post-Kyoto emissions reduction commitments, which provide uncertain precedents for basing REDD+ heavily on donor support (Phelps et al. 2011, 2012b). Lawmakers in the United States have stalled over climate legislation, and the European Union does not currently plan to integrate REDD+ based carbon credits into its existing emission trading system until after 2020. In fact, enforceable international commitments to reduce emissions through the UNFCCC are a crucial financial condition for full-scale REDD+ implementation, as it will otherwise be an immense challenge to recruit necessary voluntary investment (Phelps et al. 2011).

Even when credible demand for emission reduction credits exists, adequate regulation of market-based finance represents another condition for effective implementation. For example, there is potential for oversupply of forest-based credits in carbon markets that could reduce carbon prices and compromise mechanism function (Hare and Macey 2007; Livengood and Dixon 2009; Phelps et al. 2011). This implies that regulated market supply and a 'price floor' for carbon might be necessary to ensure financial sustainability. Similarly, rigorous standards for forest-based carbon credits would be needed to ensure the reliability of market-based trading. Additionally, long-term demand for forest-based emission credits is also a requisite, although there is a danger that these will be used only as a temporary mitigation approach while developing low-carbon technologies (Piris-Cabezas and Keohane 2008; Streck and Parker 2012). It follows that the role of forestry-sector carbon credits could decline if they ceased to be the most economically efficient form of mitigation (Ghazoul et al. 2010a; Phelps et al. 2011).

Fund-based financing also entails an alternative or parallel approach. As in the case of the UNFCCC Green Climate Fund, it can be funded through voluntary commitments from Annex countries but could also be supported through mandated contributions from polluting countries and industries. As with markets and public finance, these would be more robust if based not on year-to-year voluntary support, but on reliable investments to offset emissions in accordance with international commitments. Reliable, long-term Annex contributions and financial management that provide participants with financial certainty are crucial. Especially in the context of carbon market immaturity and uncertainty about long-term donor support, fund-based finance could ensure reliable and more predictable REDD+ payments (Hare and Macey 2007) and could help circumvent market volatility (see Phelps et al. 2011).

REDD+ financial resilience would also benefit from an expanded approach to ecosystem services, combining or expanding REDD+ payments for carbon emissions reductions with payments for other ecosystem services such as water quality, recreation, biodiversity, erosion control or flood protection (Scholz and Schmidt 2008). Combining multiple financing streams could help to ensure that other services are not overlooked by single-commodity payments, and could potentially increase REDD+ funding. This would be particularly helpful in areas with high opportunity costs, where REDD+ might not otherwise be financially competitive (Phelps et al. 2010b).

Crucially, however, resources and donor support for the forestry sector must also address the distribution of (financial) benefits. At present, it is not clear how REDD+ funds will be channelled within countries. The UNFCCC is not likely to promulgate rules about how the carbon revenues are to be shared within recipient countries, since this is a matter of subsidiarity (Balderas Torres and Skutsch 2012). There are also strong arguments to suggest that state management of REDD+ finances could fail to fairly distribute financial rewards and benefits from REDD+ (Peskett et al. 2008). For example, a majority of submissions for World Bank financial support of REDD+ have not adequately addressed governance challenges related to benefit sharing (Davis et al. 2009). In cases where REDD+ is implemented on public land, there is a need to design benefit-sharing mechanisms that avoid locally powerful actors from reaping disproportionate benefits (elite capture, Grieg-Gran et al. 2005; McElwee 2012; Corbera and Pascual 2012). This will require empowering local users with the authority and resources to negotiate revenue sharing (Phelps et al. 2010a), as further discussed in Sect. 8.4.3.

8.4.3 *Forest Governance and Institutions: Donor Engagement with Multiple Levels of Governance*

In the context of many tropical developing countries, forest governance reform is a condition for improving sustainable forest management and for maximizing associated foreign aid (Ostrom 2010; Agrawal et al. 2008; Kanowski et al. 2011; Nasi et al. 2011). Yet, Davis et al. (2009) note that a review of 25 of the Forest Carbon Partnership Facility REDD country participants revealed that their proposals lacked adequate conservation of governance factors related to law enforcement, land tenure clarity, benefit-sharing mechanisms or transparency and accountability of forest monitoring systems, potentially limiting forest protection (Bisson et al. 2003) and related emissions reductions (Phelps et al. 2010a), as well as negatively affecting forest-dependent communities. Indeed, there are considerable and diverse risks associated with overlooking the governance complexities (Phelps et al. 2010a, b; Nagendra and Ostrom 2012; Korhonen-Kurki et al. 2012).

'Good' governance may be a condition for some types of funding, as some private-sector and donor actors may both lack confidence to invest in states with weak governance structures (Balderas Torres and Skutsch 2012), which could jeopardize forestry sector finance (see Karsenty and Ongolo 2011). While much ODA and forestry-sector financing occurs through national-level negotiations, we highlight the importance of donors engaging with the forestry sector through multi-level interventions that recognize the complexity of forest governance systems and need to engage with actors beyond the national level (Nagendra and Ostrom 2012).

REDD+ necessitates collaboration among the diverse stakeholders with interests in the forestry sector, and large-scale improvements to forestry-sector conservation and management cannot be addressed through interventions at any single level of governance alone (Nagendra and Ostrom 2012). International donors, whose engagement is often at the national level, are challenged to consider forestry-sector issues, processes and decisions at other scales. For example, there is scope for donor engagement at the project level, especially in the context of a 'nested' REDD+ approach that allows for site-based and subnational funding and implementation (Angelsen 2008). Such an approach has considerable support,[2] yet also requires strategic engagement to ensure regional and national scaling up and integration. Indeed, subnational REDD+ efforts highlight the polycentric nature of REDD+ governance, and the need to accommodate stake, regional and community-based management (Corbera et al. 2010).

Similarly, REDD+ interventions have the potential to increase centralized management of forest resources, at the cost of local management rights (Phelps et al. 2010a; Sandbrook et al. 2010; Kanowski et al. 2011). In addition, REDD+ implementation might place further demands on national forest managers, favouring a more centralized system of forest control for monitoring and coordination (Phelps et al. 2010a). This may seriously affect indigenous and forest-dependent people's rights over their traditional lands and resources (Brown et al. 2008; Schroeder 2010; Kanowski et al. 2011), increasing tenure insecurity (Cotula and Mayers 2009). As such, ensuring that REDD+ reforms address issues such as local land tenure, resource access, autonomy or participation and benefit distribution are essential conditions for consideration among donors.

8.4.4 *Considering Multiple Performance Trade-offs*

Finance mechanisms for REDD+ and their associated governance and institutional architecture strongly condition potential trade-offs associated with forest management decisions. To date, REDD+ donor financing is largely focused on forests' carbon sequestration, which has largely defined resource management priorities. Nevertheless, REDD+ has the potential to enhance other co-benefits such us biodiversity conservation and other ecosystem services, poverty alleviation and adaptive capacities. However, such co-benefits seldom occur automatically. We highlight here the trade-offs between carbon sequestration and (1) biodiversity conservation, (2) economic and human development, and (3) climate change adaptation. These trade-offs need to be candidly addressed as a condition for REDD+ implementation to avoid undesirable outcomes, maximize donor resources, and to ensure the permanence of conservation actions (Stickler et al. 2009; Kanowski et al. 2011; Blom et al. 2010; Ghazoul et al. 2010b; Harvey et al. 2010; Hirsch et al. 2010; Phelps et al. 2012a).

8.4.4.1 Biodiversity Conservation

While some authors recognize REDD+ potential to jointly address declines in forest-based carbon storage and biodiversity (Miles and Kapos 2008;

Venter et al. 2009), approaches for linking biodiversity conservation to climate change mitigation strategies are not straightforward, and REDD+ mechanisms do not guarantee biodiversity co-benefits automatically (Venter et al. 2009; Phelps et al. 2012b). In fact, one major risk is that REDD+ finance could overlook support for biodiversity conservation priority sites with low carbon stocks (Grainger et al. 2009; Miles and Kapos 2008; Phelps et al. 2011). Moreover, in the definition of 'forests', REDD+ does not distinguish between natural forests and plantations, and therefore could create incentives for forest degradation in favour of less biodiversity-rich strategies associated with commercial tree planting (Sasaki and Putz 2009; Lindenmayer et al. 2012). Additionally, REDD+ policies could displace deforestation pressures (leakage) into other, unprotected high-biodiversity forests, both within the same country (Angelsen et al. 2009; Harvey et al. 2010) and internationally (Mudiyarso et al. 2008; Strassburg et al. 2009).

REDD+ planning has focused predominantly on identifying high-carbon and high-deforestation countries like Indonesia, though there is a need for much broader country participation in order to account for a greater range of emissions sources, avoid international leakage and account for future pressures on forests (Mudiyarso et al. 2008; Strassburg et al. 2009; TCG 2008). In this sense, not all countries are ensured participation in a REDD mechanism, because of low forest cover, low deforestation rates and high opportunity costs (Miles and Kapos 2008; Venter et al. 2009).

Since donors are primarily interested in economies of scale to tackle global warming cost effectively (Butler and Laurance 2008), some large tropical areas within specific countries that account for a significant share of global tropical deforestation (e.g. Brazil, Peru, Indonesia and the Democratic Republic of Congo) become key REDD+ targets. But as Campell (2009) points out, REDD+ initiatives could look beyond the humid tropics to target other biodiversity hotspots, even if REDD+ donors might need to pay a premium price for carbon credits (Venter et al. 2009). If biodiversity co-benefits were to be integrated into REDD+, financial backers, including donors, would need to acknowledge the degree to which reduced carbon benefits would be acceptable in order to enhance biodiversity outcomes (Venter et al. 2009; Minteer and Miller 2011). Even where biodiversity co-benefits are possible, unless the additional costs of biodiversity conservation, monitoring and reporting are incorporated into REDD, carbon–biodiversity trade-offs may persist (Hirsch et al. 2010; Phelps et al. 2012b).

8.4.4.2 Economic and Human Development

REDD+ involves a number of development trade-offs, both in terms of larger-scale economic trade-offs (see Box 8.4), as well as a range of economic and non-economic human development trade-offs (Box 8.5). Both types relate to forgone activities in order to allow for forest conservation, as REDD+ interventions often involve restrictions on forest access, use and harvest, thereby involving foregone revenues (e.g. from timber, crops and livestock) and/or limiting subsistence and smaller-scale resource use (Shepherd 2004; Ghazoul et al. 2010a). Economic trade-offs at both scales are of immediate interest to donors because they inform the scale of financing necessary to offset alternative land uses.

Box 8.4 REDD+ and Large-Scale Economic Trade-Offs

Restrictions on resource use associated with REDD+ policies inherently involve economic trade-offs, which are significant within the industrial timber sector and related and downstream industries (Ghazoul et al. 2010a). REDD+ must be economically competitive with these types of high-value alternative land uses, in order to incentivize conservation. For example, opportunity costs associated with the paper and pulp industry in the Indonesian province of Riau have been estimated at ~US$5 billion in 2006 (ibid.). Although this figure is debatable, it provides an indication of the relative value of industrial forestry extraction when compared to forecasts of REDD+ financing for the entire country of Indonesia (US$3.8–15 billion per year). Furthermore, the forestry sector in Indonesia directly employs around 350,000 people and about 3.1 million people in broader forestry-related businesses (ibid.). Malaysia's oil palm industry offers a comparable example. It contributes 5–6% of Malaysian GDP and provides direct employment for 570,000 people, while employing 830,000 in downstream activities. It generates US$10.1 billion annually in foreign exchange, which has been essential for human development associated with basic service provision, such as piped water, electricity, communications, roads, schools and healthcare (MPOC 2012).

Source: Compiled by the authors.

Perceived restrictions that conflict with powerful national economic interests and development agendas are likely to result in delays and opposition to REDD+. Similarly, REDD+ programmes that conflict with local livelihoods and human development goals can face opposition, thereby potentially compromising the efficiency and effectiveness of such foreign aid (Corbera and Pascual 2012; Beymer-Farris and Bassett 2012). Some studies address how REDD+ could involve trade-offs in local agricultural production, food security, resource access and local hardships (McElwee 2012; Stickler et al. 2009), although there remains considerable debate over whether REDD+ should also be pro-poor and make active contributions to rural livelihoods, or merely be designed to not harm the poor. This decision is likely to depend on others' willingness to pay for additional social benefits—including the willingness of international donors.

8.4.4.3 Adaptation to Climate Change

Climate change policies have tended to pay disproportionately more attention to mitigation than to adaptation, although adaptation is receiving increasing political attention (Pielke et al. 2007; Jerneck and Olsson 2008; Parry et al. 2009). However, it was not until the 2010 UNFCCC 16th Conference of Parties that countries reached the first global agreement on adaptation, including foreign aid commitments. But that agreement still overlooked the role of forest ecosystems in enhancing social and ecological resilience through the provisioning of key ecosystem services (Locatelli et al. 2011).

Successful long-term forest management requires that ecosystems are able to adapt to climate change (ibid.). Communities and industries that depend on forest resources also face pressures to adapt to climate change. Thus, managing for climate change adaptation potentially adds a new critical dimension of forest management. Joint mitigation-adaptation efforts in forested landscapes can be illustrated through mangrove restoration, which can simultaneously contribute to carbon stocks (Donato et al. 2011), while also protecting coastal areas from erosion and storm surges particularly associated with climate change and accelerated sea level rise (Das and Vincent 2009; Turner et al. 2009). Traditional forest-based agricultural systems may also successfully link adaptation and mitigation (see Box 8.5).

Box 8.5 REDD+ and Human Development Trade-Offs in the Context of Swidden Agriculture

Swiddening, also known as shifting cultivation, is an extensive agricultural system that involves rotating amongst forest plots—clearing forestland using fire, cultivation, fallow and a return to previously farmed sites. Importantly, it is a land use type heavily targeted for transformation under a number of REDD+ policies: REDD+ policymakers across the tropics are proposing that REDD+ carbon finance can be used to provide incentives for swidden farmers to transition to other land uses (Ziegler et al. 2012; Pirard and Belna 2012). However, this type of cultivation is the main source of livelihood or an important source of supplementary income for millions of people worldwide (Cramb et al. 2009; Mertz et al. 2009). Although the proposed land use shifts have the potential to increase food security and farm incomes, this equally exemplifies how climate change mitigation efforts may compromise communities' adaptive capacity. A recent meta-analysis of more than 250 studies supports a reassessment of policies that encourage land-cover conversion away from (especially long-fallow) swidden systems (Ziegler et al. 2012). There is little evidence to suggest that transitions from swidden agriculture to most other land uses will directly or reliably produce positive carbon gains. On the contrary, many transitions—including the replacement of various types of swidden agriculture with oil palm, rubber or some types of agroforestry systems—may negatively impact other ecosystem services, food security and local livelihoods, such that the entire carbon and non-carbon benefits stream should also be taken into account before prescribing transition with ambiguous carbon benefits (ibid.). While extensive agriculture might not necessarily be a viable management strategy in all contexts or for the bulk of global food production (Ghazoul et al. 2010b), long-fallow rotations within swidden agriculture can serve as effective stores of carbon while preserving traditional livelihoods and maintaining adaptive capacity (Ziegler et al. 2012). Notably, REDD+ policies should not preclude the option of maintaining or rehabilitating traditional, intermediate and long-fallow swidden and agroforestry systems within the broader forest landscape. From a long-term carbon perspective, intermediate and long-fallow swidden systems could conceivably represent optimal land-use options in some situations (Ziegler et al. 2011, 2012).

Source: Compiled by the authors, based on material cited above.

8.5 What Is Scalable?

8.5.1 Financial and Geographic Scaling Up

This section considers how REDD+ as a foreign aid instrument for SFM is serving to mainstream and scale up lessons learned from previous forest conservation instruments. We specifically consider how this scaling up, while technologically possible, represents a complex, political process that involves stakeholders at multiple levels of governance. We focus on the role of donors in identifying priorities within the forestry sector and helping to mediate decision-making.

REDD+ represents not only a dramatic and unprecedented increase in foreign aid for the forestry sector (Sect. 8.3.3), but a geographic scaling up when compared with previous forestry-sector initiatives. Crucially, REDD+ implementation generally involves a highly centralized approach to forest management policy, as the mechanism requires national-level carbon accounting to demonstrate net emissions reductions and avoid leakage (Phelps et al. 2010a).[3] As a result, REDD+ is serving to mainstream landscape and national-level land use planning and cross-sectoral planning (e.g. across forestry sector, agriculture, transport sectors) (Wunder 2008; CBD 2010; Phelps et al. 2011). Multi-lateral initiatives such as the FCPF are supporting forest conservation efforts in about 40 developing countries to prepare REDD+ national strategies by 2015, and to conduct landscape-level planning to target interventions and ensure policy harmonization. This scaling up extends beyond the national level and also includes international cooperation to reduce leakage. In fact, dozens of tropical developing countries are planning or implementing REDD+ activities, most supported by international public finance (see Kshatriya and Sills 2010; VRD 2011). The national- and international-level nature of current forest sustainability initiatives reflects a much larger scaling of forest sustainability initiatives than traditionally represented by a project or site-based approach to conservation and forest management.

In recognition of the extraordinary challenges associated with this broadened scope (see Sect. 8.4.3), a bulk of ODA for REDD+, including Norway's bilateral support to tropical developing countries, has been targeted towards capacity building, early planning efforts and demonstration activities with an aim to scaling up (Creed and Nakhooda 2011). Indeed, many existing national REDD+ strategies include explicit plans for scaling up from local REDD+ pilot projects to regional and national-level forestry-sector reforms (e.g. Philippines, PNRPS 2010).

8.5.2 Scalability and Decision-Making

Not all of the relevant stakeholders will perceive efforts to significantly scale up SFM and sustainability principles across sectors, landscapes and countries in the same way. We have addressed key considerations for effective donor engagement within the forestry sector in terms of mechanism financing conditions, governance and institutional conditions, and in terms of acknowledging social–ecological trade-offs (Fig. 8.1). Additionally, there are considerable disagreements among stakeholders regarding mechanism design and financing (McDermott et al. 2012; Korhonen-Kurki et al. 2012; Phelps et al. 2012a). Moreover, PES schemes such as REDD+ can have different impacts on different stakeholders, with particular potential to negatively harm local service providers (Pagiola et al. 2005). In fact, the cascading socioeconomic effects of REDD+ policies and the diversity of associated stakeholders make REDD+ policy development an especially complex case of environmental governance (Ghazoul et al. 2010a; Corbera and Schroeder 2011; McDermott et al. 2012; Nagendra and Ostrom 2012; Korhonen-Kurki et al. 2012). Donors have a significant role to play in identifying and negotiating priorities and helping to mediate decision-making.

REDD+ policy development is generally centred on the formal UNFCCC process, in which a future global REDD+ mechanism remains under negotiation. However, REDD+ policies are simultaneously being formed at multiple other levels. We can identify eight levels at which stakeholders are engaging with SFM and REDD+ policies, and which stand to shape the outcomes of foreign aid for climate change. These stakeholders range from actors operating principally within international level negotiations to local implementers and resource users (Box 8.6 below).

Naturally, many of these levels overlap, adding complexity to REDD+ stakeholder architecture. For instance, parties engaged in bi- and multilateral REDD+ initiatives are also active in UNFCCC processes; individual REDD+ sites fit into national-level carbon accounting and REDD+ planning; local civil society actors are influencing global negotiations; third party verification schemes will run in parallel with government standards and international safeguards.

The ways in which REDD+ will address financial needs and deliver financial flows are dependent on which actors are able to define mechanism priorities and the associated trade-offs, which in turn are shaped by the interplay among diverse stakeholders (Corbera and Schroeder 2011;

Box 8.6 Eight Levels of Stakeholders Engaging in SFM and REDD+

1. Non-UNFCCC multi-lateral initiatives (e.g. Forest Carbon Partnership Facility) and bilateral initiatives (e.g. Norway-Indonesia) are supporting country partners to develop national REDD+ policies, pilot projects and 'readiness' in preparation to engage with a future REDD+ mechanism.
2. Participating governments are shaping global policy while domestically establishing pilot projects, developing and beginning to implement national REDD+ strategies.
3. Local and regional government actors are engaging with national and local counterparts, to implement national strategies and, in some cases, to establish more autonomous local and regional initiatives, such as in Aceh (Indonesia).
4. A wide range of civil society initiatives are contributing to, contesting and negotiating REDD+ policies at numerous scales (e.g. Friends of the Earth Indonesia).
5. The private sector and industry actors are both funding and independently developing site-based REDD+ projects for voluntary carbon markets and in anticipation of emissions caps and future carbon trading (e.g. Carbon Conservation).
6. Third party private and non-governmental organizations are setting up verification schemes and standards for REDD+ (e.g. Community, Conservation and Biodiversity Alliance).
7. Actors within academia, NGOs, government agencies and the private sector are developing carbon quantification and monitoring technologies to help operationalize REDD+.
8. NGOs and community groups are developing site-specific REDD+ projects (e.g. Cambodia's Oddar Meanchey REDD project) but in other cases also struggling against externally imposed initiatives (e.g. Rufiji Delta in Tanzania).

Source: Compiled by the authors.

McDermott et al. 2012; Ribot and Larson 2012). In fact, the way in which SFM objectives are defined, prioritized and scaled up depends largely on such interplay.

Among key stakeholders' interests, we specifically highlight the roles of local resource users and Annex donor nations in defining REDD+ objectives and designing the future financial mechanism, notably because (1) these groups represent opposite ends of the payment-for-ecosystems relationship, and reflect differences in scale: local versus global; (2) local resource users will be among the most directly and heavily affected by REDD+ policies, yet are among the most marginalized in the REDD+ policy processes (Thompson et al. 2011; Ribot and Larson 2012), and (3) Annex nations (donors) are among the best positioned stakeholders to help ensure that the identified challenges are meaningfully addressed and that foreign aid is effectively leveraged as well as being responsible for the bulk of carbon emissions and investments into REDD+ and its policy development.

Annex countries are central to the REDD+ policy development and financing processes, not only through the UNFCCC process but via the various multilateral processes and through direct bilateral engagement with forested countries (Agrawal et al. 2011; McDermott et al. 2012). The UNFCCC negotiations may eventually result in new agreements on international emission abatement targets. This would vastly extend to role of Annex finance, as industrialized nations purchase carbon credits in a future market, or finance a carbon fund in order to finance REDD+ based emissions reductions. However, as in any PES scheme, service beneficiaries exert significant power because they help to define terms of purchase/finance, and are likely to approach REDD+ with considerable expectations (see Clements 2010).

Likewise, as Creed and Nakhooda (2011) suggest, some donor nations have already demonstrated preference for bilateral finance arrangements because they allow them to exert even greater control over how REDD+ finance is spent. It potentially also allows financers to impose conditions that shape how REDD+ addresses issues such as resource governance, tenure reforms, equitable benefit distribution, integration of biodiversity into REDD+ planning, and the monitoring of social and environmental safeguards (Phelps et al. 2010a; Phelps et al. 2011). For example, Norway has targeted foreign aid towards Indonesia in the interests of maximizing investments to reduce emissions but has also placed conditions on its financing, including a prioritization of biodiversity conservation alongside emissions reductions.

At the opposite end of the international PES relationship, there are serious concerns that REDD+ policies could deny local resource users and traditional landholders their territories and livelihoods, especially because these actors generally lack commensurate agency within REDD+ decision-making (Brown et al. 2008; Schroeder 2010; Thomson et al. 2011;

Lyster 2011; Peskett et al. 2008). Many existing REDD+ decision-making processes are also characterized by 'deep inequities', notably biased against local forest resource users, whose participation is heavily limited by 'class, ethnic and other social inequities and economic hurdles' (Ribot and Larson 2012). A rapid scaling up of efforts in the forestry sector could profoundly impact on dependent communities, potentially much more than a site-based approach. Notably, a scaled up, national-level approach to REDD+ implementation has the potential to centralize control over forest resources, reducing local decision-making (Phelps et al. 2010a; Sandbrook et al. 2010).

This clearly illustrates the trade-offs and multi-scalar nature inherent to REDD+ decisions. As discussed, while many of the benefits of tropical forest conservation are global, including climate change mitigation benefits, the costs of protecting forests are largely borne by local communities, especially those that depend on forest resources (Balmford and Whitten 2003; Naidoo et al. 2008). The 'passive costs' of conservation can have disproportionate impacts on local communities (Balmford and Whitten 2003; e.g. Pagiola et al. 2005), which further highlights the importance of equitable distribution and engaging local actors in decision-making. A similar disparity of scales is evident when comparing climate change mitigation and adaptation, as mitigation interests and interventions are generally global in nature, while adaptation concerns are more local (Locatelli et al. 2011).

Broadening recognition of the local social implications of REDD+ implementation has given rise to mounting social safeguards within the UNFCCC process, a number of participating countries and third party groups (Rutt 2012). However, safeguards within the UNFCCC have been criticized as inadequate because they are voluntary and unenforceable, and because even if local stakeholders have the right to benefit from REDD+, they may lack the ability to benefit due to such inadequate engagement and representation (Ribot and Larson 2012).

Industrialized nations, as donors and potentially as buyers of emissions credits, are positioned to continue to shape the ways in which the voices of service providers—local communities responsible for forest resource conservation and management on the ground—are heard within international and national policy decisions. Further addressing the relationships between international donors/service buyers and local service providers, and the differences and trade-offs in what they consider priorities within REDD+ (e.g. carbon trade-offs with human development goals) remains a key challenge. Leveraging effective foreign aid through REDD+ critically relies on matching interests at these two scales.

8.6 Conclusion

This chapter identifies key conditions for maximizing the effectiveness of foreign aid into the forestry sector, particularly in the context of climate change mitigation and balancing multiple objectives via SFM. It has focused on the potential for transformational changes as a result of the new, large-scale donor-based climate change finance for REDD+ policies.

The chapter identifies the potential for SFM principles to help bolster sustainability with the forestry sector, including climate change mitigation and adaptation. It draws on historical developing country experiences with sustainable forest management and conservation to explore how contemporary foreign aid can serve to promote greater sustainability within tropical forests management. Moreover, it considers how the scale and design of new REDD+ policies make the initiative unique and distinct from previous forestry-sector experiences, with the potential to overcoming barriers that have historically hindered forest conservation and the implementation of improved management.

Through a conceptual framework of the intertwined financial conditions, governance conditions and policy trade-offs that shape REDD+, the chapter discusses how greenhouse gas emissions mitigation objectives need to be harmonized with a diversity of other forestry-sector objectives, embodied by the United Nations SFM principles. We concur with a broadening consensus that REDD+ as a foreign aid instrument should contribute to biodiversity conservation, sustainable rural development and improved forest management. However, we also stress that win-win outcomes for carbon emissions, human development and other ecosystem services are not automatic, and involve interplay of complex finance and governance architecture.

We have also specifically examined how REDD+ financing, catalysed by official development aid, has the potential to move beyond previous SFM efforts. Likewise, we warn that the mechanism still faces uncertainty over the long-term sustainability of financing. This drives us to highlight that financing for the forestry sector is not a stand-alone issue that can be addressed independently of governance considerations and without proper acknowledgement of a range of trade-offs associated with REDD+.

We note that leveraging and materializing sufficient foreign aid for the sustainability of the forestry sector remains a major challenge. Notwithstanding the need to engage with different types of finance (public and private, market and non-market-based), there are particular political challenges associated with mandating the emissions reductions needed to effectively generate REDD+ finance that is both sustainable and resilient. This is critically associated with the complexity of multiple stakeholder interests, and the role of Annex nations in shaping REDD+ policy. Furthermore, the role of ODA in leveraging REDD+ finance must be properly taken into account, as different options have the potential to compromise the scarce existing development aid.

We have also emphasized international donors' role, via engagement with participating countries, foreign aid and conditionality, in not only promoting carbon conservation and enhancement but also addressing many of the underlying governance issues and policy trade-offs. In this regard, we suggest that within a polycentric approach to governance, livelihood impacts and local legitimacy of REDD+ must be understood and considered within REDD+ negotiations and planning.

Given the crucial role of forests to both climate change mitigation and adaptation, REDD+ policy development and the increase in foreign aid/ODA for the forestry sector is promising but confronts complex future challenges. Hence, however novel, we stress that these initiatives need to draw lessons and best practices from decades of donor investment into developing tropical forests. This includes recognition of the financing conditions and governance complexities.

Last but not least, while donor support for the forestry sector is heavily focused on carbon, REDD+ efforts need to consider the broader diversity of services provide by forest ecosystems, and the profound impacts that forest reform will have on national development and local-level livelihoods, particularly among forest-dependent local communities.

Annex

Table 8.2 Financial flows in REDD+ (in US$ million)

Fund	*Pledged*[a]	*Deposited*[b]	*Approved*[c]	*Disbursed*[d]	*Donors*	*Projects*
Amazon Fund						
	1032.22	57.49	141.59	42.48	• Germany • Norway • Petroleo Brasileiro S.A. Petrobras	• Brazil

(*continued*)

Table 8.2 (continued)

Fund	*Pledged*[a]	*Deposited*[b]	*Approved*[c]	*Disbursed*[d]	*Donors*	*Projects*
Congo Basin Forest Fund (CBFF)						
	165.0	165.0	75.05	12.1	• Norway • United Kingdom	• Burundi • Cameroon • CAR • Chad • Guinea Equatoriale • Congo DRC • Gabon • Congo • Rwanda • Sao Tomas et Principe
Forest Carbon Partnership Facility—Carbon Fund (FCPF-CF)						
	204.5	179.3	0.57	0.20	• BP • CDC Climate • EC • Germany • IFCI • Norway • Switzerland • TNC • United Kingdom • United States	• Global
Forest Carbon Partnership Facility—Readiness Fund (FCPF-RF)						
	229.6	229.6	27.24	9.14	• AFD (France) • Canada • Denmark • European Commission • Finland • Germany International Forest Carbon Initiative (IFCI) • Italy • Japan • Netherlands • Norway's International Climate and Forest Initiative • Spain • Switzerland • UK's International Climate Fund • United States	• Cameroon • Colombia • Congo DRC • Costa Rica • El Salvador • Ethiopia • Ghana • Indonesia • Kenya • Lao • Liberia • Nepal • Nicaragua • Congo • Uganda • Vanuatu .../.

(*continued*)

Table 8.2 (continued)

Fund	*Pledged*[a]	*Deposited*[b]	*Approved*[c]	*Disbursed*[d]	*Donors*	*Projects*
Forest Investment Programme (FIP)						
	644	459	50.96	3.18	• Denmark • International Forest Carbon Initiative (IFCI) • Japan's Fast Start Finance • Norway's International • Climate and Forest Initiative • Spain • UK's International Climate Fund • United States	• Brazil • Indonesia
Global Climate Change Alliance (GCCA)						
	226.12	224. 6	24. 70	0.00	• Cyprus • Czech Republic • Denmark • EC Fast Start Funding • European Community • European Development Fund • International Forest Carbon Initiative (IFCI) • Ireland • Japan's Fast Start Finance • Norway's International Climate and Forest Initiative • Spain • Sweden • UK's International Climate Fund • United States	• Bangladesh • Belize • Cambodia • Ethiopia • GCCA Support Facility • Guyana • Intra ACP • Jamaica • Maldives • Mali • Mauritius • Mozambique • Nepal • Pacific Region • Rwanda • Seychelles • Solomon Islands • Tanzania

(continued)

Table 8.2 (continued)

Fund	*Pledged*[a]	*Deposited*[b]	*Approved*[c]	*Disbursed*[d]	*Donors*	*Projects*
Indonesia Climate Change Trust Fund (ICCTF)						
	18.61	8.81	1.25	1.19	• Australia • Denmark • International Forest Carbon Initiative (IFCI) • Japan's Fast Start Finance • Norway's International Climate and Forest Initiative • Spain • Sweden • UK's International Climate Fund • United Kingdom • United States	• Indonesia
UN-REDD						
	119.9	118.9	117.56	95.36	• Denmark • Japan's Fast Start Finance • Norway's International Climate and Forest Initiative • Spain	• Bolivia • Cambodia • Congo DRC • Indonesia • Nigeria • Panama • Papua New Guinea • Paraguay • Philippines • Solomon Islands • Tanzania • United Nations • Vietnam • Zambia

Source: Compiled by authors based on the following sources: UN-REDD data from www.mptf.undp.org/factsheet/fund/CCF00; other facts from www.climatefundsupdate.org/data and UN-REDD. There is also information on the countries of red funding destiny with approved funding and figures of each countries amount. All data in the table refer to REDD+ funds: defined in the database as countries' efforts to reduce emissions from deforestation and forest degradation and to foster conservation sustainable management of forests and enhancement of forest carbon stocks (UNFCCC 2013)

[a]Pledges: represent verbal or signed commitments from donors to provide financial support for a particular fund. All pledges are cumulative

[b]Deposits: represent the funds that have been transferred from the donor into the account(s) of the fund. Also known as committed funds. All deposits are cumulative

[c]Approved: represents funds that have been officially approved and earmarked to a specific project or programme. All approvals on figures are cumulative

[d]Disbursed: represents those funds that have been spent, either through administrative means or directly to an implementation programme or project, with proof of spend. All disbursements on figures are cumulative

Notes

1. According to the World Bank, more than 1.6 billion people directly depend on forest resources.
2. For example, de Gryzde and Durschinger (2010); Swickard and Carnahan (2010) in Balderas Torres and Skutsch (2012).
3. Leakage is when interventions to reduce deforestation or degradation at one site simply displace pressures and increase emissions elsewhere.

References

Adams, W.M., R. Aveling, D. Brockington, B. Dickson, J. Elliott, J. Hutton, et al. 2004. Biodiversity Conservation and the Eradication of Poverty. *Science* 306: 1146–1149.

Agrawal, A., and A. Angelsen. 2009. Using Community Forest Management to Achieve REDD+ Goals. In *Realising REDD+: National Strategy and Policy Options*, ed. A. Angelsen et al. Bogor: Center for International Forestry Research (CIFOR).

Agrawal, A., A. Chhatre, and R. Hardin. 2008. Changing Governance of the World's Forests. *Science* 320: 1460–1462.

Agrawal, A., D. Nepstad, and A. Chhatre. 2011. Reducing Emissions from Deforestation and Forest Degradation. *Annual Reviews* 36: 373–396.

Angelsen, A., ed. 2008. *Moving Ahead with REDD: Issues, Options, and Implications*. Bogor: CIFOR.

Angelsen, A., M. Brockhaus, M. Kanninen, E. Sills, W.D. Sunderlin, and S. Wertz-Kanounnikoff. 2009. *Realising REDD+: National Strategy and Policy Options*, 390. Bogor: CIFOR.

Arndt, C., and C. Bach. 2011. *Foreign Assistance in a Climate-Constrained World.* WIDER Working Paper 2011/66. Helsinki: UNU-WIDER.

Arnstein, S.R. 1969. A Ladder of Citizen Participation. *Journal of the American Institute of Planners* 35: 216–224.

Asen, A., M. Bosolo, R. Carrillo, K. van Dijk, C. Nordheim-Larsen, S. Oystese, et al. 2011. Unlocking National Opportunities. In *New Insights for Financing Sustainable Forest and Land Management*. Rome: FAO.

Balderas Torres, A.B., and M. Skutsch. 2012. Splitting the Difference: A Proposal for Benefit Sharing in Reduced Emissions from Deforestation and Forest Degradation (REDD+). *Forests* 3: 137–154.

Balmford, A., and T. Whitten. 2003. Who Should Pay for Tropical Conservation, and How Could the Costs Be Met? *Oryx* 37: 238–250.

Bennett, E.L. 2011. Another Inconvenient Truth: The Failure of Enforcement Systems to Save Charismatic Species. *Oryx* 45: 476–479.

Berry, N.J., O.L. Phillips, S.L. Lewis, J.K. Hill, D.P. Edwards, N.P. Tawatao, et al. 2010. The High Value of Logged Tropical Forests: Lessons from Northern Borneo. *Biodiversity and Conservation* 19: 985–997.

Beymer-Farris, B.A., and T.J. Bassett. 2012. The REDD Menace: Resurgent Protectionism in Tanzania's Mangrove Forests. *Global Environmental Change* 22: 332–341.

Bisson, J., E.S. Guiang, P. Walpole, and D. Tolentino. 2003. *Better Governance Critical to Reversing Forest Degradation Trends in Southeast Asia*. Paper Presented at XII World Forestry Congress, 21–28 September, Quebec City. www.fao.org/DOCREP/ARTICLE/WFC/XII/0837-A4.HTM

Blom, B., T. Sunderland, and D. Murdiyarso. 2010. Getting REDD to Work Locally: Lessons Learned from Integrated Conservation and Development Projects. *Environmental Science & Policy* 13: 164–172.

Boccucci, M., R. Boer, A. Contreras-Hermosilla, J. Leitmann, N. Masripatin, and J. Spears. 2008. REDD and Investment. In *Financing Sustainable Forest Management*, ed. J. Holopainen and M. Wit. Wageningen: Tropenbos International.

Brown, D., F. Seymour, and L. Peskett. 2008. How Do We Achieve REDD Co-benefits and Avoid Doing Harm? In *Moving Ahead With REDD. Issues, Options, and Implications*, ed. A. Angelsen, 107–118. Bogor: CIFOR.

Butler, R.A., and B. Laurance. 2008. New Strategies for Conserving Tropical Forests. *Trends in Ecology and Evolution* 23 (9): 469–472.

Butler, R.A., L.P. Koh, and J. Ghazoul. 2009. REDD in the Red: Palm Oil Could Undermine Carbon Payment Schemes. *Conservation Letters* 2: 67–73.

Campbell, B. 2009. Beyond Copenhagen: REDD+, Agriculture, Adaptation Strategies and Poverty. *Global Environmental Change* 19: 397–399.

Canadell, J.G., and M.R. Raupach. 2008. Managing Forests for Climate Change Mitigation. *Science* 320 (5882): 1456–1457.

CBD (Convention on Biological Diversity). 2010. Global Expert Workshop on Biodiversity Benefits of REDD in Developing Countries, 20–23 September, Nairobi. www.cbd.int/doc/?meeting=EWREDD-01

Cerbu, G.A., B.M. Swallow, and D.Y. Thompson. 2011. Locating REDD: A Global Survey and Analysis of REDD Readiness and Demonstration Activities. *Environmental Science and Policy* 14: 168–180.

Chan, K.M.A., R.M. Pringle, J. Ranganathan, C.L. Boggs, Y.L. Chan, P.R. Ehrlich, et al. 2007. When Agendas Collide: Human Welfare and Biological Conservation. *Conbio* 21: 59–68.

Chhatre, A., and A. Agrawal. 2009. Trade-offs and Synergies between Carbon Storage and Livelihood Benefits from Forest Commons. *Proceedings of the National Academy of Sciences* 106 (42): 17667–17670.

Chipeta, M.E., and M. Joshi, eds. 2001. *Financing Sustainable Forest Management*. Report of the International Workshop of Experts, 22–25 January 2001, Oslo, Norway. Bogor: CIFOR, 109p.

Clark, P., J. Chaffin, and J. Blass. 2013. EU Carbon Prices Crash to Record Low. *Financial Times*, 23 January. Online. www.ft.com/cms/s/0/77764dda-6645-11e2-b967-00144feab49a.html#axzz2RYAYPM4m

Clements, T. 2010. Reduced Expectations: The Political and Institutional Challenges of REDD+. *Oryx* 44: 309–310.

Contreras-Hermosilla, A., H.M. Gregersen, and A. White. 2008. *Forest Governance in Countries with Federal Systems of Government: Lessons and Implications for Decentralization*. Bogor: CIFOR. www.cifor.org/publications/pdf_files/Books/BContreras-Hermosilla0701.pdf

Corbera, E., and U. Pascual. 2012. Ecosystem Services: Heed Social Goals. *Science* 335: 655–656.

Corbera, E., and H. Schroeder. 2011. Governing and Implementing REDD+. *Environmental Science and Policy* 14: 89–99.

Corbera, E., M. Estrada, and K. Brown. 2010. Reducing Greenhouse Gas Emissions from Deforestation and Forest Degradation in Developing Countries: Revisiting the Assumptions. *Climatic Change* 100: 355–388.

Cotula, L., and J. Mayers. 2009. Tenure in REDD—Start-Point or Afterthought? *IIED Natural Resource Issues*, 15. London: International Institute for Environment and Development. www.iied.org/pubs/pdfs/13554IIED.pdf

CPF. 2012a. *Study on Forest Financing*. Advisory Group on Finance Collaborative Partnership on Forests, June.

CPF (Collaborative Partnership in Forestry). 2012b. SFM and REDD+. *SFM Fact Sheet*, No. 5. www.cpfweb.org/76228/en/

Cramb, R.A., C.J.P. Colfer, W. Dressler, P. Laungaramsri, T.L. Quang, E. Mulyoutami, et al. 2009. Swidden Trans-formations and Rural Livelihoods in Southeast Asia. *Human Ecology* 37: 323–346.

Creed, A., and S. Nakhooda. 2011. REDD+ Finance Delivery: Lessons from Early Experience. In *Climate Finance Policy Brief*. London: ODI.

Das, S., and J.R. Vincent. 2009. Do Mangroves Provide an Effective Barrier to Storm Surges? *Proceedings of the National Academy of Sciences USA* 106: 7357–7360.

Davis, C., F. Daviet, S. Nakhooda, and A. Thuault. 2009. A Review of 25 Readiness Plans. In *Idea Notes from the World Bank Forest Carbon Partnership Facility*. Washington, DC: World Resources Institute. www.wri.org/gfi

Dellasala, D.A., J.M. Fitzgerald, B.-G. Jonsson, J.A. Mcneely, B.D. Dovie, M. Dieterich, et al. 2012. Priority Actions for Sustainable Forest Management in the International Year of Forests. *Conservation Biology* 26: 572–575.

Donato, D.C., J.B. Kauffman, D. Murdiyarso, S. Kurnianto, M. Stidham, and M. Kanninen. 2011. Mangroves among the Most Carbon-Rich Forests in the Tropics. *Nature Geoscience* 4: 293–297.

EC (European Commission). 2008. *Addressing the Challenges of Deforestation and Forest Degradation to Tackle Climate Change and Biodiversity Loss*. Communication from the Commission to the European Parliament, The Council, The European Economic and Social Committee and the Committee of the Regions, Brussels.

Eliasch, J. 2008. *Climate Change: Financing Global Forests: The Eliasch Review*. London: Earthscan.

Engel, S., S. Pagiola, and S. Wunder. 2008. Designing Payments for Environmental Services in Theory and Practice: An Overview of the Issue. *Ecological Economics* 65: 663–674.

FAO. 2010. *Global Forest Resources Assessment*. FAO Forestry Paper 140. Rome: FAO.

FAO (Food and Agriculture Organization). 2000. *Global Forest Resources*. Rome: FAO.

Friess, D., J. Phelps, E. Garmendia, and E. Gómez Baggethun. 2015. Payments for Ecosystem Services (PES) in the Face of External Biophysical Stressors. *Global Environmental Change-Human and Policy Dimensions* 30: 31–42. https://doi.org/10.1016/j.gloenvcha.2014.10.013

Fischlin, A., and G.F. Midgley. 2007. Ecosystems, Their Properties, Goods, and Services. In *Climate Change 2007: Impacts, Adaptation and Vulnerability*, Contribution of Working Group II to the Fourth Assessment Report of the Intergovernmental Panel on Climate Change, ed. M.L. Parry et al., 211–272. Cambridge: Cambridge University Press.

Garcia-Fernandez, C., M. Ruiz-Perez, and S. Wunder. 2008. Is Multiple-Use Forest Management Widely Implementable in the Tropics? *Forest Ecology and Management* 26: 1468–1486.

Garzia-Ruiz, J.M., D. Regüés, B. Alvera, N. Lana-Renault, P. Serrano-Muela, E. Nadal-Romero, et al. 2008. Flood Generation and Sediment Transport in Experimental Catchments Affected by Land Use Changes in the Central Pyrenees. *Journal of Hydrology* 356: 245–260.

GCF (Green Climate Fund). 2013. United Nations Framework Convention on Climate Change. www.unfccc.int/cooperation_and_support/financial mechanism /green_climate_fund/items/5869.php

GEF (Global Environmental Facility). 2010. The GEF Incentive Mechanism for Forests a New REDD+ Multilateral Finance Programme. www.thegef.org/gef/pubs/SFM_REDD_2010

Geist, H.J., and E.F. Lambin. 2002. Proximate Causes and Underlying Driving Forces of Tropical Deforestation. *Bioscience* 52: 143–150.

Ghazoul, J., R.A. Butler, J. Meteo-Vega, and L.P. Koh. 2010a. REDD: A Reckoning of Environment and Development Implications. *Trends in Ecology and Evolution* 25: 396–402.

Ghazoul, J., L.P. Koh, and R.A. Butler. 2010b. A REDD Light for Wildlife Friendly Farming. *Conservation Biology* 24: 644–648.

Global Witness. 2012. Safeguarding REDD+ Finance. Ensuring Transparent and Accountable International Financial Flows. www.globalwitness.org/library/safeguarding-redd-finance-ensuring-transparent-and-accountable-international-financial-flows

Grainger, A., D.H. Boucher, P.C. Frumhoff, W.F. Laurance, T. Lovejoy, J. McNeely, et al. 2009. Biodiversity and REDD at Copenhagen. *Current Biology* 19: R974–R977.

Grieg-Gran, M., I. Porras, and S. Wunder. 2005. How Can Market Mechanisms for Forest Environmental Services Help the Poor? Preliminary Lessons from Latin America. *World Development* 33: 1511–1527.

de Gryzde, S., and L. Durschinger. 2010. *An Integrated REDD Offset Program (IREDD) for Nesting Projects under Jurisdictional Account*. San Francisco: Terra Global Capital.

Hamilton, K., M. Sjardin, M. Peters-Stanley, and T. Marcello. 2010. *Building Bridges: State of the Voluntary Carbon Markets 2010*. Ecosystem Marketplace & Bloomberg New Energy Finance.

Hare, B., and K. Macey. 2007. *Tropical Deforestation Emissions Reduction Mechanism*. Greenpeace Discussion Paper. Amsterdam: Greenpeace. www.greenpeace.org/raw/content/international/press/reports/TDERM.pdf

Harris, N., S. Brown, S.C. Hagen, S.S. Saatchi, S. Petrova, W. Salas, et al. 2012. Baseline Map of Carbon Emissions from Deforestation in Tropical Regions. *Science* 336: 1573–1576.

Harvey, C.A., B. Dickson, and C. Kormos. 2010. Opportunities for Achieving Biodiversity Conservation through REDD. *Conservation Letters* 33: 53–61.

Hirsch, P.D., W.M. Adams, J.P. Brosius, A. Zia, N. Bariola, and J.L. Dammert. 2010. Acknowledging Conservation Trade-Offs and Embracing Complexity. *Conservation Biology* 25: 259–264.

Holopainen, J., and M. Wit, eds. 2008. *Financing Sustainable Forest Management*. Wageningen: Tropenbos International.

Houghton, R.A. 2008. *Carbon Flux to the Atmosphere from Land-use Changes: 1850–2005*. Oak Ridge: Carbon Dioxide Information Analysis Center, US Department of Energy.

Hutton, J., W. Adams, and J. Murombedzi. 2006. Back to Barriers? Changing Narratives in Biodiversity Conservation. *Forum for Development Studies* 32: 342–370.

IPCC (Intergovernmental Panel on Climate Change). 2007. *Climate Change 2007: The Fourth Assessment Report of the Intergovernmental Panel on Climate Change*. Geneva: IPCC.

Jerneck, A., and L. Olsson. 2008. Adaptation and the Poor: Development, Resilience and Transition. *Climate Policy* 8 (2): 170–182.

Kanowski, P.J., C.L. McDermott, and B.W. Cashore. 2011. Implementing REDD+: Lessons from Analysis of Forest Governance. *Environmental Science and Policy* 14: 111–117.

Karsenty, A., and S. Ongolo. 2011. Can "Fragile States" Decide to Reduce Their Deforestation? *Forest Policy and Economics* 18: 38–45.

Kindermann, G., M. Obersteiner, B. Sohngen, J. Sathaye, K. Andrasko, E. Rametsteiner, et al. 2008. Global Cost Estimates of Reducing Carbon Emissions through Avoided Deforestation. *PNAS* 105 (30): 10302–10307.

Kinzig, A.P., C. Perrings, F.S. Chapin, S. Polasky, V.K. Smith, D. Tilman, et al. 2011. Paying for Ecosystem Services-Promise and Peril. *Science* 334: 603–604.

Korhonen-Kurki, K., M. Brockhaus, A.E. Duchelle, S. Atmadja, and P.T. Thuy. 2012. Multiple Levels and Multiple Challenges for REDD+. In *Analysing REDD+: Challenges and Choices*, ed. A. Angelsen et al. Bogor: CIFOR. www.cifor.org/online-library/browse/view-publication/publication/3805.html

Kshatriya, M., and E. Sills. 2010. *Global Database of REDD+ and Other Forest Carbon Projects*. Bogor: CIFOR. www.forestsclimatechange.org/redd-map/

Lindenmayer, D.B., K.B. Hulvey, R.J. Hobbs, M. Colyvan, A. Felton, et al. 2012. Avoiding Bio-Perversity from Carbon Sequestration Solutions. *Conservation Letters* 5: 28–36.

Livengood, E., and A. Dixon. 2009. *REDD and the Effort to Limit Global Warming to 2C: Implications for Including REDD Credits in the International Carbon Market*. Greenpeace International and KEA3.

Locatelli, B., V. Evans, A. Wardell, A. Andrade, and R. Vignola. 2011. Forests and Climate Change in Latin America: Linking Adaptation and Mitigation. *Forests* 2 (1): 431–450.

Lyster, M. 2011. REDD+, Transparency, Participation and Resource Rights: The Role of Law. *Environmental Science and Policy* 14 (2): 118–126.

MA. 2005. *Millennium Ecosystem Assessment, Ecosystems and Human Well-Being*. Washington, DC.

McDermott, C.L., A. O'Carroll, and P. Wood. 2007. *International Forest Policy: The Instruments, Agreements and Processes that Shape It*. New York: United Nations Forum on Forests. www.un.org/esa/forests

McDermott, C.L., L. Coad, A. Helfgott, and H. Schroeder. 2012. Operationalizing Social Safeguards in REDD+: Actors, Interests and Ideas. *Environmental Science and Policy* 21: 63–71.

McElwee, P. 2012. Payments for Environmental Services as Neoliberal Market-Based Forest Conservation in Vietnam: Panacea or Problem? *Geoforum* 43: 412–426.

McShane, T.O., P.D. Hirsch, T.C. Trung, A.N. Songorwa, A. Kinzig, B. Monteferri, et al. 2011. Hard Choices: Making Trade-Offs between Biodiversity Conservation and Human Wellbeing. *Biological Conservation* 144: 966–972.

Mertz, O., C. Padoch, J. Fox, R.A. Cramb, S.J. Leisz, N.T. Lam, et al. 2009. Swidden Change in Southeast Asia: Understanding Causes and Consequences. *Human Ecology* 37: 259–264.

Miles, L., and V. Kapos. 2008. Reducing Greenhouse Gas Emissions from Deforestation and Forest Degradation: Global Land-Use Implications. *Science* 320: 1254–1255.

Minteer, B.A., and T.R. Miller. 2011. The New Conservation Debate: Ethical Foundations, Strategic Trade-Offs, and Policy Opportunities. *Biological Conservation* 144: 945–947.

MPOC (Malaysian Palm Oil Council). 2012. Factsheets on Malaysian Palm Oil. www.mpoc.org.my/pubs_view.aspx?id=cd34de06-b843-48ca-90e3-b873fb812499

Mudiyarso, D., M. Skutsch, M. Guariguata, M. Kanninen, C. Luttrell, and P. Verweij, et al. 2008. How Do We Measure and Monitor Forest Degradation? In *Moving Ahead with REDD: Issues, Options and Implications*, ed. A. Anglesen. Bogor: CIFOR. www.cifor.cgiar.org/publications/pdf_files/Books/BAngelsen0801.pdf

Nagendra, H., and E. Ostrom. 2012. Polycentric Governance of Multifunctional Forested Landscapes. *International Journal of the Commons* 6: 104–133.

Naidoo, R., A. Balmford, R. Costanza, B. Fisher, R.E. Green, B. Lehner, et al. 2008. Global Mapping of Ecosystem Services and Conservation Priorities. *PNAS* 105: 9495–9500.

Nakhooda, S., A. Caravani, A. Wenzel, and L. Schalatek. 2011. The Evolving Global Climate Finance Architecture. In *Climate Change Fundamentals, Brief 2*. London: ODI and Heinrich Böll Stiftung North America. www.odi.org.uk/resources/docs/7468.pdf

Nasi, R., F.E. Putz, P. Pacheco, S. Wunder, and S. Anta. 2011. Sustainable Forest Management and Carbon in Tropical Latin America: The Case for REDD+. *Forests* 2: 200–217.

Nepstad, D., B.S. Soares-Filho, F. Merry, A. Lima, P. Moutinho, J. Carter, et al. 2009. The End of Deforestation in the Brazilian Amazon. *Science* 326: 1350–1351.

Norway (Government of Norway Ministry of the Environment and Ministry of Foreign Affairs). 2011. Norwegian Climate and Forest Initiative. *Oslo*. www.regjeringen.no/upload/MD/Vedlegg/Klima/klima_skogprosjektet/mai2010.pdf

Oliveira, P., G.P. Asner, D. Knapp, A. Almeyda, R. Galván-Gildemeister, S. Keene, et al. 2007. Land-use Allocation Protects the Peruvian Amazon. *Science* 317: 1233–1236.

Ostrom, E. 2010. Beyond Markets and States: Polycentric Governance of Complex Economic Systems. *American Economic Review* 100: 641–672.

Pascual, U., J. Phelps, E. Garmendia, K. Brown, E. Corbera, A. Martin, E. Gomez-Baggethun, and R. Muradian. 2014. Social Equity Matters in Conservation Payments. *Bioscience*. https://doi.org/10.1093/biosci/biu146

Pascual, U., I. Palomo, W. Adams, K. Chan, T. Daw, E. Garmendia, E. Gómez-Baggethun, et al. 2017. Off-Stage Ecosystem Service Burdens: A Blind Spot for Global Sustainability. *Environmental Research Letters* 12. http://iopscience.iop.org/article/10.1088/1748-9326/aa7392

Pagiola, S., A. Arcenas, and G. Platais. 2005. Can Payments for Ecosystem Services Help Reduce Poverty? An Exploration of the Issues and Evidence to Date from Latin America. *World Development* 33: 237–253.

Parker, C., J. Brown, and J. Pickering. 2009. *The Little Climate Finance Book*. Oxford: Global Canopy Programme.

Parker, C., M. Cranford, N. Oakes, and M. Leggett, eds. 2012. *The Little Biodiversity Finance Book*. Oxford: Global Canopy Programme.

Parry, M., J. Lowe, and C. Hanson. 2009. Overshoot, Adapt and Recover. *Nature* 458: 1102–1103.

Pattanayak, S.K., S. Wunder, and P.J. Ferraro. 2010. Show Me the Money: Do Payments Supply Environmental Services in Developing Countries? *Review of Environmental Economics and Policy* 4: 254–274.

Persha, L., A. Agrawal, and A. Chhatre. 2011. Social and Ecological Synergy: Local Rulemaking, Forest Livelihoods, and Biodiversity Conservation. *Science* 33: 1606–1608.

Peskett, L., D. Huberman, E. Bowen-Jones, G. Edwards, and J. Brown. 2008. Making REDD Work for the Poor. *Poverty Environment Partnership*. www.cmsdata.iucn.org/downloads/making_redd_work_for_the_poor_final_draft_0110.pdf

Phelps, J., M.C. Guerrero, D.A. Dalabajan, B. Young, and E.L. Webb. 2010a. What Makes a "REDD" Country? *Global Environmental Change* 20: 322–332.

Phelps, J., E.L. Webb, and A. Agrawal. 2010b. Does REDD+ Threaten to Recentralize Forest Governance? *Science* 328: 312–313.

Phelps, J., E.L. Webb, and L.P. Koh. 2011. Risky Business: An Uncertain Future for Biodiversity Conservation Finance through REDD+. *Conservation Letters* 4: 88–94.

Phelps, J., D.A. Friess, and E.L. Webb. 2012a. Win-win REDD+ Approaches Belie Carbon-Biodiversity Trade-Offs. *Biological Conservation* 154: 53–60.

Phelps, J., E.L. Webb, and W.M. Adams. 2012b. Biodiversity Co-Benefits of Policies to Reducing Forest-Carbon Emissions. *Nature Climate Change* 2: 497–503.

Phelps, J., A. Dermawan, and E. Garmendia. 2017. Institutionalizing Environmental Valuation Into Policy: Lessons from 7 Indonesian Agencies. *Global Environmental Change-Human and Policy Dimensions* 43: 15–25.

Pielke, R.J., G. Prins, S. Rayner, and D. Sarewitz. 2007. Lifting the Taboo on Adaptation. *Nature* 445: 597–598.

Pirard, R., and K. Belna. 2012. Agriculture and Deforestation: Is REDD+ Rooted in Evidence? *Forest Policy and Economics* 21: 62–70.

Piris-Cabezas, P., and N. Keohane. 2008. *Reducing Emissions from Deforestation and Forest Degradation: Implications for the Carbon Market*. Washington, DC: Environmental Defense Fund.

PNRPS (Philippines National REDD Plus Strategy). 2010. www.unredd.net/index.php?option=com_docman&task=doc_download&gid=4191&Itemid=53

Putz, F.E., P.A. Zuidema, T. Synnott, M. Peña-Claros, M.A. Pinard, et al. 2012. Sustaining Conservation Values in Selectively Logged Tropical Forests: The Attained and the Attainable. *Conservation Letters* 5: 296–303.

Rametsteiner, E., and M. Simula. 2003. Forest Certification: An Instrument to Promote Sustainable Forest Management? *Journal of Environmental Management* 63: 87–98.

Ribot, J., and A.M. Larson. 2012. Reducing REDD Risks: Affirmative Policy on an Uneven Playing Field. *International Journal of the Commons* 6: 233–254.

Rosendal, G.K., and S. Andresen. 2011. Institutional Design for Improved Forest Governance through REDD: Lessons from the Global Environment Facility. *Ecological Economics* 70: 1908–1915.

Rutt, R. 2012. *Social Protection in REDD+ Initiatives: A Review*. Responsive Forest Governance Initiative, University of Illinois, International Union for the Conservation of Nature. Mimeo. www.theredddesk.org/resources/reports/social_protection_in_redd_initiatives_a_review

Sandbrook, C., F. Nelson, W.M. Adams, and A. Agrawal. 2010. Carbon, Forests and the REDD Paradox. *Oryx* 44: 330–334.

Sasaki, N., and F.E. Putz. 2009. Critical Need for New Definitions of "Forest" and "Forest Degradation" in Global Climate Change Agreements. *Conservation Letters* 2: 226–232.

Scholz, I., and L. Schmidt. 2008. REDD: Opportunities for SFM. In *Financing Sustainable Forest Management*, ed. J. Holopainen and M. Wit. Wageningen: Tropenbos International.

Schroeder, H. 2010. Agency in International Climate Negotiations: The Case of Indigenous Peoples and Avoided Deforestation. *International Environmental Agreements: Politics, Law and Economics* 10 (4): 317–332.

Shepherd, G. 2004. Poverty and Forests: Sustaining Livelihoods in Integrated Conservation and Development. In *Getting Biodiversity Projects to Work*, ed. T.A. McShane and M.P. Wells, 340–371. New York: Columbia University Press.

Sidle, R.C., A.D. Ziegler, J.N. Negishi, A.R. Nik, R. Siew, and F. Turkelboom. 2006. Erosion Processes in Steep Terrain: Truths, Myths, and Uncertainties. *Forest Ecology and Management* 224 (1): 199–225.

Simula, M. 2010. Analysis of REDD+ Financing Gaps and Overlaps. *REDD+ Partnership*. www.reddpluspartnership.org/25159-09eb378a8444ec 149e8ab 32e2f5671b11.pdf

Smith, R.J., and M.J. Walpole. 2005. Should Conservationists Pay More Attention to Corruption? *Oryx* 39: 251–256.

Stephenson, S. 2011. Does ODA Grow on Trees? A Legal Analysis of REDD-ODA Finance. *European Journal of Legal Studies* 4 (1): 81–101.
Stern, S.N. 2006. *The Economics of Climate Change*. Cambridge: Cambridge University Press.
Stickler, C.M., D.C. Nepstad, M.T. Coe, D.G. McGrath, H. Rodrigues, W.S. Walker, et al. 2009. Potential Ecological Cobenefits of REDD. *Global Change Biology* 15: 2803–2824.
Strassburg, B., R.K. Turner, B. Fisher, R. Schaeffer, and A. Lovett. 2009. Reducing Emissions from Deforestation: The "Combined Incentives" Mechanism and Empirical Simulations. *Global Environmental Change* 19: 265–278.
Streck, C. 2012. Financing REDD+: Matching Needs and Ends. *Current Opinion in Environmental Sustainability* 4: 628–637.
Streck, C., and C. Parker. 2012. Financing REDD+. In *Analysing REDD+: Challenges and Choices*, ed. A. Angelsen et al. Bogor: CIFOR.
Sunderlin, W.D., A. Angelsen, B. Belcher, P. Burgers, R. Nasi, L. Santoso, et al. 2005. Livelihoods, Forests, and Conservation in Developing Countries: An Overview. *World Development* 33 (9): 1383–1402.
Swickard, N., and K. Carnahan. 2010. Integrating Project and National REDD+: The Importance of the Private Sector. In *Pathways for Implementing REDD+*, ed. X. Zhu et al. Roskilde: UNEP Risø Centre.
TCG (Terrestrial Carbon Group). 2008. How to Include Terrestrial Carbon in Developing Nations in the Overall Climate Change Solution. www.terrestrialcarbon.org/default.aspx
Thompson, I., B. Mackey, S. McNulty, and A. Mosseler. 2009. Forest Resilience, Biodiversity, and Climate Change. A Synthesis of the Biodiversity/Resilience/Stability Relationship in Forest Ecosystems. In *Technical Series*, vol. 43. Montreal: Secretariat of the Convention on Biological Diversity.
Thompson, M.C., M. Baruah, and E.R. Carr. 2011. Seeing REDD+ as a Project of Environmental Governance. *Environmental Science and Policy* 14: 100–110.
Tomaselli, I. 2006. *Brief Study on Funding and Financing for Forestry and Forest-Based Sector*. Report to the United Nations Forum on Forests Secretariat. New York, NY: UN.
Turner, W.R., M. Oppenheimer, and D.S. Wilcove. 2009. A Force to Fight Global Warming. *Nature* 428: 278–279.
UN (United Nations). 2008. *62/98 Resolution on Non-legally Binding Instrument on All Types of Forests*. New York, NY: UN. www.daccess-dds-ny.un.org/doc/UNDOC/GEN/N07/469/65/PDF/N0746965.pdf?OpenElement
UNESCO (United Nations Educational, Scientific and Cultural Organization). 2007. *World Heritage Forests: Leveraging Conservation at the Landscape Level*. World Heritage Report, 21. Paris: UNESCO. whc.unesco.org/uploads/activities/documents/activity-43-8.pdf

UNFCCC. 2010. Methodological Guidance for Activities Relating to Reducing Emissions. Decision 4/CP.15. www.unfccc.int/documentation/decisions

———. 2013. Report to the Conference of Parties on its Eighteenth Session. Document FCCC/CP/2012/8/Add.1. www.unfccc.int/resource/docs/2012/cop18/eng/08a01.pdf

UNFCCC (United Nations Framework Convention on Climate Change). 2007. *Investment and Financial Flows to Address Climate Change*. Bonn: UNFCCC.

Upton, C., and S. Bass. 1995. *The Forest Certification Handbook*. Wallingford: CABI Publishing.

Venter, O., and L.P. Koh. 2012. Reducing Emission from Deforestation and Forest Degradation (REDD+): Game Changer or Just Another Quick Fix? *Annals of the New York Academy of Science* 1249: 137–150.

Venter, O., W.F. Laurance, T. Iwamura, K.A. Wilson, R.A. Fuller, and H.P. Possingham. 2009. Harnessing Carbon Payments to Protect Biodiversity. *Science* 326 (5958): 1368.

Verchot, L.V., and E. Petkova. 2009. *The State of REDD Negotiations: Consensus Points, Options for Moving Forward and Research Needs to Support the Process*. Background Document for the UN-REDD Sponsored Support to Regional Groups.

Vickers, B. 2008. REDD: A Steep Learning Curve. In *Financing Sustainable Forest Management*, ed. J. Holopainen and M. Wit. Wageningen: Tropenbos International.

VRD (Voluntary REDD+ Database). 2011. REDD+ Partnership. www.reddplus-database.org/

WCED. 1987. *Report of the World Commission on Environment and Development: Our Common Future*. United Nations.

Wiersum, K.F. 1995. 200 Years of Sustainability in Forestry: Lessons from History. *Environmental Management* 19 (3): 321–329.

Winterbottom, R. 1990. *Taking Stock: The Tropical Forestry Action Plan after Five Years*. Washington, DC: World Resources Institute.

Wunder, S. 2008. How to Deal with Leakage? In *Moving Ahead With REDD. Issues, Options, and Implications*, ed. A. Angelsen, 65–76. Bogor: CIFOR.

Ziegler, A.D., J.M. Fox, E.L. Webb, C. Padoch, S. Leisz, R.A. Cramb, et al. 2011. Rapid Land Use Changes and the Role of Swiddening in SE Asia. *Conservation Biology* 25: 846–848.

Ziegler, A.D., J. Phelps, J.Q. Yuen, D. Lawrence, W.L. Webb, T.B. Bruun, et al. 2012. Carbon Outcomes of Major Land-Cover Transitions in SE Asia: Great Uncertainties and REDD+ Policy Implications. *Global Change Biology* 18: 3087–3099.

CHAPTER 9

The Evolution, Paradigm Shift and Guidelines for Foreign Aid in Forestry

Pekka E. Kauppi

9.1 Introduction

Forests cover about one third of the global land area, and are important to rural development in most countries worldwide. Forests provide raw material for construction, industry, and energy, and uphold the majority of terrestrial biodiversity. Forests grant also many other benefits. The relative importance of the various ecosystem services depends on local conditions, because there are large differences in climate, deforestation history, population density, and so forth. For example, the role of forests in the protection of devastating floods is very important in the densely populated lowlands of China (Niu et al. 2012). Flood protection is less important in other environments, such as arid highlands.

Forestry is a large economic sector, which mainly evolves unrelated to the efforts of foreign aid. Forest ecosystem services provide win-win benefits within the forestry sector and between other sectors, but there are also trade-offs. It is important to appreciate the long time horizon of forest-related action; time lags until the materialization of benefits are characteris-

P.E. Kauppi (✉)
Department of Environmental Sciences, University of Helsinki, Helsinki, Finland

Y. Huang, U. Pascual (eds.), *Aid Effectiveness for Environmental Sustainability*, https://doi.org/10.1007/978-981-10-5379-5_9

tic of forestry policies. This calls for careful preparation of aid programmes, commitment, and persistence. Forestry aid must be tailored to the specific ecological, economic, and cultural characteristics of the recipient region.

The richer countries are urged to donate 0.7% of their gross national product (GNP) to official development assistance (ODA). The United Nations General Assembly Resolution first pledged to this target in 1970. Since then, several international agreements over the years have referred to this goal. For example, the International Conference on Financing for Development in 2002, in Monterrey, Mexico, and the World Summit on Sustainable Development held in Johannesburg later that same year, reaffirmed this demanding and altruistic target.

The combined gross domestic product (GDP) of the Organisation for Economic Co-operation and Development (OECD) countries was US$43,000 billion in 2011. A share of 0.7% would imply a global aid target of approximately US$300 billion. The OECD nations actually provided US$133.5 billion of net official development assistance, which did not reach the defined target but nevertheless equates to a considerable international funding. Most of the aid support will be reserved to other areas, such as education, healthcare, agriculture, sanitation, electricity infrastructure, and so on. If 1% of foreign aid is allocated to promoting sustainable forestry, approximately US$1.3 billion can be made available for forestry projects. This is adopted here as an order of magnitude estimate of forestry aid. Hence the question is what can be achieved in forestry with the best possible allocation of foreign aid in the order of US$1.3 billion annually?

This chapter seeks to identify guidelines for foreign aid in forestry in the short, mid-, and long term. It first describes the evolution and paradigm shifts of forestry aid in the past. From these experiences, it then elaborates on questions posed by the UNU-WIDER institute, which are as follows.

Regarding foreign aid and sustainable forestry:

- what works?
- what could work?
- what is scalable?
- what is transferable?

These questions are addressed, noting that foreign aid operates in the real world where several forces interact, and aid projects alone cannot solve development problems. Foreign aid is a piece of a large puzzle, with which good governance is maintained in rural regions of the world.

The forestry sector has special characteristics, which call for particular attention. Firstly, forests tend to be located apart from the main population centres in remote rural areas. The population density is often low in forested areas, the infrastructure is weak, and human skills and capacities are not always easy to find. Secondly, forests rarely produce goods with high and immediate market value. Forest goods and services are diverse, and many important services cannot directly be measured in monetary terms. For example, forests regulate the hydrological cycle, with mountain forests protecting lowland areas from devastating floods. Such services are enormously valuable but direct funding mechanisms are lacking, which would encourage the management of upstream forests to provide this ecosystem service. Thirdly, forests in many cases are considered of secondary importance compared with agriculture and animal husbandry in rural areas of the developing world. Forests grow on residual land. This is often the case in both subsistence farming and cash crop systems. Solving farmland problems is sometimes the best way of promoting sustainable forestry.

The international scientific community has gradually learnt to understand and appreciate the important role of forests in the global carbon budget. Science has made progress, documented as follows:

- prospects have been presented that the era of deforestation may come to an end in the process called 'forest transition' (Mather and Needle 1998);
- historically, losses of biomass in deforestation has released carbon dioxide into the atmosphere (Houghton 2003). In contrast, greening of the planet would help remove higher amounts of carbon dioxide from the atmosphere;
- the removal process already dominates in large areas of the world, where forest biomass is expanding (Goodale et al. 2002);
- woody debris and forest soils accumulate carbon, and amplify the impact of biomass expansion (Pan et al. 2011);
- cropland area of the world is near a steady state, even though widespread cultivation of bioenergy crops triggers land conversion from forest to non-forest. The consumption patterns of food, energy, and fibre have improved, thus avoiding wasteful use of the land resources (Ausubel et al. 2013).

The importance of forests in the global carbon cycle has received significant public and political attention, especially in donor countries. This importance has been a key argument in support of aid projects in forestry, adding a novel dimension to aid policies of the twenty-first century. Forestry projects have moved up in aid project ranking, when assessed from the donor perspective.

9.2 What Has Worked in the Past?

Tree planting in general has worked in many places, both as a component of aid programmes and unrelated to foreign aid. Wangari Maathai, who was awarded the Nobel Peace Prize in 2004, appreciated the feasibility and multiple benefits from tree planting (Michaelson 1994): 'I think that when you look at a tree you planted and see it grow, it is like looking at a child grow. You develop a relationship that is very pleasant. You get to love the tree.'

A target of planting 100 million trees by 2017 was introduced in the Rio 20+ conference in 2012, as a means to promote sustainable forestry. This specific Rio 20+ programme involves primary schools all over the world, connected to one another via the internet and the social media (for more detail see Vanhanen 2012). This programme has an educational element extending way beyond 2017, and perhaps preparing children to work as future professionals within the forestry sector. However, the programme cannot reach remote rural areas, where the internet and electricity are not available, and even school systems may be missing. Yet, the programme is well buffered against corruption risks.

Preconditions of success have been analysed using the history of Swedish forestry as an example (Persson 2003). Wood raw material in Sweden became valuable when the demand for timber increased in response to the evolution of wood-working and forestry industries. At the same time, Swedish farming methods improved, and food production from the existing cropland became sufficient to meet the demands of the rural population. Therefore, land clearance for agriculture came to an end. Growth of Swedish cities and large-scale emigration to North America helped to lower the population pressure of the rural regions. The land use patterns became stable and predictable. Sustainable management and protection of forests became a shared value of a large majority of citizens. The economic profit from Swedish forests and forestry was sufficiently high to provide the basis of reinvestments to the forestry sector. Forestry became sustainable, not as a response to implementing measures within the forestry sector but due to many changes in the society and to development in general. The requirements for sustainable forestry were met. Persson (2003) also describes the evolution of global aid paradigms in the field of forestry as follows.

Phase 1: Industrial Forestry (1960s)

Forestry was viewed as an engine of modernization and economic progress in the developing countries. Aid projects contributed to establishing planted forests and constructing sawmills, and even pulp and paper factories in some cases. Nordic countries were important donors.

Phase 2: Social Forestry (1970s)

The rise of the environmental movement in the 1970s and droughts in the Sahel region prompted the birth of the next paradigm: forestry for local community development. Persson (2003) uses terms such as farm forestry, social forestry, community forestry, and village forestry, in describing this era of foreign forestry aid.

Phase 3: Environmental Forestry (1980s)

Foreign aid in the 1980s was largely motivated by the notions of saving the rainforests and halting deforestation. The concept of biological diversity—biodiversity for short—became widely known in science in the 1980s (Soulé and Wilcox 1980; Wilson and Peter 1988). The biodiversity issue was swiftly placed on the aid agenda.

Phase 4: Sustainable Management of Renewable Natural Resources (1990s)

Poor rural people in particular were the primary focus of aid programmes during the 1990s, rather than nature per se. This was a response to the notion that biodiversity cannot be protected successfully unless local communities cooperate.

Phase 5: Reducing Emissions from Deforestation and Forest Degradation (REDD+, Since 2005)

A new phase of forestry aid began following the publication of Persson (2003). Climate assessments emphasized the importance of tropical forests as a global climate system component. Carbon sequestration and the production of renewable biomass became a priority (Angelsen 2009).

Economic productivity was the main focus of early aid, but ecological, social, and cultural issues have gradually moved to the forefront of the programmes. Nevertheless, the economic dimension was never abandoned. All aid, after all, contains a funding element.

This is a broad account of the emphasis of development in forestry aid, although elements of the five phases have co-existed over the past decades. Foreign aid has not always worked very well (Persson 2003). The frequently remote location of forests and lack of infrastructure have meant logistical problems. Land tenure has been unclear on forested land. Forestry is a slow economic sector, where investors need patience. Sustainable development refers to seeking a balance between ecological, economic, social, and cultural dimensions of human action. It has not proven easy in general, let alone with forest projects in particular, to implement the Swedish path of improving the preconditions of human life, and of development more broadly.

While paradigms and the emphasis of foreign aid have shifted over time, forests and forestry of the world have evolved simultaneously, mainly unrelated to aid contributions. Fighting poverty remains as a main ultimate goal of all foreign aid, including forestry aid. Learning from the history of the nineteenth and twentieth centuries, it is possible to see positive trends and success stories regarding what has worked in the past. They have often been driven by the private sector responding to growing demand for forest products. Forestry investments have been triggered and maintained by the high and improving value of the wood raw material. Nevertheless, ecological and social concerns remain.

Kauppi et al. (2006) showed that forest area has been sustained in nations where GNP exceeds a minimum level of about 5000 US$ per capita. This indicates that development and the evolution of socioeconomic structures improves forestry as a process, which, in most countries has not been driven by aid contributions. Transition that is a shift from forest contraction to forest expansion has occurred (Mather and Needle 1998; Mather 2007).

Tree planting has worked in the past as an element of forestry aid, as it has been integrated into the broad patterns of rural development. Plantation forestry has expanded mainly unrelated to aid programmes. A tree planting programme based on aid funding has been a small contribution to a large and powerful global trend. The area covered by planted forests grew from 17.8 million hectares in 1980 to 264 million hectares in 2010 (UN/FAO 2000, 2010). In China, forested area expanded from 139.3 to 155.6 million hectares between 1990 and 2007, largely in response to tree planting (Pan et al. 2011). The contribution of aid to tree planting programmes in China in general was relatively small.

This expansion of tree planting was mainly driven by private companies, which were interested in growing industrial raw material, and by government policies for the promotion of services from forest ecosystems (Sedjo 1999). A combination of private-sector action and government initiatives has promoted the rapid expansion of tree plantations in Chile, for example. Hence, several forces interacted: private-sector interests, the action of local and regional government, and the development of rural infrastructure, largely unrelated to forestry. Forestry aid worked when successfully integrated into the general transition of the society.

Regarding forests specifically, the concept of 'forest transition' refers to a shift from shrinking to expanding forests (Mather and Needle 1998). The development of forestry and the rural landscape responds to universal changes of lifestyles and technologies (Rudel et al. 2005; Ausubel et al. 2013).

Wikipedia in 2012 explains the process of forest transition as follows:

Forest transition refers to a geographic theory describing a reversal or turnaround in land-use trends for a given territory from a period of net forest area loss (i.e., deforestation) to a period of net forest area gain.

Forest recovery resulting in net increases in forest extent can occur by means of spontaneous regeneration, active planting, or both.

Studies of forest transitions have been conducted for several nations, as well as sub-national regions. Territories reported to have experienced forest transitions after the onset of industrialization include: Bangladesh, China, Costa Rica, Cuba, Denmark, Dominican Republic, El Salvador, France, Gambia, Hungary, Ireland, Morocco, New Zealand, Portugal, Puerto Rico, Rwanda, Scotland, South Korea, Switzerland, the United States, and Vietnam. Furthermore, forest-transition dynamics have been documented for regions within Brazil, Ecuador, and Mexico.

The environmental effects of these forest transitions are very variable, depending on whether deforestation of old-growth forests continue, the proportions and types of tree plantations versus natural regeneration of forests, and the location and spatial configuration of the different types of forests.

The findings of returning forests in these widespread studies raise questions about the prospects of a worldwide forest transition. In other words, can the global extent of forests be expected to reach a turning point in the future, reversing the current trend of overall forest decline towards overall forest expansion? Studies showed that given an increased competition for productive land between different land uses, a global forest transition would require major policy and technological innovations, as well as shifts in demands for fiber, fuel, and food, and that these changes cannot be taken for granted.

The United Nations (UN) and Food and Agriculture Organization (FAO) statistics (UN/FAO 2010) show that, despite forest planting and forest transition in many parts of the world, the forested area of the world keeps shrinking, albeit at a decelerating rate (see Pan et al. 2011). A general development goal is to reach a balance and a steady state in terms of the global land cover that is forest, and to bring to an end the expansion of croplands and pastures. Later, global forest cover may start expanding and returning to areas where forests have been lost. It is important to assess aid projects in this broad perspective. Drivers of forest transition have been analysed in the literature (Mather and Needle 1998; Lambin et al. 2001; Meyfroidt and Lambin 2008; DeFries et al. 2010; Saikku et al. 2012; Ausubel et al. 2013). An integration of aid projects into the general pattern of forest transition has been critical to past successes. Tree planting serves as an example of such a good integration and of an approach that has worked in the past.

9.3 How to Encourage Successful Forestry Aid?

9.3.1 Selecting Goals

Both the goals and means of forestry matter when addressing what kind of aid *could work*, which UNU-WIDER posed as the second question for this study. Regarding goals, the concept of ecosystem services refers to the benefits people obtain from ecosystems. For example, halting deforestation and reducing forest degradation both help to sustain ecosystem services such as fuelwood production, landslide prevention, and biodiversity preservation. Maintaining and improving ecosystem services is the goal of sustainable forestry. An agreement of clear goals is the first step to forestry aid, in terms of what could work and bring results.

The concept of ecosystem services and its classification into regulatory, supportive, provisioning, and cultural services was promoted in 2005 by the UN's Millennium Ecosystem Assessment (MEA), a four-year study involving more than 1300 authors worldwide. Forest ecosystems generally provide numerous and diverse benefits. Many different services are available from a given forest at the same time. Therefore, so-called win-win opportunities exist, and a variety of ancillary benefits are available, when foreign aid is directed to the forestry sector. For example, protecting forest biodiversity automatically preserves carbon stocks and vice versa. However, notable harmful side effects also exist, and the trade-offs must be assessed.

A yardstick is needed in analysing the pros and cons of forestry aid projects, with reference to ecosystem services' sensitivity to side effects and adverse impacts, and its aptness for co-benefits. A cost–benefit analysis is useful. However, not all the impacts can be easily quantified in monetary terms, especially in poor rural regions, where subsistence economy prevails. An intensive effort is presently under way internationally for assessing forest ecosystem services. Poor rural regions of China, for example, assist the social and economic development of the lowlands, particularly by protecting Chinese lowland plains from flood damage (Niu et al. 2012).

There are risks that international interest in the protection of the global climate does not fully acknowledge the multiple benefits from forests at local and regional level. Carbon dioxide emissions from forests have been the main focus in REDD+ programmes, and rightly so. This is relevant, because forests greatly affect the carbon dioxide concentration in the atmosphere (Pan et al. 2011). Co-benefits and positive side effects of the REDD+ project, however, can sometimes exceed its climate benefits. Noting the complexity and diversity of forests, a comprehensive assessment of forest ecosystem services is important, in order to gain the full benefit of the project.

Aid, by definition, must promote poverty abatement one way or another. A foreign aid REDD+ project aims at managing the carbon sequestration of a specified forest, by directing financial recourses for that specific purpose. For example, as carbon sequestration is the intended aid goal, fuelwood service is recognized as a co-benefit.

In selecting aid goals, it is important to set geographical priorities. Forests cover about one third of all the land area of the world, affecting the planet's regulating systems, such as the global hydrological and carbon cycles. These services benefit all people. Forest provisioning services, such as food, bioenergy, and wood-based industrial products, interest smaller groups of people, namely those directly involved in gleaning such benefits. Forests provide cultural, aesthetic, and spiritual services, which are essential to the wellbeing of people. Biodiversity has economic value by supporting ecotourism and providing pharmaceutical potential. Biodiversity is also a good yardstick of ecosystem naturalness.

Forests extend to different regions, and there exist large variations between forest systems and the services they provide. It is important to note the large variability in forest ecology, population density, affluence, infrastructure, and social and cultural factors between regions of the world. As the largest biome, the tropical forests are particularly diverse. One small part of this huge domain is interesting in this context—the remote rural fraction, where people are short of arable land and short of wood material to be used in construction and as fuel. This region, mainly on the African continent, must remain as the geographical prime focus of forestry aid—an area, where aid is very much needed and aid projects could work in the future.

9.3.2 Selecting Means

Forestry is a large enterprise and a slow system, where time horizons are long and patience is required for obtaining returns. All programmes, large and small, need a preparatory phase before launching. The search for partners, preparation of project plans and budget negotiations, as well as reaching and signing agreement always take time. Immediate results are not possible. Some initiatives can provide relatively near-term benefits, while others require more persistent and patient support. Projects and programmes with a long time horizon pose unique challenges but are important too in promoting sustainable development. A horizon of several decades, with the dynamics of forest transition, is sometimes necessary for a full-scale realization of forestry impacts. Capacity building in particular takes time—decades rather than years.

A minimum time horizon for a focused tree planting project is approximately three years, given the time required for preparation and implementation. On the other hand, capacity building takes much longer.

A comparison with the education system is salutary here: directing a young person through primary school to higher education and then on to a PhD typically takes 25 years. Establishing a high-quality research university can take even longer, up to 50–100 years. Thus, feasibility assessments must cope with the fact that only a certain kind of aid project can provide results within a short timespan, say of 3–10 years. Short-term programmes are rarely the ones that remain in history as the best success stories.

Foreign aid in forestry is potentially successful if it adopts modern and realistic aid paradigms, seeks to promote co-benefits and ecosystem services in collaboration with the best local expertise, integrates into the general development of forests and forestry, adopts a realistic time horizon, and comes with a long-term time perspective. The means of aid projects must also be chosen in a way that the risks of misuse and corruption are minimized. Successful forestry aid ultimately contributes to poverty reduction. Examples are given in the next section of aid programmes that show promise at various time horizons.

9.4 Opportunities for Scaling Up

9.4.1 Short Term

A potential area for scalable measures is in shifting away from wasteful consumption of forest-based food, energy, and fibre. Inefficient patterns of material flows are common in all countries. Ausubel et al. (2013) elaborate on the significant progress that has been achieved in the past in terms of efficiency improvements. Such examples encourage scaling up this progress by means of forestry aid. As an example project, fuel-efficient cooking stoves can be introduced to foster better development in rural areas which are not yet connected to the electric grid.

The wood-fired stove has an analogy in solar energy systems, consisting of solar panels connected to a battery. Both systems can operate in rural conditions outside the electricity grid. Solar cells and the battery are industrial products, which are best produced in the donor country, and then transferred to poor regions of the recipient country. The cell captures solar energy, which is stored in the battery for later use—in the same way that tree leaves capture solar energy, produce electrons in photosynthesis, and the energy is stored in the wooden 'battery'. Energy can then be released in the burning process, providing the necessary heat for cooking hot meals (Fig. 9.1).

Fig. 9.1 A simple and robust cooking stove. Note: The stove quickly produces a hot flame using a small amount of wood. The idea is simple: cooking with an efficient stove spares fuelwood and saves forests. Source: Photo by the author

A stove, unlike solar electricity, is a single-purpose item, which applies to cooking only. But for cooking purposes, the analogy to solar energy holds. A modern stove made of high-quality steel is an industrial product like solar cell and battery, which needs to be produced in a factory and be transported to rural regions. The stove is simple and robust compared with the solar cell-battery system. No maintenance is needed, if the construction is decent. The lifetime of a stove is up to ten years and the unit cost is affordable. Therefore, the system is easily scalable.

A forestry objective of the fuel-efficient stove is to spare trees. As an efficient stove replaces inefficient wood combustion, the demand for wood is reduced per cooking event. The energy output from a harvested tree improves, as more meals can be cooked per unit of wood. The impact is amplified, as tree planting becomes more profitable and socially attractive.

In the win-win mode, an efficient cooking stove supports social development goals. Working time is spared, both in cooking and in the effort of collecting firewood. Cooking with an efficient stove enables the preparation of five meals with 20 kilos of wood, which has been carried home by a household member. The same amount of wood is only enough for the preparation of two meals when using inefficient combustion of wood.

Manufacturing and distributing fuel-efficient stoves as a forestry aid project is an interesting and scalable approach, which improves consumption patterns, preserves forest resources, and promotes social development goals.

As with tree planting, a campaign could be launched with the goal of distributing 100 million fuel-efficient wood-fired cooking stoves to rural regions of the world, which are not yet connected to the electricity grid. A robust cooking stove is an example of a transferable device, which can be distributed to unconnected, poor rural areas, including the most remote ones. Manufacturing and distributing stoves is possible with a reasonable and affordable level of expenditure using foreign aid. Thus, a distribution programme of 100 million stoves could be launched within the current budget constraints of foreign aid allocated to forestry.

As less wood is needed, the harvest pressure is decreased and harvest levels become more sustainable. The benefit is distributed among different groups of people and among the different ecosystem services. As a co-benefit, the wellbeing of the household is improved, as fuelwood collection becomes less laborious. Planting trees near the village and thus shortening the walk to collect wood can further enhance the benefit obtained from the stove. If one cooking stove spares 1000 trees over its lifetime of ten years, a programme of distributing 100 million stoves has a potential of sparing of 100 billion trees. This would make a significant contribution to halting deforestation and encouraging forest transition at continental, and even the global level.

If an average family using one stove consists of five people, and 100 million stoves are distributed successfully, the lives of 0.5 billion people would improve in the least developed rural areas of the world. This would be feasible within a relatively short time, by 2018, and at costs which are within the budget constraints of forestry aid at present. A 100 million stove programme is relatively robust against corruption. However, large-scale manufacturing, transportation, distribution, and end-use of the devices do imply certain risks of misconduct.

9.4.2 Mid-Term

Improving forestry databases is both feasible and very important. Information on forest attributes is crucial for policymakers in defining baselines and setting development goals for policies on rural development. Forest inventory skills can be transferred to recipient countries, with a high potential of building critical capacity and creating innovation centres of future.

A system has recently been tested and applied in Tanzania's National Forest Resources Monitoring and Assessment programme (NAFORMA). Tanzania's trees have been counted to determine their age, size, and species, with the work carried out in collaboration with national and international experts.[1] Publication of results is not available at present, and so project evaluation is not currently possible, but the fieldwork has been completed successfully. The programme is expected to replace outdated statistics and help this East African country assess the services provided by forest ecosystems and then allow the raising of REDD+ funding.

The NAFORMA project is not as easily scalable and transferable as the short-term project proposed in Sect. 9.4.1. But in Tanzania's dry tropical forests, the trees grow in a relatively sparsely, making it easy for a measurement team to penetrate and thus favourable for carrying out systematic observations. Moreover, remote sensing methods are effective when the forest canopy is relatively open. Similar methods do not currently exist for the monitoring of tropical rainforests. The costs of NAFORMA are modest, in the order of US$10–20 million, but implementing such programmes is science-intensive, and the technical and scientific skills required are not easily available. Patience will be needed, as the measurements must be repeated after a few years to detect any possible trends in biomass, biodiversity, and carbon. Despite scaling obstacles, a measurement programme such as NAFORMA is useful and cost-effective for promoting sustainable forestry and for serving policy design and implementation in the mid-term of 10 to 20 years. However, the approach is prone to corruption and misuse, because national and international organizations are required with staff, equipment, and vehicles. Tanzania's example nonetheless shows that corruption risks can be circumvented.

Forest monitoring programmes can be scalable; yet, they require a mid-term time horizon of 5–15 years, and can be applied only to certain areas of the tropical biome. Methods of forest monitoring exist and are available, but their application requires highly qualified staff. What may succeed in Tanzania may not be scalable to tropical rainforests, with their dense vegetation and multi-layer canopies.

Scalable measures are available also in shifting away from the wasteful consumption of agricultural products. If food consumption is wasteful, and farming practices inefficient, farms expand excessively, and forests suffer as a consequence. Increasing yields has been a dominant and very valuable trend in global land use since the 1960s (Ausubel et al. 2013).

Forestry benefits scale up if the pressure of land conversion from forest to non-forest can come to an end. Investing in forests that soon will disappear logically cannot support forest ecosystem services. In other words, it is unlikely that scaling up will be successful by working against the great wave of forest transition. During the phase when forest cover is diminishing, it is important to improve farming practices and consumption patterns, with the objective of reaching steady-state land cover as soon as possible. After forest transition, when forests can again expand, investing in forest improvement and management becomes more rewarding. It is crucial, at all times, that farming practices improve, food material is not wasted, and bioenergy development is reconciled with other ecosystem services from farmland, pastures, and forests. Population growth is bound to continue and the demands are high for improving nutrition, especially of the poor.

9.5 Transferring Lessons: The Role of Universities

A functioning educational system is crucial to sustainable development in the long term. It has been acknowledged across the industrial world that universities serve as embryos of economic prosperity and social progress; and this is also the case regarding the development of sustainable forestry. Harvard University, founded in 1638, and the numerous other excellent universities in the eastern United States are the foundation of the country's wealth. California's blooming economy is also based on universities as centres of innovation and higher education. China is now following the USA's path, with excellent universities moving up international ranking lists, based on substantial national funding and giving support to the best talent.

The Swedish example, to which Persson (2003) refers, is indicative. Sweden has developed and maintained scientific and multi-disciplinary approaches to forests and forestry based on university research. Primary and secondary schools are also important, but the role of universities is crucial. An educational system needs good teachers to work and collaborate with young people. University staff, including professors, refer to the latest results of international scientific research in their teaching. After graduation, some former university students accept jobs as teachers and distribute their knowledge to pupils at all levels of the school system.

The case of Africa is different: here the university network is sparse and weak. This presents an obstacle to all development, and in particular, improving farming and providing forestry practices with their basic needs. Hayward (2012) writes:

> Higher education in Africa in the 1960s and 1970s pictured excitement, creativity, and pride—given that faculty members dedicated to teaching were involved in innovative research, and many helped lay the foundations for governance and development. Quality was high, and universities held in great esteem. Most students were eager scholars, exhilarated by their good fortune, and certain they were destined for leadership roles. And a start was made on graduate programmes. By the early 1980s, the picture was different for most universities—including budget shortfalls in declining national economic circumstances, repression, curtailed academic freedom, civil unrest, and loss of status. Donor interest shifted to primary education, and external funding declined from US$103 million annually as late as 1994, dropping to an average of US$30.8 million from 1995 to1999.

The United Nations Educational, Scientific and Cultural Organization (UNESCO) is active in finding support for African universities and trying to increase funding for higher education on the continent. The organization refers to the difficulties Africa must face in overcoming the challenges to good higher education. These include the rapid increase in the number of young people reaching the college age, uncontrolled brain-drain, financial shortcomings, and hence the the low quality of teaching and research, as well as general difficulties in governance structures. There are prestigious universities in South Africa, Egypt, and elsewhere, but the lack of universities is striking in many African countries, given that the population of the continent is twice as large as that of Europe. There is far too little research in Africa focusing on domestic issues such as forests and forestry.

Universities typically work internationally, but in addition, they respond to local and national interests, select national issues, and local themes as research topics. The ultimate goal of higher education is to detect talented individuals and pave their way to professions where they can work to improve the wellbeing of themselves and their fellow citizens. African universities need loyal and persistent support and a clear focus on forestry, agriculture, and rural development. The continent is very diverse ecologically and culturally. Therefore, it is important that each country and region can support research and university education, which responds to local needs using the best local talent and expertise.

The shortage of higher education is a root cause of the development crisis in Africa. UNESCO's Association for African Universities has 173 university members in 34 African countries. None of the African universities is listed among the 100 best universities in the world in any of the international ranking lists. Half a dozen African universities rank among the 500 best universities in the world, but the high-quality universities are mostly located in just one country, South Africa.

While the education system is weak, the number of young people is growing. In 2012, nine out of ten countries in the world with the highest fertility rate were located in Africa. Bright children born in 2012 will need college education in 2030–35. Any aid programme must seriously address the growing imbalance in Africa between the number of young people and the lack of higher education.

The lack of universities in Africa is a general problem, only partly related to the shortage of research and teaching of forestry. However, agriculture and forestry are key components of future development in Africa. Patient international support to the faculties of sustainability science and forestry in African universities appears as scalable and transferable action available to foreign aid donors. Certain risks of misconduct and corruption do exist, but most of them can be avoided if support is directed to the academic staff, and if their immediate needs for infrastructure are satisfied. The recruitment procedure of academic staff must follow best international practices, with the priority in assessing the scientific publication record of professorial candidates.

9.6 Discussion and Conclusions

The broad development of forests and forestry is affected by social and economic drivers. World population keeps growing, fortunately at a decreasing pace but still with a momentum that will generate billions of new adults in the next few decades. Improving lifestyles and diets, combined with the growing population, create pressure on land ecosystems. This is the big picture, into which forestry aid must merge.

People convert forests to croplands in order to obtain food for the growing population. Farming methods improve and average yields rise, while urbanization, industrialization, and the development of the service sector replace subsistence farming in many parts of the world. The fundamentals of human livelihood change. This sort of development can rescue forest ecosystems, which otherwise would become converted to arable lands, in order to feed the expanding population.

Ausubel et al. (2013) refer to 'peak farmland', noting that the area of cultivated agricultural lands has ceased to grow. More precisely, it would have ceased to grow in absence of bioenergy expansion, which covered about 20 million hectares of croplands in 2012. Despite bioenergy threats, the big picture is surprisingly positive in the world on average. Improving yields has been the main single contributor to this positive development.

Meanwhile, the gap between rich and poor has increased. Foreign aid is meant to improve the livelihood of the poorest people and the most disadvantaged regions. More than a billion people are still not connected to the electricity grid, and many of them live in or near forests.

Significant foreign aid has been allocated to sustainable forestry since the 1960s. The paradigm of aid contributions has evolved over time. Too little attention was given in the early decades to scaling delays, and to balancing economic, social, and ecological objectives, as well as circumventing corruption and appreciating the diversity of forest ecosystem services. Even at present, there is not enough awareness of the general process of forest transition, which frames all projects. Concurrently, though, the process called 'forest transition' made surprising progress, deforestation in many nations ceased, and forests started to expand. The forest biomass of the world is no longer a source but a sink of atmospheric carbon (Mather and Needle 1998; Kauppi et al. 2006; Rautiainen et al. 2011; Pan et al. 2011; Ausubel et al. 2013).

Foreign aid has made modest contributions to the promotion of sustainable practices in forestry (Persson 2003). The positive development of forest transition has been associated with social and economic development and with an alleviation of poverty (Lambin et al. 2001; Lambin and Meyfroidt 2011). The evolution of approaches to forests and forestry has primarily been unrelated to foreign aid. Foreign aid dedicated to forestry has rarely, if ever, triggered, or even significantly promoted, forest transitions. This is no surprise, as the volume of forestry aid is far smaller economically than that of the forestry-sector system as a whole.

Thomas Rudel, in his recent book, describes the motivation of people to protect the environment as 'defensive' or 'altruistic' (Rudel et al. 2013). By defensive environmentalism, he refers to interests of people to protect the domestic, local environment. Promoting sustainable forestry far away from home is primarily a global 'altruistic' effort. The sense of 'defensiveness' may evolve and intensify as globalization increases contact between people.

The success of an aid project is measured in terms of improvements achieved in forests in the recipient country. The perspective foregrounded in this chapter is that of the recipients, who are poor people in the rural areas of the world. Donors appreciate sustainable forestry, which seeks to balance biodiversity protection, carbon sequestration, creation of income to rural areas, improving social relations through respect, and enhancement of local cultures. Examples have been elaborated in this chapter that have the potential to meet these goals.

The risks of corruption have particularly been emphasized regarding aid expenditures in the least-developed economies (e.g. Moyo 2006). These risks further call for integrating forestry aid policies with development policies more broadly. Integration of forestry with agriculture, energy, transport, mining, rural development, education, governance, and law enforcement is important. Ultimate success depends on enhancing the electric grid, road and rail networks, improving healthcare, providing clean drinking water, and improving the food chain from farming practices to cooking instruments. Development liberates rural populations to help themselves in creating small businesses and finding a livelihood, which is benign to the nearby forest vegetation. In other words, sustainable forestry needs a favourable general setting, where forest transition can be reached and passed.

Carbon sequestration as an ecosystem service is becoming valued in monetary terms. Citizens in donor countries are concerned about climate. A 'close-to-home' appeal is attached to climate mitigation. If *they* do not take good care of their lands, *our* climate will become adversely affected (Rudel et al. 2013). If REDD+ projects are integrated into plans for providing ecosystem services at the global scale, this holds promise to sustainable forestry. Monitoring REDD+ is a challenge. Deforestation of dry tropical forests can be accurately measured with modern technologies, but forest degradation is difficult to observe in tropical rainforests with multi-layer canopies.

Forests, by definition, are rural. Rural cultures tend to be more diverse than urban ones. Projects and approaches are recommended in this chapter for the fostering of sustainable forestry by means of foreign aid. In this context, tree planting can be promoted and cooking methods can be improved. Forest observation systems can be improved. Universities can be supported, especially in Africa, focusing on forestry, agriculture, and rural development.

In conclusion, sustainable forestry can be promoted by foreign aid in rural areas, assisted with a broad spectrum of measures. A minimum of three to five years is needed to implement any major project, and a long-term commitment is required in most cases. In due course, forestry must be organized by domestic actors. Therefore, the importance of developing the education system cannot be overemphasized.

Notes

1. Personal correspondence with Erkki Tomppo.

References

Angelsen, A. 2009. Policy Options to Reduce Deforestation. In *Realising REDD+: National Strategy and Policy Options*, ed. A. Angelsen. Bogor: CIFOR.

Ausubel, J.H., I.K. Wernick, and P.E. Waggoner. 2013. Peak Farmland and the Prospect for Land Sparing. *Population and Development Review* 38 (Suppl): 217–238.

Defries, R.S., T. Rudel, M. Uriarte, and M. Hansen. 2010. Deforestation Driven by Urban Population Growth and Agricultural Trade in the Twenty-first Century. *Nature Geoscience* 3 (3): 178–181.

Goodale, C.L., M. Apps, R. Birdsey, C. Field, L. Heath, R. Houghton, J. Jenkins, G. Kohlmaier, W. Kurz, S. Liu, G. Nabuurs, S. Nilsson, and A. Shvidenko. 2002. Forest Carbon Sinks in the Northern Hemisphere. *Ecological Applications* 12 (3): 891–899.

Hayward, F.M. 2012. Graduate Education in Sub-Saharan Africa: Prospects and Challenges. *International Higher Education* 66: 21–22.

Houghton, R.A. 2003. Revised Estimates of the Annual Net Flux of Carbon to the Atmosphere from Changes in Land Use and Land Management 1850–2000. *Tellus B* 55: 378–390.

Kauppi, P.E., J.H. Ausubel, J. Fang, A.S. Mather, R.A. Sedjo, and P.E. Waggoner. 2006. Returning Forests Analyzed with the Forest Identity. *Proceedings of the National Academy of Sciences* 103 (46): 17574–17579.

Lambin, E.F., and P. Meyfroidt. 2011. Global Land Use Change, Economic Globalization, and the Looming Land Scarcity. *Proceedings of the National Academy of Sciences of the United States of America* 108 (9): 3465–3472.

Lambin, E.F., B.L. Turner, H.J. Geist, S.B. Agbola, A. Angelsen, J.W. Bruce, O.T. Coomes, et al. 2001. The Causes of Land-use and Land-cover Change: Moving Beyond the Myths. *Global Environmental Change* 11 (4): 261–269.

Mather, A.S. 2007. Recent Asian Forest Transitions in Relation to Forest-transition Theory. *International Forestry Review* 9: 491–502.

Mather, A.S., and C. Needle. 1998. The Forest Transition: A Theoretical Basis. *Area* 30: 117–124.

Meyfroidt, P., and E.F. Lambin. 2008. The Causes of the Reforestation in Vietnam. *Land Use Policy* 25 (2): 182–197.

Michaelson, M. 1994. Wangari Maathai and Kenya's Green Belt Movement: Exploring the Evolution and Potentialities of Consensus Movement Mobilization. *Social Problems* 41 (4): 540–561.

Moyo, D. 2006. *Dead Aid. Why Aid Is Not Working and How There Is a Better Way for Africa*. New York: Farrar, Straus and Giroux.

Niu, X., B. Wang, S. Liu, C. Liu, W. Wei, and P.E. Kauppi. 2012. Economical Assessment of Forest Ecosystem Services in China: Characteristics and Implications. *Ecological Complexity* 11: 1–11.

Pan, Y., R.A. Birdsey, J. Fang, R. Houghton, P.E. Kauppi, W.A. Kurz, O.L. Phillips, A. Shvidenko, S.L. Lewis, J.G. Canadell, P. Ciais, R.B. Jackson, S. Pacala, A.D. McGuire, S. Piao, A. Rautiainen, S. Sitch, and D. Hayes. 2011. A Large and Persistent Carbon Sink in the World's Forests. *Science* 333: 988–983.

Persson, R. 2003. *Assistance to Forestry: Experiences and Potential for Improvement*. Bogor: CIFOR.

Rautiainen, A., I. Wernick, P.E. Waggoner, J.H. Ausubel, and P.E. Kauppi. 2011. A National and International Analysis of Changing Forest Density. *PloS One* 6 (5): e19577. https://doi.org/10.1371/journal.pone.0019577.

Rudel, T.K., O.T. Coomes, E. Moran, F. Achard, A. Angelsen, J. Xu, and E. Lambin. 2005. Forest Transitions: Towards a Global Understanding of Land Use Change. *Global Environmental Change* 15: 23–31.

———. 2013. *Defensive Environmentalists and the Dynamics of Global Reform*. New York and Cambridge: Cambridge University Press.

Saikku, L., S. Soimakallio, and K. Pingoud. 2012. Attributing Land-use Change Carbon Emissions to Exported Biomass. *Environmental Impact Assessment Review* 37: 47–54.

Sedjo, R.A. 1999. The Potential of High-Yield Plantation Forestry for Meeting Timber Needs: Recent Performance, Future Potentials, and Environmental Implications. *New Forests* 17: 339–359.

Soulé, M.E., and B.A. Wilcox. 1980. *Conservation Biology: An Evolutionary-Ecological Perspective*. Sunderland: Sinauer Associates.

UN/FAO. 2000. *Global Forest Resources Assessment 2000. Main Report*. Forestry Paper 140. Rome: Food and Agriculture Organization of the United Nations.

———. 2010. *Global Forest Resources Assessment 2010: Main Report*. Forestry Paper 163. Rome: Food and Agriculture Organization of the United Nations.

Vanhanen, M. 2012. ENO's Final Breakthrough in Rio+ Summit. http://www.uncsd2012.org/content/documents/790Summary%20of%20Voluntary%20Commitments%20Registered%20at%20Rio20%20v6.pdf. Accessed 17 October 2013.

Wilson, E.O., and F.M. Peter, eds. 1988. *Biodiversity*. Washington, DC: National Academy Press.

CHAPTER 10

Financing Sustainable Agriculture under Climate Change with a Specific Focus on Foreign Aid

Jikun Huang and Yangjie Wang

10.1 Introduction

Hundreds of millions of people suffer from hunger and food insecurity. According to estimates, the total number of undernourished people in the world in 2009 was 1.023 billion, although this was expected to decline by 2010 to 925 million (FAO 2010a). But in 2010, the actual number of hungry people was higher than the level which had existed when world leaders at the World Food Summit in 1996 agreed to reduce these numbers by half (ibid.). Most of the world's hungry live in the developing countries.

Global food security is likely to face even greater challenges in the coming decades. According to estimates from the Food and Agriculture Organization (FAO 2009a), global food production must increase by at least 70% to meet the growing food demands of a world population that is expected to surpass 9 billion by 2050 (ibid.). Furthermore, the growth rate of agricultural productivity has been falling; for example, average annual

J. Huang (✉)
School of Advanced Agricultural Sciences, Peking University, Beijing, China

Y. Wang
Business School of Central South University, Changsha, China

Y. Huang, U. Pascual (eds.), *Aid Effectiveness for Environmental Sustainability*, https://doi.org/10.1007/978-981-10-5379-5_10

growth rate of cereal yields has decreased from about 2–3% in the 1970s and 1980s to 1–2% in the 1990s and early 2000s (World Bank 2007).

Agriculture and food security may face even greater difficulties under climate change. Despite some existing uncertainties, increasing evidence indicates that the earth's climate is experiencing significant changes. According to projections (FAO 2009a), owing to continued and rising global warming, by 2050 developing countries may experience a decline of between 9 and 21% in overall agricultural productivity. In addition to long-term effects, global and regional weather conditions are also expected to become more varied, with increases in the frequency and severity of extreme events such as cyclones, floods, hailstorms, and droughts (Easterling et al. 2000; IPCC 2007a, 2012). Such extreme weather conditions will bring larger fluctuations to crop yields and local food supplies as well as higher risks of food insecurity (FAO 2008a, 2009a; IPCC 2012).

How severely climate change will affect agriculture depends on whether these impacts can be countered by investments in agriculture. The amount of investment needed for sustainable agriculture in the developing countries is tremendous even without taking climate change into consideration and must be greatly increased to address food insecurity issues (FAO 2009a). When climate change is included in the equation, even greater efforts will be necessary in the coming decades. However, investment and foreign aid in agriculture have either fallen or not grown appropriately. Current investments and commitments fall far short of the requirements necessary to meet the growing needs, especially in the developing world (Islam 2011). In addition, there has been a decline in the share targeted to the agricultural sector in aggregate foreign aid. For example, while the share of aid to agriculture in total aid increased from 13.0% in 1973–5 to 23% in 1979–81, it has declined since the mid-1980s (Table 10.1).

The international community has called for measures for climate change adaptation to be incorporated into national development plans (World Bank 2010). Climate change adaptation is defined by IPCC (2001) as 'adjustment in natural or human systems in response to actual or expected climatic stimuli or their effects, which moderates harm or exploits beneficial opportunities'. This would mean learning to manage new risks by preparing to deal with shocks and by strengthening resilience in the face of change. In the food and agriculture sector, FAO (2007) has already highlighted some measures for climate change adaptation but these need large amounts of investment for implementation in the developing countries.

Table 10.1 Average annual bilateral and multilateral agricultural and total aid

	1973–5	*1979–81*	*1991–3*	*2000–2*	*2003–5*	*2006–8*
Agriculture commitments:	In US$ billion (constant 2007 price)					
– Bilateral	3.4	6.7	5.4	3.0	4.0	3.4
– Multilateral	2.1	4.7	2.4	2.0	2.3	2.1
– Bilateral plus multilateral	5.5	11.4	7.8	5.1	6.3	5.5
Total aid to all sectors	42.5	50.5	69.7	92.9	104.8	42.5
Agriculture commitments:	In percentage (%)					
– Bilateral	7.9	13.2	7.8	5.4	3.3	3.8
– Multilateral	5.0	9.3	3.4	2.8	2.2	2.2
– Bilateral plus multilateral	12.9	22.5	11.2	8.1	5.4	6.0
Total aid to all sectors	100.0	100.0	100.0	100.0	100.0	100.0

Source: Based on OECD/DAC and OECD/CRS (various years) and Islam (2011)

While agriculture is the most sensitive and vulnerable sector to climate change, it is also one of the major contributors of greenhouse gas (GHG) emissions (IPCC 2007b; FAO 2008b). Projections indicate that these emissions will increase if agricultural development is continued under the 'business-as-usual' model. According to recent data by IPCC (ibid.), agriculture accounts for 13.5% of global GHGs or about 6.8 Gt of carbon dioxide equivalent (e) in 2004. The World Resources Institute (Herzog 2006) indicates that the energy sector's emissions attributable to the use of fossil fuels by agricultural and food processing account for another 2.4% of GHG emissions. Agriculture is also the largest producer of both methane and nitrous oxide, which together make up about 22% of global emissions (Baumert et al. 2005). Agricultural nitrous oxide emissions are projected to grow by 35–60% by 2030 owing to increases in both nitrogen fertilizer use and animal manure production (IPCC 2007b). As about 74% of total agricultural GHG emissions originate in the developing countries, their mitigation is also important for slowing down climate change.

Obviously, however, mitigation and adaptation need investment. With the recently rising awareness of the consequences of climate change, this is likely to find its way into international and national action climate change plans, but the design and implementation of effective mitigation and adaptation strategies in agriculture are still in its infancy. It is still not clear what measures are needed to ensure the implementation of such plans or action. A series of questions exists that need to be investigated:

- How can agricultural mitigation and adaptation plans be funded?
- What is the role of foreign aid?
- How can the funds be used most effectively?
- What measures work?
- What evidence exists for foreign aid effectiveness for agriculture?
- What projects could work? What types of foreign aid practices have the potential to work for agriculture?
- What measures are scalable? What types of foreign aid (projects) that have delivered on a small scale can be scaled up, and what needs to be done to deliver foreign aid on a bigger scale?
- What is transferable? What aid experiences can successfully be transferred from one region to another?

The overall goal of this chapter is to examine how finance, particularly foreign aid, can be used to achieve the joint objectives of development, mitigation, and adaptation in agriculture in the developing world. The analysis is based on existing literature and case studies. The chapter is organized as follows. The next section provides an overview of the financing needed for sustainable agriculture under climate change, with a specific focus on foreign aid. Sections 10.3–10.6 examine the role of foreign aid in financing the mitigation of and adaptation to climate change in agriculture by examining each of the following four questions: what works, what could work, what is scalable, and what is transferable. The last section concludes and discusses policy implications.

10.2 Financing and Aid to Agriculture under Climate Change

10.2.1 Overall Financing and Aid to Agriculture

Agriculture has been largely underfunded. A global assessment of agricultural development by the World Bank (2007) concludes that insufficient investment in agriculture has been one of the primary causes of falling agricultural productivity since the 1980s. The lack of incentive, largely owing to low agricultural prices and market failure, is apparent in both public and private sectors.

Over the past two decades, investment in agriculture through foreign aid has also experienced a falling trend until the recent global food crisis.

According to Organisation for Economic Co-operation and Development (OECD) statistics, while average annual foreign aid to agriculture, including bilateral and multilateral aid measured in constant 2007 prices, increased from US$5.5 billion in 1973–5 to US$11.4 billion in 1979–81, it decreased to US$7.8 billion in 1991–3 and US$5.5 billion in 2006–8 (Table 10.1). Measured in relative terms, the fall in agricultural aid was even larger. Table 10.1 shows that aid to agriculture accounted for 22.5% of total aid to all sectors in 1979–81, declining to 11.2% in 1991–3 and 5.4% in 2003–5. Despite a slight recovery when aid to agriculture increased during the global food crisis (2006–8), its share in total aid was still only 6% (Table 10.1). Financing agriculture is going to face much greater challenges in the future. To increase global production by 70% to feed the world's forecasted 9.1 billion people in 2050, it is estimated that net investments to agriculture must top US$83 billion per year. This is about 50% more than current levels (FAO 2009a).

10.2.2 Overall Climate Change Funds

In facing the challenges of climate change, international communities have initiated several global-wide funds raised through both bilateral and multilateral channels for 'climate finance' for the developing countries (OECD 2009). Under the United Nations Framework Convention on Climate Change (UNFCCC), three multilateral funds address climate-related needs and are managed by the Global Environment Facility (GEF). They are the Adaptation Fund, the Least Developed Countries Fund (LDCF), and the Special Climate Change Fund (SCCF) (Oxfam 2009; OECD 2009). The SCCF was created to fund projects in capacity building, adaptation, technology transfer, and climate change mitigation. The LDCF is designed to help the poorest countries cover the costs of preparing and implementing their national adaptation programmes of action (NAPAs). The newest fund, the Adaptation Fund, was established to finance adaptation projects and programmes in developing countries that are parties to the Kyoto Protocol in order to protect vulnerable communities from the impacts of climate change. A summary of the multilateral adaptation funding channels is provided in Table 10.2.

Moreover, climate change funding initiatives outside the UNFCCC (non-convention funds) are also rising. These funds are used either in general or specific areas (e.g. forests) to address both adaptation and miti-

Table 10.2 Multilateral adaptation funding channels

Funding stream	*Institution*	*Pledged US$ million*	*Received US$ million*	*Disbursed US$ million*
Least Developed Countries Fund (LDCF)	GEF	176.5	135.0	31.4
Special Climate Change Fund (SCCF)	GEF	121.0	100.5	46.9
GEF Trust Fund's Strategic Priority for Adaptation	GEF	N/A	50.0	50.0
Kyoto Protocol Adaptation Fund	Adaptation Fund Board	Increasing to 300 p.a.	18.5	
Pilot Programme for Climate Resilience (PPCR)	World Bank	546.0	95.8	N/A
Total		843.5	399.8	128.3

Source: Compiled from Oxfam (2009)

gation issues. The total of these funds, if the sums can be raised in full, is indeed impressive (Table 10.3).

Despite the recent funds that have emerged, they are still far from the amounts needed to effectively mitigate and adapt to climate change. As of 2010, the combined climate finance from the UNFCCC, multilateral and bilateral sources including the CDM, GEF Trust Fund, Adaptation Fund, World Bank Climate Investment Funders and others amounted to US$8 billion per annum (UN-HAGCCF 2010). In addition, by November 2011, approximately US$450 million had been pledged to the LDCF and US$250 million to the SCCF (Schalatek et al. 2011) but are nevertheless much less than the estimated additional investment needed to finance climate change (Table 10.4; World Bank 2009b).

10.2.3 Climate Change Funds for Agriculture

Climate finance offers an opportunity to strengthen food security and promote climate change mitigation and adaptation in the developing countries. However, current climate-related financial flows specifically targeted to agriculture in the developing nations cover only a tiny fraction of the total climate change funds (Climate Focus 2011). By 2008, the average

Table 10.3 Summary of existing climate change funding initiatives outside UNFCCC (non-convention funds)

Fund	*Pledged amount, US$*	*Administrator*	*Short description*
CIF Climate Investment Fund that encompasses SCF Strategic Climate Fund	6.3 billion	World Bank	Piloting new approaches or scaling-up activities in developing: (1) the SCF for increasing climate resiliency; reducing emissions from deforestation and forest degradation (under consideration); and scaling-up renewable energy. (2) the CIF for demonstrating and transferring low-carbon technologies.
FCPF Forest Carbon Partnership Facility	165 million	World Bank	Focusing on mitigation through Reducing Emissions from Deforestation and Forest Degradation (REDD).
CPF Carbon Partnership Facility	470 million	World Bank	Supporting developing countries towards lower carbon development paths. Components of the Investment Framework for Clean Energy and Development (CEIF).
CBFF Congo Basin Forest Fund	200 million	FFDB	Promoting biodiversity conservation, natural resource management and mitigation through REDD.
SPA Strategic Priority on Adaptation	50 million	GEF	A three-year pilot programme for adaptation planning.
UN-REDD Programme	35 million	UNDP	Focusing on mitigation through REDD.
Millennium Development Goals Achievement Fund: Environment and Climate Change window	90 million	UNDP	Focusing on adaptation and general mitigation.
Global Climate Change Alliance European Union (EU)-Global Climate Change Alliance	300 million (€220 m)	UN	Focusing on adaptation, general mitigation and REDD.

(*continued*)

Table 10.3 (continued)

Fund	*Pledged amount, US$*	*Administrator*	*Short description*
Cool Earth Initiative (Japan)	10 billion	Japan	Focusing on adaptation and mitigation activities.
Environmental Transformation Fund (UK)	1.2 billion (£800 m)	UK	Focusing on adaptation and mitigation with some components administered by World Bank and the African Development Bank (Congo Basin Forest Fund).
International Climate Initiative (Germany)	170 million (€120 m)	Germany	Focusing on adaptation and general mitigation.
International Forest Carbon Initiative (Australia)	180 million (AUD 200 m)	Australia	Focusing on mitigation through REDD.

Source: Adapted from a report prepared for the Financing for Development Conference on Climate Change, Kigali, 21–2 May 2009

Table 10.4 Estimated additional investment and financial flows needed for adaptation in 2030 (US$)

Sector	*Investment flow, US$/year*	*In developing countries, US$/year*	*Africa, US$/year*
Agriculture, forestry & fisheries	14 billion	7 billion (50%)	1000–2000 million
Water resources	11 billion	9 billion (80%)	2788–2913 million
Coastal zones	11 billion	5 billion (45%)	528–612 million (2030) 1197–1319 million (2080)
Human health	5 billion	5 billion (all)	2166–3328 million
Infrastructure	8–130 billion	2–41 billion	22–371 million
Total	49–171 billion	28–67 billion (57–39%)	7173–9931 million

Source: UNFCCC (2007)

total aid on agriculture was less than US$6 billion (Table 10.1), yet when considering the annual investment needs for agricultural adaptation are about US$7 billion (Nelson et al. 2010), climate finance is unlikely to meet most of the developing countries requirements for mitigation and adaptation. On the other hand, it is estimated that the potential increase in global investment flows to agriculture, forestry, and fishery sectors will reach US$14 billion per year by 2030, of which US$7 billion/year is assumed to go developing countries (SEI 2008). However, according to projections, mitigation costs in agriculture will reach about US$20 billion by 2030 (Table 10.5). The amount of investment flows on agriculture will be significantly less than the expected costs for agricultural mitigation and adaptation (Louis 2007).

10.2.4 Financing Measures to Battle Climate Change in Agriculture

Financing sustainable agriculture under climate change includes financing both the mitigation of and adaptation to climate change. The extent of reduction of agricultural GHG emissions depends on the potential and marginal cost of limiting emissions. Table 10.6 summarizes four major areas of potential support to mitigate the effects of climate change in agriculture. Based on a review of literature and various adaptation programmes and practices, we summarize four broad categories for agricultural adaptation to climate change in Table 10.7. In the following sections, we base our analysis on four questions. Which measures work? What could work? Which procedures are scalable? Which are transferable?

10.3 What Will Work?

Financing mitigation of and adaptation to climate change in agriculture through foreign aid is an international movement that has been in existence only for a short period, but a number of successful experiences have emerged. This section discusses some of the foreign aid projects in developing countries that have worked well in terms of areas financed and successful outcomes.

10.3.1 Major Areas of Involvement

In financing the mitigation of climate change in agriculture, foreign aid-supported measures have worked well in locations where GHG emissions are significant but relatively easy to reduce with appropriate technologies. Agricultural GHG are caused by nitrous oxide emissions from soils, methane from ruminants and paddy fields, carbon dioxide emissions from soil

Table 10.5 Estimate of the reductions of emissions from non-carbon dioxide and soil carbon GHGs ($MtCO_2e$) and the investment needed to achieve these reductions (US$ billion) between 2000 and 2030 at a cost of US$30 tCO_2e (2000$)

	2000		*2010*		*2020*		*2030*	
Subsector	Reductions	Cost	Reductions	Cost	Reductions	Cost	Reductions	Cost
Cropland	172	7.74	183	5.48	168	5.04	180	5.39
Rice	200	6.0	226	6.79	238	7.14	243	7.3
Livestock	131	3.93	143	4.28	158	4.73	175	5.26
Total	529	15.88	596	17.89	631	18.92	684	20.5

Source: Based on Verchot (2007)

Table 10.6 Potential areas for financing mitigation of climate change in agriculture in developing world through foreign aid

	Potential areas	Foreign aid			
		Works	*Could work*	*Scalable*	*Transferable*
1	Reducing nitrous oxide emissions from soils:				
	For example, by improving efficiency of fertilizer use through better technology extension service and training		***	***	**
2	Reducing methane from ruminants and paddy fields:				
	Reducing emissions from ruminants by reducing animal number in degraded grassland		*	*	*
	Reducing emissions from paddy fields through better farm management	**		***	**
3	Reducing carbon dioxide emission from soil:				
	Soil carbon sequestration through injection		**	*	*
	Soil carbon sequestration through land-use conversion or conservation	**		**	*
	Reducing carbon dioxide emissions by changing farming practices such as less or zero tillage, alternative fallow and tillage periods	**		**	**
4	Reducing carbon dioxide emissions through energy-saving technology:				
	For example, saving energy use through water-saving technology, less land preparation		**	**	**

Note: *, **, and *** indicate the level of probability (minimal, normal, or high, respectively) for 'what works', 'what could work', 'what is scalable', and 'what is transferable'

Source: Author's analysis

and from the energy used in agricultural production (IPCC 2007b; Table 10.6). The two major areas where foreign aid has worked well (Table 10.6: column 1) are reductions in paddy field methane emissions and in carbon dioxide soil emissions such as soil carbon sequestration through land-use conversions or from changes in farm practices (e.g. less or

Table 10.7 Potential areas for financing adaptation to climate change in agriculture in developing world through foreign aid

	Potential areas	*Foreign aid*			
		Works	*Could work*	*Scalable*	*Transferable*
1	Investment in water conservation infrastructure				
	Develop/improve irrigation infrastructure	***		***	***
	Water transfer or diversion projects within a country	*		*	*
	Land contouring, terracing, water storage, etc.	**		**	**
	Development of integrated drainage systems	**		***	***
2	Investment in agricultural science and technology				
	Investing in research for a better understanding of climate change impacts and vulnerability	**		***	***
	Developing new crop varieties, e.g. drought-resistant or flood-tolerant varieties	**		***	***
	Facilitating international technology transfer and local technology extension service	**		***	***
	Others (e.g. biotech, water saving technology, ecological and organic agriculture in some areas)	**		***	***
3	Investment in capacity-building programme				
	Capacity to develop/implement adaption plans by national and local government	**		***	***
	Community planning and management capacity		**	**	**
	Improving farmers' capacity through farmers' associations (e.g. water users associations and cooperatives) and training	**		***	**
4	Investment in risk management				
	Subsidized agricultural insurance		**	**	**

(*continued*)

Table 10.7 (continued)

Potential areas	*Foreign aid*			
	Works	*Could work*	*Scalable*	*Transferable*
Natural disaster release and food aid programme	***		***	*
Early warning and information systems to provide timely weather predictions and forecasts	**		***	***
Restore the natural capacity to buffer climate impacts	**		**	**

Notes and source: as given in Table 10.6

zero tillage, alternative periods of fallow and tillage). An example from the Philippines is presented later to show how foreign aid has worked to reduce paddy field methane emissions, and a grassland carbon sequestration project for controlling carbon dioxide emissions is presented in Sect. 10.6.

With regard to agricultural adaptation to climate change, foreign aid has been present to finance nearly every area listed in Table 10.7 (column 1). These include investments: (1) in water conservation infrastructure; (2) in agricultural science and technology, (3) in capacity-building programmes, and (4) in risk management. The wide coverage of the finances for climate change adaptation in agriculture may be because the aim of many of these measures is to further enhance agricultural and rural development. Irrigation is a priority area that has received considerable attention from foreign financing. This is not surprising, given that irrigation infrastructural development has been targeted by many international development and financing agents. In the next subsection, to illustrate the importance of mainstreaming climate change adaptation into agricultural development, we examine the role of foreign aid in China with respect to irrigation and its successful outcome.

10.3.2 Successful Experiences

In this subsection, we examine several cases where foreign aid has had a prominent role in the successful financing of sustainable agriculture under climate change.

Box 10.1: Mitigating Methane Emissions Through New Irrigation Schemes in Rice Production in the Philippines

Methane emission is a major component of overall GHG emissions and has been rising over time (Oberthür and Ott 1999; Tyler et al. 1999). Paddy fields are a primary source of methane emissions and are also one of the few anthropogenic sources where management of methane is possible (Wassmann et al. 2009). A more integrated approach to rice paddy irrigation and fertilizer application can substantially reduce methane emissions but it requires modifications to farm management, such as changes in the mid-season drainage of rice paddies and intermittent irrigation.

This box summarizes measures to mitigate methane emissions in rice production in Bohol Island (the Philippines) based on the report by Wassmann et al. (2009) and FAO (2010b). Bohol Island, one of the country's biggest rice-growing areas, in the Visayas region, has experienced declining productivity because of defective existing irrigation systems. Before completion of the Bohol Integrated Irrigation System (BIIS) in 2007, two older reservoirs (Malinao and Capayas Dam) were beset by problems, unable to ensure sufficient water for the second crop (November to April), especially for farmers farthest downstream from the dam. The problem was aggravated by unfair water distribution practices and water wastage through continuous flooding to irrigate rice crops.

In the face of declining rice production and ineffective water management, the National Irrigation Administration (NIA) created the BIIS action plan in 2007, with the overall goal to improve the efficiency of water management, which would also achieve simultaneous benefits of mitigating methane emissions, and increasing rice productivity in Bohol. The project included construction of a new dam (Bayongan Dam; funded by a loan from the Japan Bank for International Cooperation) and implementation of a water-saving technology known as Alternate Wetting and Drying (AWD) which was developed by the International Rice Research Institute (IRRI) in cooperation with national research institutes. The visible success achieved with AWD in pilot farms as well as specific training programmes for farmers dispelled the widely held perception of possible yield losses from non-flooded rice fields.

Wide adoption of AWD improved irrigation water usage, so that crop intensity could be increased from *c.* 119% to *c.* 160% (compared to the 200% maximum of double-cropping systems). Moreover, based on the revised Intergovernmental Panel on Climate Change (IPCC) methodology (IPCC 2006), modifications to the water regime have the potential to reduce methane emissions by 48% over the traditional method of continuous flooding of rice fields.

The AWD project therefore generated multiple benefits with regard to methane emission reductions (mitigation), decreased water use (adaptation where water is scarce), and increased productivity, thereby contributing to food security (author, based on Bouman et al. 2007).

A Project for Reducing Methane Emissions in the Philippines
Methane emission constitutes an important component of global GHG emissions. The general consensus is that the potential to reduce methane emissions at its major source—rice fields—is high (FAO 2010b), but the problem is how to incorporate a methane emission reduction objective into farming practices while also maintaining or even improving agricultural productivity.

Box 10.1 describes one such case of successful foreign aid intervention in the Philippines where emission reductions were achieved through new irrigation schemes. Investment also benefitted the farmers who participated in the programme in Bohol Island, one of the biggest rice-growing areas of the Philippines. As this case indicates, a programme aimed at reducing GHG emissions can be successful if it is incorporated into the agricultural development agenda, provides incentives for farmers to participate, and attracts the interest of major stakeholders.

A Project for Financing Measures for Climate Change Adaptation in Agriculture
Faced with the reality of global warming, adaptation to climate change through appropriate measures and investment is essential. As Table 10.7 shows, four major categories of investment in the agricultural sector for adaptation to climate change could produce successful outcomes. Here, we introduce a foreign aid scheme in China that underlines the importance of mainstreaming climate change adaptation into a national development programme (Box 10.2).

Box 10.2: Mainstreaming Climate Change Adaptation in Irrigated Agriculture in 3H Basin

The Huang-Huai-Hai River Basin (3H Basin), a region with a population of 425 million is beset with challenging climatic risks. Primarily an agricultural region, producing about 50% of China's grain, it is heavily dependent on irrigation water. But the region's per capita water availability is only one-third of the country's average, and available resources are already fully allocated and often overexploited, making the region highly vulnerable to climate change. Higher temperatures and higher crop evapotranspiration further aggravate the problem.

To ease the water shortage, the World Bank, supported by the SCCF of the Global Environment Facility, implemented a project in 2008–10 on mainstreaming climate change adaptation measures in irrigated agriculture. The project consisted of three phases, each with a specific target. The first phase identified and prioritized different adaptation measures; the second phase constituted demonstration and implementation of the measures; while the third component was to mainstream adaptation into the national comprehensive agricultural development (CAD) programme and institutional strengthening.

The activities included a series of measures to promote capacity building, technical assistance, knowledge sharing, public awareness, and the preparation of a national climate change adaptation plan for CAD. The procedure for integrating and mainstreaming climate change adaptations into the national plan also engaged officials from the National Development and Reform Commission, the Ministry of Finance, and provincial government, and scholars from the Chinese Academy of Sciences and the Chinese Academy of Agricultural Sciences.

Through the efforts of SCCF and Third Irrigated Agriculture Intensification Project (IAIL3) projects, communities are currently better informed about climate threats but, more importantly, their ability to sustain and perhaps even improve that knowledge and use it to guide future coping choices has increased. Equipped with a toolkit of immediate instruments, the communities are better prepared to protect their livelihoods, and to expand the toolkit in accordance with changing climatic circumstances and increased knowledge. This represents the beginning of an adaptive capacity that rural communities across the developing world will need to safeguard their livelihoods against the effects of global warming.

The project created the first line of defence in five provinces across the 3H Basin by exploring and demonstrating how the achievements of IAIL3 and other CAD initiatives can be used to safeguard against climate change. More detailed information on the project is given in World Bank (2012) and Conrad and Li (2012).

Source: Author, based on reports by World Bank (2012) and Conrad and Li (2012).

The Mainstreaming Climate Change Adaptation in Irrigated Agriculture is a project supported by the GEF-managed SCCF and focused on the Huang-Huai-Hai River Basin (3H Basin) in the northern plains of China. All project objectives were fulfilled: to introduce, demonstrate and implement specific adaptation measures in selected demonstration areas, adjust and integrate appropriate adaptation measures into the implementation of the Third Irrigated Agriculture Intensification Project (IAIL3), and to reduce vulnerability to climate change in the 3H Basin (Conrad and Li 2012). The project was successful in increasing local ability to react to changing circumstances. For example, more than 1000 water users associations, 209 farmer associations, and 20 specialized farmer cooperatives were established under the overall IAIL3 project. According to interviews with national officials, the project also generated a general framework and approach for the Office of the National Comprehensive Agricultural Development (CAD), the Ministry of Finance on integrating and mainstreaming climate change adaptation into the national CAD programme.

A Project to Invest in Agricultural Technology

Box 10.3 illustrates the successful results of an investment in the research for drought-tolerant maize for Africa. This case shows that investment in research to develop relevant technology is a priority area for financing agriculture under climate change.

Within a short period, this project has demonstrated high investment returns in science and technology by the international public research organizations and the importance of agricultural technology in mitigating the impacts of climate change in developing countries. Maize productivity is increasing, and the adoption rate of drought-tolerant maize varieties can

Box 10.3: The DTMA Project for Drought-Tolerant Maize
A typical programme of foreign aid research and development, the Drought Tolerant Maize for Africa (DTMA) project, was launched in 2006 and was jointly funded by the Bill and Melinda Gates Foundation, the Howard G. Buffett Foundation, USAID, and the UK Department for International Development. Coordinated by the International Maize and Wheat Improvement Center and the International Institute for Tropical Agriculture, the current ten-year phase of DTMA covers the period 2006–16 and focuses on 'expanded use by farmers of certified, drought-tolerant maize seed, and should enable delivery of enough seed to benefit 30–40 million people in sub-Saharan Africa and provide added grain worth US$160–200 million each year in drought-affected areas' (DTMA 2012).

Recent studies suggest that the return to investment is impressively high. By 2012 farmers in the 15 participating countries already had access to 34 drought-tolerant seed varieties and hybrids (DTMA 2012). Yields of drought-tolerant maize over normal varieties, depending on the seriousness of actual drought conditions, have improved by 3–34%, which has significantly increased farmer income, household food security, and local food supply. An impact assessment reported by La Rovere et al. (2010) shows that 'at the most likely rates of adoption, based on several recent studies and expert advice, drought tolerant maize can generate US$0.53 billion from increased maize grain harvests and reduced risk over the study period, assuming conservative yield improvements'. The report also estimated the likely impacts of the project under a more optimistic yield-gain scenario, and concluded that the economic benefit could reach as high as US$0.88 billion in the 15 African countries covered in this project.

Sources: Author based on La Rovere et al. (2010).

be expected to be high. Should wide adoption of drought-tolerant varieties materialize, it is estimated that over 4 million people will 'escape poverty and many millions more will be able to improve their livelihoods' (La Rovere et al. 2010). Moreover, the impacts of the project are expected to continue after the conclusion of the first phase in 2016, as non-participating countries can also benefit from technology spillovers.

Other Schemes for Investing in Agricultural Technology
Other major foreign aid investment areas in agricultural technology include biotechnology, water saving technology, and technologies supporting ecological agriculture (Table 10.7). Recent investment in biotechnology by the Bill and Melinda Gates Foundation in Consultative Group on International Agricultural Research (CGIAR), Africa and South Asia for improving food security and poverty reduction in the less developed countries has been impressive. Kostandini et al. (2009) document the ex ante impact of transgenic research for mitigating drought in rain-reliant production areas for maize, rice, and wheat in Asia and Africa.[1] Their results show that the biotech drought-tolerant crops are 'very promising for the millions of poor in the more marginal rain-fed agricultural areas of developing countries'. Water saving technology is also an area that has often attracted foreign aid from the World Bank and several regional development banks worldwide, particularly in Africa and Asia (World Bank 2009a, 2010; Howden and Meinke 2003). In Senegal, in responding to increasing desertification from climate change, the International Fund for Agricultural Development (IFAD) supported a successful project on drip irrigation (World Bank 2010). Some programmes are also aimed at developing ecological farming and organic agriculture (EU focus 2010; Tirado and Cotter 2010). For example, the EU-supported small projects facility has helped Philippines farmers adopt organic agriculture, thereby increasing their export potential for to European markets (EU focus 2010). In Burkina Faso, the IFAD-supported sustainable rural development programme is encouraging the adoption of more environmentally friendly technologies, such as soil and water conservation techniques and agroforestry (IFAD 2010).

10.4 What Measures Could Work

In this section, we present an examination of foreign aid-supported measures which could work but where foreign aid is underrepresented, although it has the potential to produce results (column 2 of Tables 10.6 and 10.7). We also highlight a few examples to show how foreign aid could work in these areas.

10.4.1 *Major Areas*

Foreign aid can do more to mitigate the effects of climate change on the agricultural sector. As was mentioned earlier, primary areas where foreign financing has been used for this specific purpose include the reduction of methane emissions from paddy fields, increased carbon sequestration through land use changes, and limiting carbon dioxide emissions from the soil. But there are other areas that are also potential targets for foreign finance in developing countries. These concern the control of nitrous oxide emissions from crop production or limiting carbon dioxide emissions through energy-saving technology.

In financing agricultural adaptation to climate change, most areas identified in Table 10.7 have often involved foreign aid projects that have produced good results (column 1). Two additional areas should also be considered as possible target areas where foreign aid could work or work better: investment in community planning and management capacity as well as subsidized agricultural insurance. Capacity building covers a wide range of activities, as the lack of capacity is a compelling problem in developing countries in their fight against climate change; this includes improving community capacity to adapt to global warming. So far, little experience has been gained from this type of financing because community capacity building is a complicated task. There are a lot of open questions with respect to the impacts of climate change at the local level, and large diversity exists among local communities. The performance of a foreign aid project in this field could be improved with more information and understanding of the consequences of climate change at the local level and actual needs of specific communities in their adaptation approach. With respect to subsidized agricultural insurance, creating an enabling environment for foreign investment in agricultural insurance and closely working with the local government are critical for the success of foreign aid project, as is indicated by the case study presented later in this section.

10.4.2 Examples of Potential Areas for Foreign Aid Investment

In this subsection, we review three schemes to illustrate possible areas of foreign aid financing that could achieve sustainable agriculture under climate change.

Project to Control Nitrous Oxide Emissions from Soils: China

Nitrous oxide emissions can be effectively reduced through the increasingly efficient use of nitrogen, thus limiting its application. While this major transition in nitrogen use is common in many developed countries, it has not occurred in the developing world. For example, the overuse of synthetic nitrogen fertilizers was common in the UK in the 1970s and early 1980s, with significant nitrous oxide emissions and other serious environmental consequences. Since then regulatory changes and investment have brought about improvements in nutrient management and agricultural technology that have allowed the increase in application rates to stop or decline slightly, whilst crop yields have risen (SAIN 2010). In the developing countries a typical problem is that farmers are credit constrained, which limits sufficient use of synthetic nitrogen fertilizer. But nitrogen fertilizer overuse is also common in many emerging countries, such as China (Huang et al. 2008).

China has considerable potential for reducing nitrous oxide emissions in crop production. The manufacture and use of synthetic nitrogen fertilizer is estimated to account for about 10% of the fossil energy used by the industrial sector, contributing to nearly 5% of China's total GHG emissions (SAIN 2010). While chemical fertilizers play an important role in increasing agricultural production and ensuring food security, farmers in China use at least 30% more per hectare than farmers in many other countries (Huang et al. 2012; SAIN 2010). If appropriate technology for improved nitrogen fertilizer use could be adopted, the resulting decrease of overuse could reduce China's total GHG emissions by more than 1% and nitrous oxide emissions by 30% or more (SAIN 2010).

While little foreign aid has been aimed at measures to reduce the overuse of nitrogen fertilizer in developing countries, several pilot experiments in China funded by both the Chinese government and international donors show that this could be an area where more financing, foreign aid included, is needed. For example, a series of training programmes shows that delivering information and knowledge on the efficiency of nitrogen fertilizer can significantly lower its use in grain production by 15–30% without adverse effects on crop yields (Hu et al. 2007; Huang et al. 2008, 2012; Peng et al. 2010). Improved nitrogen fertilizer management is a clear win–win situation with economic and environmental benefits. However, to reduce nitrous oxide emissions through more efficient use of fertilizer implies better capacity building and training programmes for the farmers. Consequently, substantial investment, including foreign aid, in agricultural extension is needed to educate hundreds of millions of small farmers in the developing countries. A brief discussion on this topic is provided in Box 10.4.

Box 10.4: Reducing Nitrogen Fertilizer Use in China Through Training Programmes

Low carbon agriculture can make a significant contribution to the overall reduction of GHG emissions. There is indisputable evidence from recent studies that the overuse of nitrogen fertilizer in China is serious and that application rates could be cut 20 to 30% (Hu et al. 2007; Huang et al. 2008, 2012; Peng et al. 2010) or even more in grain production with no loss in crop yields or national food security (Zhang et al. 2008; Zhao et al. 2010). Such a decrease in overuse could reduce China's total GHG emissions by 1–2% and nitrous oxide emissions by 30% or more (SAIN 2010).

Two pilot experiments (rice and maize) aimed at cutting nitrogen fertilizer use without affecting crop yields were conducted China. The rice pilot experiment was conducted by China Center for Agricultural Policy (CCAP) and International Rice Research Institute (IRRI) and funded by International Development Research Center (IDRC) in 2003–5. The maize pilot project was also conducted by CCAP with its collaborators from China Agricultural University in 2009, jointly funded by the Sino-German Research Project and the China-UK Sustainable Agriculture Innovation Network (SAIN).

Rice production: The technology being transferred to farmers to reduce nitrogen fertilizer use is the site-specific nutrient management (SSNM) programme developed by IRRI. Experiments were implemented in six rice-growing villages in four provinces (Guangdong, Hunan, Hubei, and Jiangsu). A half-day training course by the local extension agent outlined the details of efficient fertilizer application to the farmers. After training, some of the farmers were randomly selected for field trials (Hu et al. 2007; Huang et al. 2008). Farmers who had attended the training session reduced nitrogen fertilizer use by 18% compared to the control group of non-participating farmers, while the field trial participants decreased their usage of nitrogen fertilizer up to 35% with no difference in yields. The study also indicated that the scheme's advantages needed first to be convincingly conveyed to the extension agents and that intensive training should be provided. Although 'getting the message right' does help, intensive efforts to promote technology were needed in order to secure maximum benefits from the increased efficiency of nitrogen fertilizer use.

Maize production: Experiments were implemented in two counties in Shandong province in 2009. A training course of one to two hours was offered to farmers on nitrogen fertilizer use in maize production by trained extension staff. The study results show that the training was instrumental in reducing overall nitrogen fertilizer use by 22%, but it is also pointed out that training China's 200 million smallholders is a challenge, and despite significant reductions in nitrogen fertilizer usage by trained farmers, its use still exceeded recommended levels. Whether China's current agricultural extension system can deliver appropriate information and knowledge on the efficiency of nitrogen fertilizer to millions of farmers is an issue that requires further study, because the current agricultural extension system also faces great difficulties in providing technology services to farmers.

Source: Author's analysis.

Investments for Reducing Carbon Dioxide Emissions from Direct Energy Use in Farm Operations

Although the role of direct investment through foreign aid in farm operations is limited, foreign aid could assist developing countries in generating energy-saving technologies in agricultural production. But as the technology has to be adopted by the farmers, cost effective technologies are the prerequisite of a successful project.

Carbon dioxide emissions can be reduced by saving energy in farm operations (e.g. mechanization, land preparation, and irrigation). Energy-saving machinery and limiting the use of machinery in land preparation by changing farm practices (e.g. zero tillage) have often been discussed in the literature. Water pumps for irrigation also consume vast amounts of energy (Lal 2004; Mushtaq et al. 2009), but this source of GHG emissions has been largely neglected to date. Yet a recent empirical study from China shows that emissions from groundwater irrigation pumps totalled 33.1 $MtCO_2e$ in the late 2000s, which was about 0.5% of the country's total emissions (Wang et al. 2012). Direct savings resulting from the controlled use of such energy sources as gasoline, diesel, and the electricity used in farm operations could be achieved through investment in energy-saving technologies in land preparation, irrigation, harvesting, storage, and transportation.

Investment in Risk Management: Subsidized Agricultural Insurance

Agricultural insurance is an area that has not attracted significant investment from foreign aid but which could produce benefits. Currently, crop and livestock insurance programmes are government subsidized and are implemented mainly in developed countries (OECD 2011; Smithers 1998). In the developing countries farmers are normally more vulnerable to natural disasters, but often receive little subsidized agricultural insurance from the government because of financial constraints. Market-based private agricultural insurance is rare in the developing world because small-scale farmers lack the resources to pay insurance premiums, and private insurance companies are not interested in operating costly schemes for millions of small farmers. Thus, financial mechanisms and public policy should be deployed strategically to leverage foreign aid and private capital, and to exploit opportunities to create enabling conditions for investment in agricultural insurance.

Box 10.5 presents one innovative insurance mechanism implemented in Ethiopia that shows how foreign aid can work to promote agricultural insurance in the developing countries.

Box 10.5: Insurance Mechanism in Ethiopia: The HARITA Model

The Horn of Africa Risk Transfer for Adaptation (HARITA) is an innovative climate change resilience project launched jointly by several international donors, non-governmental organizations (NGOs), and one insurance company. Between November 2007 and December 2009, a pilot climate risk management package was designed for poor farmers in the village of Adi Ha; this consisted of a mix of risk reduction, drought insurance, and credit. The approach consisted of three main components:

- Risk reduction and minimizing vulnerability. Farmers participating in HARITA learned how to use compost, which is important for rebuilding soil nutrients and improving soil moisture retention. They also built small-scale water harvesting structures and planted trees and grasses to promote soil and water conservation.

- Risk transfer and weather index insurance. Introduced micro-insurance to strengthen Ethiopia's productive safety net programme by addressing the non-chronic, 'unpredictable' needs not covered under the programme.
- Prudent risk-taking and credit. Supported poor producers in making optimal production decisions even in the face of uncertainty for livelihood diversification, technology adoption, and entry into more profitable lines of business.

HARITA was innovative in the sense that it allowed very vulnerable farmers to pay their premiums and benefit through risk reduction measures.

Source: Oxfam America (2009).

10.5 What Can Be Scaled Up?

Currently, most of the foreign aid projects that are known to work (see columns 1 and 2 of Tables 10.6 and 10.7) are normally implemented on a small scale. These could be scaled up to bring additional benefits, but greater efforts may be needed to deliver foreign aid on a bigger scale. In this section, we introduce several examples of projects for climate change mitigation or adaptation, and review the results of the scaling-up experiences in these cases.

10.5.1 Expanding Pilot Projects for Reducing Carbon Dioxide Emissions Through Soil Carbon Sequestration

The Kyoto Protocol recognizes that it is possible to reduce net emissions either by decreasing the rate at which GHGs are emitted to the atmosphere or by increasing the rate at which GHGs are removed from the atmosphere through sinks. Agricultural soils are among the planet's largest reservoirs of carbon and hold the potential for expanded carbon sequestration, thus providing the possibility of mitigating the growing atmospheric concentration of GHGs (FAO 2001). It is estimated that soils can sequester around 20 Pg C in 25 years, more than 10% of the anthropogenic emissions.

While cost-effective technologies for soil carbon sequestration are still to be developed, there are a number of efforts and pilot projects in effect. Some of the successful pilot projects could be scaled up in the future. For example, an international network was created in 2000—the DMC (Direct sowing, Mulch based systems, and Conservation tillage)—which already includes 60 international and national institutions. The German government established a partnership with the African tillage network. French Agricultural Research Center for International Development joined this network, and with French cooperation funding set up a plan of action in Brazil, Madagascar, Mali, Laos, and Tunisia, where different agricultural practices are tested and assessed with measurements of stocks and fluxes of carbon dioxide and nitrous oxide emissions at benchmark sites.

Box 10.6: The Three Rivers Grassland Carbon Sequestration Project in Qinghai, China

The potential of grasslands to sequester carbon is being increasingly recognized. With low levels of plant biomass compared with forests or shrub land ecosystems, grasslands form a major terrestrial carbon stock that can be increased though appropriate management (UECC 2010). Nevertheless, grassland soil carbon sequestration is significant, and its effects with respect to climate mitigation are measurable and verifiable.

The pilot stage of the Three Rivers Grassland Carbon Sequestration project was launched in 2009 in Qinghai (China) where overgrazing was a serious problem. Utilizing carbon financing, the pilot project was aimed at increasing carbon stocks through restoration of degraded grasslands and enhancing livestock productivity. The project introduced better grassland management practices such as improving summer–winter pasture rotation of grazing, limiting the timing and number of animals on degraded pastures, and restoring severely degraded lands by replanting perennial grasses and ensuring appropriate long-term management. Replacing the low-input, low-output, degradation-inducing livestock system with a high-productivity, sustainable land management system can contribute to carbon sequestration.

Herders were offered a menu of options designed to fit their specific land use. These included a combination of grassland restoration zoning and stocking rate management within an incentive-based system. Given the 45% rate of overstocking prior to project implementation, considerable reductions in animal numbers, and therefore in incomes, could be expected during the first years of implementation, for which herders were to be compensated. In subsequent years, as incomes grow in response to increased livestock productivity—and possibly from small additional business support measures—compensation decreases progressively until year ten, when it will cease altogether.

Overall, during the project's first ten years, households will have fewer but more productive livestock, after which herds can be increased beyond the initial ten-year level without risk of overgrazing. Increased availability of forage will ensure higher incomes and higher levels of production over the long run, providing a financial incentive for long-term sustainable management. In addition, the project envisions developing a number of activities aimed at enhancing the profitability of livestock rearing for improved herder livelihoods. In addition to improvements in animal production (e.g. feeding, winter housing, and breeding), the project includes the development of processing activities and marketing associations.

This project hopes to break the vicious cycle of overstocking and degradation, thus demonstrating sustainable management options while generating a reduction of approximately 500,000 tCO_2e, over a ten-year period. It also aims to address some of the key barriers to smallholder access to carbon finance, which include the lack of appropriate methodologies for accessing credit, and cost-effective monitoring, reporting, and verification.

Source: Author, based on FAO (2010b).

Box 10.6 provides a pilot case which demonstrates that a carbon sequestration project can be successfully scaled up if the mitigation objectives of foreign aid are formulated in partnership with short-run compensation and the need to increase long-run agricultural productivity. The Qinghai (China) carbon sequestration project is aimed at promoting livestock productivity, while at the same time increasing carbon stocks through the restoration of degraded grasslands. This is an interesting case because there are few operational examples of carbon finance projects that are targeted at grasslands anywhere in the developing world. The project introduces improved grassland management practices, while concurrently providing compensation to herders during the initial years of project implementation. As the livestock system changes from degradation-inducing methods to sustainable land management systems, it can contribute to carbon sequestration. Based on the experience gained from the project area, the government of China and international donors are planning to scale up this pilot project, which could also be transferred to other regions of the country and the rest of world.

10.5.2 Scaling Up Existing Conservative Management Practices Aimed at Sequestrating and Reducing Carbon Dioxide Emissions

Many conservative management practices aimed at sequestrating and reducing carbon dioxide emissions from the soil can be easily scaled up. For example, alternative fallow and tillage practices can address climate change-related moisture and nutrient deficiencies. These measures have been significantly scaled up and are now widely used in Missouri, Iowa, Nebraska, and Kansas (Easterling et al. 1993). Other examples of successful scaling-up experiments include crop rotation with legumes or grass-clover leys, application of organic fertilizers, and less or zero tillage practices in many developing countries. In recent years, the World Bank has been strongly involved in the diffusion and extension programmes that focus on direct sowing and associated practices in developing countries, particularly in Brazil (FAO 2009b). Conservation tillage in crop production has also expanded rapidly in many parts of China (Wang et al. 2010).

10.5.3 Scaling Up Climate Change Adaptation with Regard to Agricultural Irrigation

As climate change adaptation in water conservation is strongly related to agricultural productivity growth and sustainable agriculture, foreign aid can play an important role in scaling up existing successful pilot projects. Here we examine the mainstreaming of climate change adaptation within agricultural irrigation in the 3H Basin discussed earlier. But it is worth noting that there are many schemes related to irrigation and drainage infrastructure related to climate change (World Bank 2010; FAO 2010b) which could be scaled up with foreign aid.

Recently, a decision was made by the government of China and the World Bank to scale up a project on mainstreaming climate change adaptation into irrigated agriculture in China because of its successful pilot stage. The World Bank approved a loan of US$80 million (Water Conservation

Project II) to China for 2012–17 to help improve water management and to increase water productivity within agriculture, and to boost the incomes of 1.3 million farmers in the Ningxia, Hebei, and Shanxi provinces. In addition to enhancing adaptation to climate change in agriculture and irrigation water management practices in 3H Basin and northwest China, the project includes activities in awareness and capacity building, and adaptation measures. Within the 3H Basin, in Anhui alone, provincial level investment projects on climate change adaptation activities will be increased from 16 to 93 counties, bringing the number of farmers who gain to benefit from 1 million to 31 million. The SCCF project has also made great efforts to reduce uncertainty through a comprehensive analysis carried out by national and international scientists and supported by a World Bank Analytical and Advisory Activity during the pilot and scaling-up stages.

The discussions above highlight several points that need to be recognized in efforts to scale up climate change financing. These include, but are not limited to, such factors as (1) investment should be targeted to a significant specific issue which either mitigates climate change in agriculture or improves agricultural adaptation to climate change; (2) investment should be relevant to agricultural development and incorporated within regular development programmes; (3) that problems are similar in both the pilot area and the scaled-up area in terms of climate change and agricultural development issues; and (4) scaled-up projects must take into the interests of all major stakeholders, particularly farmers.

10.6 What Experiences and Concepts Are Transferrable?

This section presents evidence from four experiments to show that successful intervention through foreign aid in one country can be transferred to others. These were summarized in the last column in Tables 10.6 and 10.7.

10.6.1 Foreign Aid to Reduce Methane Emissions Through the Provision of Appropriate Technologies and Integration with Development Goals

Measures to reduce methane from rice production, as exemplified by the investments made in Bohol Island in the Philippines (Sect. 10.3), are potentially transferable to other Asian countries. Rice is fundamental for food security, involving approximately 3 billion people and approximately 144 million ha of land under cultivation each year (IRRI 2010). The waterlogged, warm soils of rice paddies make this system a large producer of methane (Corton et al. 2000). According to recent studies, methane emissions in 2000 reached 625 million metric tons (mt) of CO_2e (Wassmann et al. 2009).

The Bohol project shows that the design of appropriate technologies and incorporation of the emission reduction objective into a development programme are key elements that can be transferred to other regions or countries. However, as technologies need to adapted to local conditions, participation of local farmers, extension agents, and research institutions in technology design and dissemination is critical. The above observations are also supported by an FAO report (2010b), which emphasizes that with appropriate irrigation and other farm management practices, paddy field methane emissions could be significantly reduced. Moreover, if foreign aid is to succeed in this respect, methane reduction must be fully integrated into local development goals so that farmers become interested in participating and gaining from the programme.

10.6.2 Foreign Aid Assisting Agricultural Adaptation by Mainstreaming Climate Change Adaptation into National Development Programmes and Emphasizing Local Capacity Building

Investment in irrigation and other water conservation infrastructure is one of the primary instruments for improving agricultural productivity and a priority area for financing its adaptation to climate change. If measures for climate change adaptation are to be successfully integrated into existing national development programmes, it is important that the demand and interest originates with the beneficiary country, and that close cooperation with local government is maintained. This was clearly highlighted by the 3H Basin project (Box 10.2). In the case of China, the interest in mainstreaming climate change adaptation into the nation's agricultural development programme originated with the Chinese state office of CAD, who considered this a new development concept that should be explored through a pilot project and then extended to other major projects. But for such local involvement, capacity building and technical assistance are needed, and here foreign aid can play a unique role by bringing together both international and local experts to help design adaptation measures and improve local capacity.

Investments in Agricultural Technology

Technology will have to be the primary driver for agricultural growth in the future, as it is anticipated that in the coming decades the world will have fewer natural resources to produce much more food. According to an FAO report (2009b), 80% of the production increases are projected to come from better yields and greater cropping intensity in the first half of the twenty-first century, while in land-scarce developing countries, almost the entire production growth must be achieved through improved yields. Global food security, particularly in developing countries, is expected to face an even greater challenge under climate change. Furthermore, the average growth rate of agricultural productivity has been falling (World Bank 2007).

Agricultural productivity in the developing world can be improved through technology that has been developed locally or transferred from abroad. One successful case of improving research capacity and facilitating technology transfer in developing countries is the programme to develop drought-resistant grain varieties in Asia. Recognizing the special difficulties of generating varieties for the poor and of distributing technology to these people in unfavourable environments, the Rockefeller Foundation initiated in 1998 a multiyear, multicountry programme to support research and technology transfers of drought-tolerant rice in Asia. The Foundation supported research by China, India, and Thailand as well as by the International Rice Research Institute, and to promote technology transfer it helped to provide training and networks for scientists, capital for improved screening facilities at experiment stations, and invested in the diffusion of drought-tolerant rice. The programme generated significant improvement in both the research and technological capacity of the countries involved, and experiences from the project have been transferred to countries engaged in the network. Pray et al. (2011) show that the programme generated drought-tolerant varieties which have already been adapted by farmers in the target countries. New varieties of drought-tolerant rice are being tested in other Asian countries. If these varieties could be widely adopted by Asian producers, they could help to mitigate the risks to farmers of climate change, particularly during extreme drought conditions.

Examples of International Collaboration in Agriculture: Enhancing South-South Cooperative and Learning Experiences Among Developing Countries
The patterns of international collaboration in agricultural development have been changing. Traditionally, most agricultural development programmes, including technology transfers and capacity building, have followed a north–south cooperative framework. Agricultural technological transfers have also been arranged by international organizations such as FAO, the World Bank, and the CGIAR. While these channels of investment are important and should be enhanced in the future, the recent experiences of south–south cooperative programmes in technology transfer are encouraging.

In recent years, emerging countries such as Brazil, China, and India have strived to develop agricultural collaboration with other countries in Africa and Asia. One of these south–south cooperative programmes is the China–Africa agricultural technology scheme. Under this endeavour, China has established 14 agro-technical demonstration centres (ADCs); six other ADCs are being constructed in Africa. More than 100 senior agricultural experts were dispatched in 2002 to 33 African countries (including, among others, Morocco, Sierra Leone, and Namibia). Moreover, more than 4200 agricultural officers and experts from Africa have been trained in China over the period 2004–11. While the impact of the China–Africa agricultural technology programme is yet to be evaluated, south–south cooperation does provide a new avenue of foreign aid for improving food security in the developing world and mitigating the adverse impacts of climate change.

To facilitate this cooperation, FAO launched the South–South Cooperation (SSC) Initiative in 1996. This has played an important role in the transfer of technology and development experiences within the developing world through numerous SSC agreements, the implementation of which have supported country- and regional-level action to increase food production, reduce poverty, and improve local capacity to manage climate-related disasters. By 2012 FAO had facilitated the dispatch of more than 1500 SSC experts and technicians to demonstrate how hunger and malnutrition can be alleviated and productivity improved through the adoption of new technologies, and how to reduce year-on-year production variability due to extreme weather events.

10.7 Conclusions

While agriculture is one of the major contributors to GHG emissions, it is also the most sensitive and vulnerable sector to climate change. Agricultural development is going to face great challenges in meeting global food security and can be expected to face even greater difficulties owing to climate change. Mitigating and adapting to climate change in order to achieve sustainable agriculture needs substantial investment.

This chapter has examined how finance, with a particular focus on foreign aid, can be used to fund climate change mitigation and adaptation in agriculture in the developing world. The results have showed that agriculture is greatly underfunded and that foreign aid has not increased adequately for maintaining sustainable agriculture. Although climate change funds have recently been emerging, more funding will have to be raised. While additional climate change funds are important, their effective utilization is equally important. Recently, funding agencies and donors have tried to explore innovative approaches to the challenges posed by climate change in agriculture in developing nations.

The review of literature and case analyses has showed that there is a wide range of areas where climate change mitigation and adaptation need support from foreign aid. We identified four general categories of mitigation measures that are considered to be potential areas for financing agricultural mitigation with foreign aid. They include the reduction of nitrous oxide emissions from soils (e.g. through greater efficiency in fertilizer uses with better technology extension services and training), limiting methane from ruminants and paddy fields, soil carbon sequestration through land-use conversion or conservation, scaling down carbon dioxide emissions through modifications in farming practices (e.g. zero tillage), and through energy-saving technology.

Proposed investments through foreign aid for agricultural adaptation to climate change also cover four major categories. They include investments:

- in water conservation infrastructure (e.g. irrigation, water transfers, land terracing, water storage, and integrated drainage systems);
- in agricultural science and technology (e.g. better understanding of climate change impacts and vulnerability, new crop varieties, international technology transfers and local technology extension services, biotechnology, and water-saving technology);

- in the capacity of governments, communities, and farmers to adapt to climate change, and
- in risk management (e.g. agricultural insurance, natural disaster release and food aid programme, early warning and information systems, and restoration of natural capacity to buffer climate impacts).

In each category of financial support for either the mitigation of climate change or adaptation, we emphasized four major questions. What works? What could work? What can be scaled up? What is transferable? Reviewing several aid-supported cases that have worked in certain developing countries, this chapter shows that for successful investments in agriculture in the face of climate change, foreign aid needs to consider the multiple objectives related to agricultural development, mitigation, and adaptation as well as the interests of major stakeholders involved (e.g. government and farmers).

Major requirements for successfully financing sustainable agriculture through foreign aid should include programmes and measures that are mainstreamed into each country's national action plans on climate change, close collaboration with developing country governments, enhanced local capacity, and consideration of the needs of different stakeholders. This chapter also shows that these prerequisites are key in scaling up and transferring projects within a country or across countries. Of course, the degree to which foreign aid-funded projects can be scaled up and replicated depends on the significance and similarity of problems in terms of climate change and agricultural development within the pilot area and in other potential locations.

Notes

1. E.g. India, Indonesia, Bangladesh, The Philippines, Kenya, Ethiopia, Nigeria, and South Africa.

References

Baumert, K.A., T. Herzog, and J. Pershing. 2005. *Navigating the Numbers*. WRI Greenhouse Gas Data and International Climate Policy. Washington, DC: World Resources Institute. www.pdf.wri.org/navigating_numbers.pdf

Bouman, B.A.M., R.M. Lampayan, and T.P. Tuong. 2007. *Water Management in Irrigated Rice: Coping with Water Scarcity*. Los Baños, Philippines: IRRI.

Climate Focus. 2011. *Title Carbon Market and Climate Finance for Agriculture in Developing Countries*. Washington, DC: Climate Focus. www.climatefocus.com/documents/files/title_carbon_market_and_climate_finance_for_agriculture_in_developing_countries.pdf

Conrad, B., and Q. Li. 2012. *Building a Line of Defence against Climate Change: From Reactive Coping to Adaptive Capacity in China's Irrigated Agricultural Development*. Washington, DC: World Bank.

Corton, T.M., J. Bajita, F. Grospe, R. Pamplona, R. Wassmann, and R.S. Lantin. 2000. Methane Emission from Irrigated and Intensively Managed Rice Fields in Central Luzon (Philippines). *Nutrient Cycling Agroecosystems* 58: 37–53.

DTMA (Drought Tolerant Maize for Africa). 2012. Five Years of Addressing the Needs of Smallholder Farmers in Africa. *DTMA Brief*. www.dtma.cimmyt.org/index.php/publications/doc_view/175-dtma-brief-english-revised-

Easterling, W.E., P.R. Crosson, N.J. Rosenberg, and Others. 1993. Agricultural Impacts of and Responses to Climate Change in the Missouri-Iowa-Nebraska-Kansas (MINK) Region. *Climate Change* 24: 23–61.

Easterling, D.R., G.A. Meehl, C. Parmesan, S.A. Changnon, T.R. Karl, and L.O. Mearns. 2000. Climate Extremes: Observations, Modelling, and Impacts. *Science* 289: 2068–2074.

EU Focus. 2010. *The EU and Food Security*. Washington, DC: EU Focus. www.eurunion.org/News/eunewsletters/EUFocus/2010/EUFocus-EU&FoodSecur-Mar2010.pdf

FAO. 2007. *Adaptation to Climate Change in Agriculture, Forestry and Fisheries: Perspective, Framework and Priorities*. Report of the FAO Interdepartmental Working Group on Climate Change. Rome: FAO.

———. 2008a. *Climate Change and Food Security: A Framework Document*. Rome: FAO. www.fao.org/forestry/15538-079b31d45081fe9c3dbc6ff34de4807e4.pdf

———. 2008b. *Enabling Agriculture to Contribute to Climate Change Mitigation*. Rome: FAO. www.unfccc.int/resource/docs/2008/smsn/igo/036.pdf

———. 2009a. *How to Feed the World in 2050*. Rome: FAO. www.fao.org/fileadmin/templates/wsfs/docs/expert_paper/How_to_Feed_the_World_in_2050.pdf

———. 2009b. *How to Feed the World in 2050: The Technology Challenge*. Rome: FAO. www.fao.org/fileadmin/templates/wsfs/docs/Issues_papers/HLEF2050_Technology.pdf

———. 2010a. *The State of Food Insecurity in the World 2010*. Rome: FAO. www.fao.org/docrep/013/i1683e/i1683e.pdf

———. 2010b. *"Climate-Smart" Agriculture: Policies, Practices and Financing for Food Security, Adaptation and Mitigation*. FAO: Rome. www.fao.org/docrep/013/i1881e/i1881e00.htm

FAO (Food and Agriculture Organization). 2001. *Soil Carbon Sequestration for Improved Land Management*. FAO World Soil Resources Reports 96. Rome: FAO.

Herzog, T. 2006. *Greenhouse Gases and Where They Come From*. Washington, DC: WRI. www.wri.org/climate/topic_content.cfm?cid=4177

Howden, S.M., and H. Meinke. 2003. Climate Change: Challenges and Opportunities for Australian Agriculture. In *Proceedings of 'Climate Impacts on Australia's Natural Resources: Current and Future Challenges'*, 53–55. Queensland: Gold Coast.

Hu, R., J. Cao, J. Huang, S. Peng, J. Huang, X. Zhong, Y. Zou, J. Yang, and R.J. Buresh. 2007. Farmer Participatory Testing of Standard and Modified Site-specific Nitrogen Management for Irrigated Rice in China. *Agricultural Systems* 94: 331–340.

Huang, J., R. Hu, J. Cao, and S. Rozelle. 2008. Training Programmes and In-the-field Guidance to Reduce China's Overuse of Fertilizer without Hurting Profitability. *Journal of Soil and Water Conservation* 63 (5): 165–167.

Huang, J., C. Xiang, X. Jia, and R. Hu. 2012. Impacts of Training on Farmers' Nitrogen Use in Maize Production in Shandong, China. *Journal of Soil and Water Conservation* 67 (4): 321–327.

IFAD (International Fund for Agricultural Development). 2010. *Climate Change Strategy*. Rome: IFAD.

IPCC. 2006. *Revised Guidelines for National Greenhouse Gas Inventories*. IPCC/OECD/IEA/IGES. Vol. 4. Agriculture, Forestry and Other Land Use.

———. 2007a. Climate Change 2007. *Impacts, Adaptation and Vulnerability*. Contribution of Working Group II to the Fourth Assessment Report of IPCC. Cambridge: Cambridge University Press.

———. 2007b. *Climate Change 2007. Mitigation of Climate Change*. Contribution of Working Group III to the Fourth Assessment Report of IPCC. Cambridge: Cambridge University Press.

———. 2012. *Managing the Risks of Extreme Events and Disasters to Advance Climate Change Adaptation*. Special Report of the Intergovernmental Panel on Climate Change. Cambridge: Cambridge University Press.

IPCC (Intergovernmental Panel on Climate Change). 2001. *Climate Change 2001. Impacts, Adaptation, and Vulnerability*. Contribution of Working Group II to the Fourth Assessment Report of IPCC. Cambridge: Cambridge University Press.

IRRI. 2010. *Rice in the Global Economy: Strategic Research and Policy Issues for Food Security*. Los Baños: IRRI.

Islam, N. 2011. *Foreign Aid to Agriculture—Review of Facts and Analysis*. IFPRI Discussion Paper 01053. Washington, DC: IFPRI.

Kostandini, G., B.F. Mills, S.W. Omamo, and S. Wood. 2009. Ex ante Analysis of the Benefits of Transgenic Drought Tolerance Research on Cereal Crops in Low-income Countries. *Agricultural Economics* 40 (4): 477–492.

La Rovere, R., G. Kostandini, T. Abdoulaye, J. Dixon, W. Mwangi, Z. Guo, and M. Bänziger. 2010. *Potential Impact of Investments in Drought Tolerant Maize in Africa*. Addis Ababa: CIMMYT.

Lal, R. 2004. Carbon Emission from Farm Operations. *Environment International* 30: 981–990.

Louis, V.V. 2007. *Opportunities for Climate Change Mitigation in Agriculture and Investment Requirements to Take Advantage of These Opportunities*. A Report to the UNFCCC Secretariat. Nairobi: ICRAF. www.icraf.com/downloads/publications/PDFs/rp15435.pdf

Mushtaq, S., T.N. Maraseni, J. Maroulis, and M. Hafeez. 2009. Energy and Water Tradeoffs in Enhancing Food Security: A Selective International Assessment. *Energy Policy* 37: 3635–3644.

Nelson, G.C., M.W. Rosegrant, J. Koo, R. Robertson, T. Sulser, T. Zhu, C. Ringler, et al. 2010. *The Costs of Agricultural Adaptation to Climate Change*. WB Discussion Paper 4: Development and Climate Change. Washington, DC: World Bank.

Oberthür, S., and H.E. Ott. 1999. *The Kyoto Protocol: International Climate Policy for the 21st Century*. Berlin: Springer Verlag.

OECD. 2011. *Managing Risk in Agriculture: Policy Assessment and Design*. Paris: OECD.

OECD (Organization for Economic Co-operation and Development). 2009. *Financial Resources and Investment for Climate Change*. Paper prepared for the Special Session of the Africa Partnership Forum on Climate Change, 3 September. Addis Ababa. www.oecd.org/site/africapartnershipforum/meetingdocuments/43551000.pdf

Oxfam. 2009. *Beyond Aid: Ensuring Adaptation to Climate Change Works for the Poor*. Oxfam Briefing Paper. www.preventionweb.net/files/11158_oxfam beyondaidsep09.pdf

Oxfam America. 2009. *Horn of Africa Risk Transfer for Adaptation*. HARITA Project Brief. Washington, DC: Oxfam America.

Peng, S., R.J. Buresh, J. Huang, X. Zhong, Y. Zou, J. Yang, G. Wang, Y. Liu, R. Hu, et al. 2010. Improving Nitrogen Fertilization in Rice by Site-Specific N Management. A Review. *Agronomy for Sustainable Development* 30: 649–656.

Pray, C., L. Nagarajan, L. Li, J. Huang, R. Hu, K.N. Selvaraj, O. Napasintuwong, and R. Chandra Babu. 2011. Potential Impact of Biotechnology on Adaption of Agriculture to Climate Change: The Case of Drought Tolerant Rice Breeding in Asia. *Sustainability* 3: 1723–1741.

SAIN (the China-UK Sustainable Agriculture Network). 2010. *Improved Nutrient Management in Agriculture: A Neglected Opportunity for China's Low Carbon Growth Path*. SAIN Policy Brief 1. Beijing: SAIN.

Schalatek, L., M. Cook, S. Alboher, and E. Eggerts. 2011. *The Least Developed Countries Fund and the Special Climate Change Fund*. New York: UNDP.

www.gender-climate.org/Content/Docs/Publications/UNDP_LDCF_SCCF.pdf

SEI. 2008. *Scoping Paper for Climate Change Adaptation in Africa for the Ministerial Session and Expert Group: Segments of the Dialogue on Climate Adaptation*. Lead authors: Downing (SEI) and YoubaSokona (OSS). African Ministerial Conference on Environment. Expert consultation on adaptation: 8 June 2008. Ministerial session: 12 June 2008.

Smithers, C. 1998. *Crop Insurance and Farm Management*. MA thesis, Department of Geography, University of Guelph, Guelph.

Tirado, R., and J. Cotter. 2010. *Ecological Farming: Drought-Resistant Agriculture*. Greenpeace Research Laboratories GRL-TN 02/2010. www.greenpeace.org/international/en/publications/reports/Ecological-farming-Drought-resistant-agriculture/

Tyler, S.C., H.O. Ajie, M.L. Gupta, R.J. Cicerone, D.R. Blake, and E.J. Dlugokencky. 1999. Stable Carbon Isotopic Composition of Atmospheric Methane: A Comparison of Surface Level and Free Tropospheric Air. *Journal of Geophysical Research* 104 (D11): 13895–13910.

UECC. 2010. *Carbon Sequestration & Biodiversity*. Coastal & Marine Union (EUCC). www.eeconet.org/eaf/climate-biodiversity/Carbon-Sequestration-Biodiversity -FINAL-REPORT.pdf

UNFCCC (United Nations Framework Convention on Climate Change). 2007. *Investment and Financial Flows Relevant to the Development of an Effective and Appropriate International Response to Climate Change*. Bonn: UNFCCC.

UN-HAGCCF (High-Level Advisory Group on Climate Change Financing). 2010. *Report of the Secretary-General's High-Level Advisory Group on Climate Change Financing*. 5 November 2009. New York: United Nations. www.un.org/wcm/webdav/site/climatechange/shared/Documents/AGF_reports/AGF%20Report.pdf

Verchot, L.V. 2007. *Opportunities for Climate Change Mitigation in Agriculture*. A Report to the UNFCCC Secretariat. Financial and Technical Support Programme. unfccc.int/files/cooperation_and_support/financial_mechanism/application/pdf/verchot.pdf

Wang, J., J. Huang, L. Zhang, S. Rozelle, and H.F. Farnsworth. 2010. Why Is China's Blue Revolution so "Blue"? The Determinants of Conservation Tillage in China. *Journal of Soil and Water Conservation* 65 (2): 113–129.

Wang, J., S. Rothausen, D. Conway, L. Zhang, W. Xiong, I.P. Holman, and Y. Li. 2012. China's Water–Energy Nexus: Greenhouse Gas Emissions from Groundwater Use for Agriculture. *Environmental Research Letters* 7 (1).

Wassmann, R., S.V.K. Jagadish, S. Heuer, A. Ismail, E. Redoña, R. Serraj, et al. 2009. Climate Change Affecting Rice Production: The Physiological and Agronomic Basis for Possible Adaptation Strategies. In *Advances in Agronomy*, ed. D.L. Sparks, vol. 101, 59–122. Burlington: Academic Press.

World Bank. 2007. *World Development Report 2008*. Washington, DC: World Bank.

———. 2009a. *Monitoring and Reporting on Financial Flows Related to Climate Change*. Discussion Draft. Washington, DC: World Bank. www.climatechange.worldbank.org/sites/default/files/documents/ClimateFin-Monitoring-122909.pdf

———. 2009b. *World Development Report, 2010*. Washington, DC: World Bank.

———. 2010. *Economics of Adaptation to Climate Change: Synthesis Report*. Washington, DC: World Bank. www.climatechange.worldbank.org/sites/default/files/documents/EACCSynthesisReport.pdf

———. 2012. *China: Third Irrigated Agriculture Intensification Project and Mainstreaming Climate Change Adaptation in Irrigated Agriculture Project*. Report No: ICR500. Washington, DC: World Bank.

Zhang, F.S., J.Q. Wang, W.F. Zhang, Z.L. Cui, W.Q. Ma, X.P. Chen, and R.F. Jiang. 2008. Nutrient Use Efficiencies of Major Cereal Crops in China and Measures for Improvement. *Acta Pedologica Sinica* 45: 915–924. (in Chinese with English abstract).

Zhao, B., X. Li, X. Shi, S. Huang, B. Wang, P. Zhu, et al. 2010. Long-Term Fertilizer Experiment Network in China: Crop Yields and Soil Nutrient Trends. *Agronomy* 102 (1): 216–230.

CHAPTER 11

Foreign Aid and Sustainable Agriculture in Africa

Siddig Umbadda and Ismail Elgizouli

11.1 Introduction

The recent history of foreign aid probably begins with the Marshall Plan, in which US$13 billion flowed to Europe after the Second World War. The success of the plan led President Truman, in his inaugural speech at the foundation of the North Atlantic Treaty Organization (NATO) to propose that: 'we must embark on a bold new programme for making the benefits of our scientific advances and industrial progress available for the improvement and growth of under-developed areas'.[1] The Marshall Plan perhaps represents the most effective aid ever, judging by its outcome that Europe, in less than a decade, emerged from the ruins of a war to become a strong economic and political player in the international arena (Bovard 1986). Foreign aid since took different forms and levels (during the Cold War and thereafter) and was constantly on the agenda of international conferences and economic forums. In fact, there is a commitment by major donor countries to grant aid to countries in need, to the tune of 0.7% from national income.

S. Umbadda (✉)
Mamoun Beheiry Centre for Economic & Social Studies and Research in Africa, Khartoum, Sudan

I. Elgizouli
Higher Council for Environment & Natural Resources, Khartoum, Sudan

Y. Huang, U. Pascual (eds.), *Aid Effectiveness for Environmental Sustainability*, https://doi.org/10.1007/978-981-10-5379-5_11

In general, aid is not given for purely altruistic goals. Rather it reflects the interests of donors, which may include military, political, and/or commercial interests. More often than not it does not provide the maximum benefit to the recipient. This is especially true when aid is tied. Mark Malloch Brown, the former head of the United Nations Development Programme (UNDP), estimated in 2002 that farm subsidies in advanced countries cost poor countries about US$50 billion a year in lost agricultural exports.[2] This claim was confirmed by Oxfam (2002), which reveals that aid tied to trade liberalization by the donors, such as the European Union (EU), is becoming detrimental to developing countries. The report estimated that Latin American countries, for instance, lose US$4 billion annually owing to the EU farm subsidy policies. In their opinion, aid money, instead of being used to finance development effort in these countries, ended up being used to fund terms of trade deficit brought about by trade liberalization (ibid.).

The United States (USA), for instance, gives huge subsidies to its agricultural sector. It is shown that in recent budgets at least US$5 billion is given as subsidies to rich farmers, irrespective of crop prices or yields, and a further US$10–15 billion is considered to be trade-distorting subsidies that undermine incentives to invest in agriculture in developing countries (Elliot 2011).

Another indication of the size of the loss owing to protectionism by advanced countries comes from the World Development Report (World Bank 2008). This shows that the cost of the Organisation for Economic Co-operation and Development (OECD) countries' subsidies to farmers distorts world markets, and as a result African countries are denied trading opportunities equivalent to almost twice the size of the OECD countries' foreign assistance. At the same time it is believed that rich countries spend about US$300 billion each year on agricultural subsidies—six times more than the annual US$50 billion that rich countries put into foreign aid (UK Food Group 2008: 16). In addition, the World Development Report 2008 reports that developed country agricultural policies cost developing countries about US$17 billion per year—a cost equivalent to almost five times the current levels of overseas development assistance to agriculture (World Bank 2008). As can be seen, estimates differ but all point to the complexity of judging the value of foreign aid.

Nevertheless, it could not be denied that aid has been instrumental in assisting developing countries to overcome many of their problems, including natural and man-made disasters. Aid levels have also seen substantial increases, in nominal terms, especially to Africa. In this regard it could be noted that total aid flows increased from US$58 billion in 2000 to US$133 billion in 2011, and donor commitments of aid directed to agriculture roughly doubled from US$4 billion in the mid-2000s to just over US$8 billion in 2010.

The rest of the chapter will discuss (1) the share of aid to agriculture, (2) aid to agriculture in Africa, (3) development thinking after independence of African countries and the role of aid, (4) recent developments in donor thinking about aid, including agricultural sustainability, aid effectiveness, scaling up of aid-financed projects, and transferability across countries; and finally (5) recommendations and concluding remarks.

11.2 Share of Foreign Aid to Agriculture

The statistics of official development assistance (ODA) does not always give a clear sectoral classification. For instance, between 23–36% of aid is classified as unallocable. Furthermore, there are funds that do not constitute a transfer of resources to recipient countries but are included in aid statistics; for instance, administrative costs to donors, expenditure on refugees in donor countries, and support to international non-governmental organizations (INGOs) of donor countries. Accordingly, it is difficult to ascertain exactly the share of aid to agriculture, and the whole issue is quite complicated—as aptly put by Nurul Islam: 'the task of measuring, analyzing and evaluating aid to agriculture in all its components, ramifications, and implications remains a challenging task for researchers, policy analysts and policy makers' (Islam 2011: 41). Roughly, however, the share of aid to agriculture has hovered around 5% of total aid, although some differences in shares exist between multilateral and bilateral aid, for instance lending to agriculture by the World Bank's International Development Association (IDA) remained at 9% of its total commitments.

From the 1980s, aid to agriculture began to decline both in absolute and real terms. There are both external and internal factors that militated against increased aid to agriculture during this period. The external factors include the shift of donor resources to other sectors, such as infrastructure and the social sector, because of their proven positive effect on development. For instance, the effect of rural roads and rural electrification on income growth and poverty alleviation has been demonstrated because, among others, they facilitate access to markets for both outputs and inputs.

Equally, investment in the social sector (education and health) is also recognized for its positive effect on labour productivity and promotion of human development in general. While this is true, it is also thought that civil society groups have contributed to this shift of donor focus to the social sectors by convincing donors that aid must be people-centred, instead of sector- or activity-centred (Eicher 2003). There are also other claimants on aid resources, such as commitments for debt relief for heavily indebted poor countries and humanitarian aid. The latter increased considerably in response to both natural and man-made disasters over the last two decades or so.

A further external factor that negatively impacted the share of aid to agriculture is the abundance in food production in the 1990s which led to low food prices in the international markets. At the time this led food exporters in advanced countries to oppose increased aid to agriculture for developing countries, because that would lead to a further decline in food prices. It should also be added that there was opposition from environmental groups, which saw agriculture as a contributor to natural resource destruction and environmental pollution (World Bank 2007).

The internal factors, specific to the agricultural sector, that led to a declining share of aid included delays in completion of agricultural projects in less developed countries and the associated cost overruns, and large supervision costs that tended to reduce returns to agricultural projects. Other constraints included poor road and market infrastructure, undeveloped financial sectors, and higher weather-related and disease risks. Added to this was the weak governance and institutional capacity structures entrusted to design, administer, and implement projects in an efficient manner in these countries (World Bank 2010: xi). This meant that donors had to spend time building these institutions and there were also delays in disbursement. Together these factors led donors to shift focus to policy reforms, both sectoral and macro, thus increasing policy-based lending as against direct lending to agriculture.

11.3 Aid to Agriculture in Africa

Since 1980s, aid to Africa has quadrupled from around US$11 billion to US$44 billion. But as mentioned earlier, ODA to Africa is directed mostly to other sectors (the social sector 45% and infrastructure 15%). ODA is seen as a means of leveraging other flows to ward off aid dependency. Thus donors stress the importance of simultaneously mobilizing domestic resources, promoting international trade, and encouraging foreign direct investment (FDI).[3]

Measured by net ODA as a percentage of government expenditure many African countries are aid dependent. In at least 15 sub-Saharan African (SSA) countries aid flows constitute between 50% and 770% of their government expenditures. The list includes Liberia 771%; Guinea-Bissau 221%; Rwanda 205%; the Central African Republic (CAR) 195%; Madagascar 194%; Mozambique 167%; Malawi 164%; Sierra Leone 150%; Ethiopia 133%; Democratic Republic of Congo (DRC) 126%; Uganda 98%; Guinea 91%; Zambia 84%; Senegal 80%; and Gambia 73%. Measured by aid as a percentage of gross domestic product (GDP), many other African countries could also be classified as aid dependent. Among these are Mozambique 60%; Sierra Leone 47%; and Eritrea 31% (Spagnoli 2010).

Agriculture is important for the livelihood of most Africans, and most of the poor in general. Some 75% of the world's poor live in rural areas and depend on agriculture for all or part of their livelihoods and all or most of their food supplies. The positive relationship between improving agriculture and poverty alleviation is shown empirically to be very strong. For 42 developing countries, econometric analysis shows that for the poorest 10%, a 1% GDP growth in agriculture increases income by more than 2.5%. (Kuyvenhoven 2008). The World Development Report (World Bank 2008) underscores the importance of growth in agriculture as a critical catalyst for economic growth and poverty reduction. The report points out that GDP growth from agriculture is shown to raise incomes of the poor two to four times more than GDP growth from non-agriculture.

Despite this, donors, for different reasons, accorded a low priority to agriculture, as assistance to the sector constituted only 5% of total aid. The USA, in particular, directed only 2% of its total aid budget in 2007 to agriculture (Elliot 2011). In fact this has been the case since the early 1990s. This may be due to the increase in food aid and humanitarian assistance associated with natural or man-made disasters around the globe. However, with a growing world population, and persistent poverty in many countries, more support and investment to build local capacity to increase agricultural productivity and strengthen national and regional food systems cannot be overemphasized, if food prices are to be affordable for the most vulnerable groups.

Both the volume and the share of aid earmarked for agriculture has until recently been steadily falling below its 1980s levels. This trend has been especially strong in SSA, where bilateral agricultural aid fell by 60% from US$1.3 billion to only US$524 million between 1990 and 2001 (UK Food Group 2008: 14). The perception of donors is that agriculture and rural development projects are more risky and less profitable than other types of projects. But in the period of the post-2007 world food crisis, donors committed themselves to increase aid to agriculture. The World Bank's Robert Zoellick announced in July 2008 that the World Bank would double its agricultural lending to Africa from an average of US$450 million to US$800 million a year, in the wake of international staple food price hikes (UK Food Group 2008: 15).

11.4 Agriculture in Development Thinking after Independence in African Countries and the Role of Aid

After African countries gained independence in the 1950s and 1960s, development in agriculture was not considered a priority, because it was not regarded as an important contributor to economic growth and therefore not further pursued (Ngambeki 2003). Instead, development thinking was centred on state-led industrialization and the belief that development and economic growth can be achieved by transforming agrarian-dominated societies into modern industrial countries (Eicher 2003). Achieving economic growth was considered the main priority rather than alleviating poverty through developing the agricultural sector. It was thought that creating jobs and supporting economic growth would create a trickle-down effect and thus tackle problems of poverty, as well as improving access to health and education. This belief was supported by international financial institutions such as the World Bank. For instance, in the 1960s, the World Bank's vice-president stated: 'Given the policy instruments and administrative capacity of the less-developed countries, I would judge that the employment increases generated by high growth are the most reliable means of maximising the welfare of the lower-income groups' (Chenery 1971: 37, quoted in Eicher 2003).

Instead of pursuing the state-led industrialization model, many scholars believe that the colonial extraction model, based on international trade, included some tremendous advantages, such as organizing rural space by relying on regionalism as the organization model for agricultural research. For example, the French set up regional research stations in Senegal and

Côte d'Ivoire to generate new technology and transfer it to satellite colonies, later adapting it to local conditions. Instead, many African countries followed nationalization processes, particularly of regional research centres and private plantations rather than taking part in the international trade of agrarian products (Eicher 2003). As a result, state-led industrialization in the 1960s and 1970s focused on capital accumulation and heavy reliance on foreign aid to achieve high rates of economic growth. To give an example, during the 1960s and 1970s, the World Bank increased lending for agriculture from 6% of the Bank's total lending to over 30%. Between 1974 and 1984, agricultural lending commitments reached more than US$30 billion (ibid.).

Since the 1970s, however, interest in agricultural development has increased and focused on tackling rural poverty by improving smallholder agriculture, in particular community development. Yet most integrated rural development projects were hindered by stringent macro-economic policies and were not sustainable; that is, they did not include programmes or plans to finance social and agricultural services after donor aid was phased out. Thus, while many donors invested large amounts of money into pilot projects to ensure their success, these became too expensive to be replicated or to be maintained on a national or regional level. The Cohen report (1987) on Swedish aid mentioned that Sweden invested US$41 million into the Chilalo Agricultural Development Union (CADU) rural development project in the Arsi province of Ethiopia that ran for a period of 26 years. Other common problems related to aid are structural problems, such as lack of coordination between numerous central ministries including agriculture, health, and education. In addition, Assal (2008) argues that concepts such as participation, partnership, good governance, and empowerment are often vague and therefore not applicable in local communities. They cannot address structural problems such as poverty, which hinder the recovery and development of the countries that require improved infrastructure or improved government structures.

With the growing agricultural activities of the Green Revolution in Asia, and the optimism of applying the same model to an African context, aid to agriculture rose in the 1970s. In addition, as a result of the global food crisis of 1972–4, many donors further increased global aid to agriculture. By the early 1970s, many economists had reached the conclusion that the development plans and strategies centred on economic growth through state-led industrialization were not achieving any substantial social benefits. Consequently, many donors shifted priorities and provided direct assistance to the rural poor through basic needs programmes, integrated rural development projects, and aid to small-scale agriculture (Eicher 2003).

In the 1980s, development optimism changed to pessimism and a shift to programme aid and policy reform occurred. Harsh economic policies failed to achieve the desired trickle-down effect, and the structural adjustment programmes that were pursued led to economic crises as state intervention and safety nets were either abolished or downsized.

In the 1990s, political issues such as corruption, good governance, and decentralization became important factors for donors, and they were set as prime conditions for allocating aid. This period also saw increased importance of including social sector issues such as education, health, post-conflict aid, and the environment in development planning. Owing to the growing importance of social and political factors, aid to agricultural projects has seen a steady decline. Other reasons for this involve the perception that agricultural problems can be solved outside the sector by increasingly focusing on infrastructure including roads and electricity to rural areas. Another impact is the declining support of assistance for the agricultural sector in developing countries (Brown 2009; Coppard 2010). According to the World Bank and the UK Food Group (2008) report, the major reasons for the overall decline of foreign aid to agriculture include the fall of international commodity prices, which made it less profitable to invest in agricultural activities, and increased investments in social sectors.

At the beginning of the millennium, calls for increased aid grew stronger, particularly initiated by the World Bank followed by the EU, the Department for International Development (DFID), and private institutions, though the amount allocated to agriculture has remained static at about 9%. What has changed, though, is that the largest aid to agriculture was given to Africa. In the financial years 2002–6, aid to agriculture was 47%, compared to a share of 29% in the financial years from 1996 to 2001 (UK Food Group 2008). With trends of climate change, rising food and energy prices, and rising demands in new investment in agriculture, aid to African agriculture has been put back on the international policy agenda. Between 2000 and 2008 global agricultural ODA has increased from US$3.6 billion to US$6.3 billion. While total agricultural commitments have increased, the proportion of agricultural aid has increased overall during the period (Coppard 2010: 9).

Between 2005 and 2008, 50% of the total agricultural assistance was in three areas: agricultural policy (22%), agricultural water resources (18%), and agricultural development (13%). Agricultural policy has received the largest amount of aid and includes support for agricultural ministries and measures, programmes, and actions relating to capacity building. Agricultural water resources refer to all forms of irrigation investments, and agricultural developments include farm development initiatives and projects (Coppard 2010). What is striking is the increased investment in agricultural research, which includes investment into new technologies, moving from 6% in the period 2000–3, to 10% in the period 2005–8 (Coppard 2010: 16).

11.5 Recent Developments in Aid to Agriculture: Five Main Areas of Donor Consensus

Overall, it can be noted that current donor consensus on agricultural aid for Africa is centred on seven main areas: changing perception of agriculture; market and private sector-led agricultural growth; improved governance and political processes; social services and empowerment; aid effectiveness; scaling up; and transferability.

11.5.1 Changing the Perception of Agriculture: Agricultural Sustainability

Recent perceptions of development and agriculture have seen a shift away from agriculture as engine of growth towards realizing a 'right to food and food sovereignty' (UK Food Group 2008). The Commission for Sustainable Development noted the importance of agriculture in sustaining rural life; increasing food production and enhancing food security in an environmentally sound way. Sustainable agriculture plays a key role in tackling food insecurity especially in rural areas. According to the United Nations Development Programme (UNDP) (2012b), increases in agricultural productivity and better nutrition are important for food security and human development. They argue that increased food production will increase food security by raising food availability and lowering food prices, thereby improving access to food. In addition, higher productivity will also increase people's incomes, which has positive effects on health and education (UNDP 2012a).

Previous development strategies to increase agricultural productivity were predominantly based on the industrial agriculture model, which has often proven to be environmentally, socially and/or economically unsustainable. According to USAID, of the 11% of the world's land surface that is adequate to perform agricultural activities, 38% has become degraded by poor natural resource management practices (USAID 2012). Industrial agriculture is characteristic of the use of monoculture crops which allows reducing the costs of production, thereby reducing the price for certain commodities such as wheat, corn, and soybeans (Khan 2011). Sustainable agriculture shifts away from artificial methods of increasing yields towards focusing on the growing capacity of natural inputs. This can be achieved by using a variety of techniques without affecting the environment, such as crop rotation, soil enrichment, and natural pest predators (NEPAD 2003). Crop rotation involves growing different crops in the same field instead of planting the same crop every season. This helps to ensure the long-term health of the soil, because rotating crops with nitrogen-fixing crops replace nutrients in the soil (Khan 2011; UN 2012).

With the concept of sustainable agriculture, many development agencies have sought to combine the three factors of environmental, social, and economic sustainability. Agricultural sustainability aims to apply a systems approach to address different aspects of food security. It addresses above all the economic, social, and environmental dimensions of agricultural production. Thereby a systems approach is pursued, where a system consists of the interaction of different individuals and institutions—such as researchers, unions, retailers, consumers, policymakers—that need to be considered (Amekawa 2010; ASI 2012). In this way different causes and impacts of agriculture and food insecurity can be identified as well as addressed.

11.5.2 Market- and Private Sector-led Agricultural Growth

Market- and private sector-led agricultural growth refers to the idea that agricultural growth must be market led by reducing the role of the public sector and promoting public–private partnerships. In this context, the need for a so-called 'new green revolution for Africa' is promoted, where greater emphasis is put on agricultural research activities, in particular in science and technology (Ngambeki 2003). It also involves the use of advanced technologies. Like the 'green revolution' in Asian countries, the new green revolution for Africa involves improving and diversifying crops, improving irrigation systems, and advancing technologies (UK Food

Group 2008). This also involves strategies to achieve minimum reliance on external inputs. For example, following a sustainable agriculture perspective means addressing scope and yield stability, stable food prices, and prices of fertilizers to meet economic sustainability (Amekawa 2010). Another aspect of economic sustainability is to diversify farms to avoid monoculture, thereby mitigating the risks of economic losses and responding to extreme price fluctuations associated with changes in supply and demand (ASI 2012). Yet this also requires a commitment to changing public policies, economic institutions, and social values. This is why a systems approach is important because it requires recognizing the relationship between agricultural production and society (ibid.).

Another important focus of sustainable agriculture is at policy level; that is, enhancing or introducing policies that promote environmental health, economic profitability, and social as well as economic equity; for example, supporting commodity and price programmes to allow farmers to realize the full benefits of the productivity gains. Another strategy is to modify tax and credit policies to encourage family farms rather than corporate concentration. It is important to address these policies at local, regional, national, and global level, where the last is particularly important to facilitate international trade.

With respect to international trade, African countries' share in world agricultural exports has decreased over the past decade, namely from 8% in 1971–80 to 3.4% in 1991–2000 (NEPAD 2003). Yet promoting regional, global, and bilateral trade to achieve financial sustainability is important to avoid aid dependency. Owing to high food prices, it is cheaper to buy products from international markets that are heavily subsidized than to buy locally produced goods. As a result of not being able to produce enough domestically, many African countries rely on food imports, which undermines local and national agricultural productivity and negatively affects national GDPs. The New Partnership for Africa's Development (NEPAD) estimates that in 2000 African countries spent US$18.7 billion on food imports (2003). However, affecting this requires the change of current trade policies, especially global trade policies (UK Food Group 2008).

11.5.3 Improved Governance and Political Processes

Strategies that promote sustainable agriculture for poverty reduction must also address political processes and good governance. According to the

UK Food Group (2008), to make political progress more effective, current trends focus on small and strategic improvements in governance. Therefore, priority is given to small-scale strategies to promote good governance (ibid.). Another obstacle that many African countries face is that trade and market access requires infrastructure development, financial structures, and strong national regulatory authorities to implement information and market development (NEPAD 2003). Therefore, political processes must not only refer to national political and institutional changes but also address the political framework of international trade and policies.

Despite several programmes, such as the 'Everything but Arms' arrangement, African countries' participation in world trade is still inhibited by factors such as dependency on preferential access to a few developed country markets. 'Everything but Arms' was initiated by the EU to enable duty-free and quota-free entry for all products except arms for the least developed countries (LDCs) (ibid.). Other reasons for limited world trade participation are the subsidized products of developed countries. According to NEPAD, in 2001 OECD countries subsidized their agricultural sector to the tune of US$311 billion, giving them greater advantage in the global markets. In addition, the Strategy on Agriculture and Rural Development (World Bank 2003) reported that developed countries spent about US$300 billion on agricultural subsidies. In contrast, developed countries spent only US$50 billion on foreign aid. Therefore, the NEPAD report concludes that developed countries should change their conditions of foreign trade to facilitate developing countries' access to global agricultural trade, such as ending the hidden taxation of agriculture, increasing financial allocations to rural areas, supporting rural organizations, and modifying trade tariffs. The last is of particular importance. The World Development Report (World Bank 2008) estimates that agricultural policies set by developed countries cost developing countries about US$17 billion per year, which is equivalent to about five times the ODA provision (UK Food Group 2008). Therefore, NEPAD (2003) criticized the position of many African countries, who continued to follow the belief that dynamic and sustainable agriculture should not be based on subsidies when it is developed countries that benefit from huge amounts of subsidies to agricultural activities.

According to the UK Food Group report (2008), in order to make development sustainable, African countries must take the lead in their own development. As a vital part of this, they must be involved in global decision-making processes and discourses on development and agriculture. The 2005 Commission for Africa stated in its recommendations to the G8 countries that African countries must be more involved in decision-making processes in order to take responsibility for their own development. This also involves leading positions in the International Monetary Fund (IMF) and the World Bank that are traditionally given to European or US nationals. The issue of more equitable representation is reiterated in the Human Development Report 2013. The report emphasized that 'the major international institutions need to be more representative, transparent and accountable. The Bretton Woods institutions, the regional development banks, and even the UN [United Nations] system all risk diminishing relevance if they fail to represent all member states and their people adequately. These bodies need to respect and draw constructively on the experiences of both the South and the North and to aim for equitable and sustainable outcomes for present and future generations' (UNDP 2013: 109).

11.5.4 *Social Services and Empowerment*

In order to meet sustainable agriculture, aid programmes have increasingly put emphasis on ensuring access to social services, including safety nets, as well as empowering women and small-scale farmers (UK Food Group 2008). Thus current aid programmes put greater emphasis on food production and food security, as well as agricultural and rural development. This includes ensuring adequate working and living conditions for farmers, especially those associated with health, by reducing pesticide use, for instance, as well as taking measures to protect the natural environment (ASI 2012). It also includes ensuring access to education. Examples of social sustainability are insurance, employment protection, food assistance or subsidies, and social transfers (UNDP 2012b). Human development goes beyond ensuring adequate incomes and commodities. It also addresses human choices and people's capabilities: 'their freedoms to be and do what they value' (Sen 1985, 1989). An important aspect of human development is empowerment of women, minority groups, and smallholder farmers through better education and health, and a greater share in decision-making processes. Empowering women, who make up almost half of the agricultural labour force in SSA, is important as they play a significant role in food security (UNDP 2012a).

One prominent institution that is predominantly devoted to gender empowerment and supporting small-scale farmers in rural communities is the International Fund for Agricultural Development (IFAD). IFAD recognizes that one of the main causes of food insecurity and famine are structural problems related to poverty, and the fact that most poverty is concentrated in rural areas in developing countries. Therefore, IFAD focuses on tackling rural poverty and empowering women and minority groups, including small-scale farmers, fishermen, rural poor women, landless workers, craftsmen, nomads, and indigenous people, to increase food production, raise their incomes, and thus maintain food security (IFAD 2012c, d). This means that to make aid sustainable, IFAD focuses on increasing people's access to financial services, markets, and technologies, as well as land and other natural resources.

Women's involvement in agricultural activities ranges from 20 to 70%, yet it is also noted that their participation in agriculture-related activities is increasing in developing countries. The International Assessment of Agricultural Knowledge, Science and Technology for Development (IASSTD 2008), recommends four steps for supporting women's activities in agriculture. These include supporting public services, particularly in rural areas, to improve women's living and working conditions; creating or modifying policies targeted at technological development that recognize and address women's knowledge-enhancing skills and experience in food production; and assessing and reducing negative effects of farming practices and technologies that pose risks to women's health.

Culturally, sustainable agriculture is more knowledge-based and is based on more intensive labour (Kassie and Zikhali 2009). Therefore, the understanding of ecological processes and problems is vital. It also requires greater farmer participation, fair treatment of workers, and farmer-to-farmer extension to achieve farmer empowerment. To make aid more effective and sustainable, local communities must be integrated in the design and implementation of initiated programmes to reflect their needs and constraints (UK Food Group 2008).

11.5.5 Aid Effectiveness

Recently, and in particular since the food crisis of 2007–8, there has been a growing interest in agriculture from donors, driven by food security issues and climate change challenges. This comes at a time when there is also a renewed interest in agricultural development owing to population growth and diversion of crops for energy. Although aid to agriculture still represents only around 5% of total aid, donors have begun to show keen interest in its quality. Quality of aid or aid effectiveness are difficult to measure, in particular in agriculture, mainly because of the small volume of ODA for agriculture and the fact that data used to assess aid quality is not available at sectoral level.

Donors' interest in aid effectiveness has been triggered mainly by the growing budget pressures on the donor. In 2005 both donors and recipient countries agreed on a set of principles: the Paris Declaration on Aid Effectiveness. This position was reinforced in the Accra Agenda for Action 2008 (Elliot 2012: 3).

Both declarations were, in fact, intended to respond to growing criticism that aid was not helping and might even be damaging for developing countries. The initiative on aid effectiveness revolves around a set of principles for more effective aid and a peer review process to encourage implementation. These principles include maximizing efficiency; fostering institutions in the recipient country (country ownership and alignment); reducing burden on recipient countries associated with management of aid (harmonization); and transparency and learning (mutual accountability) (Elliot 2012; Elliot and Collins 2012). However, these principles, which are used as measures for quality of ODA, are indicators of donor efforts to improve the quality of their aid and are not direct measures of effectiveness. The latter needs more effort from both donors and recipients to evaluate the actual impact of aid.

There is, however, a major element of aid effectiveness that does not seem to concern donors, namely policy-related aid. It is a fact that a considerable part of current aid to agriculture is assigned to policy and administrative management and agricultural development (41%) (Islam 2007). At the same time, the percentage of aid that goes to the production of food is quite small, currently 10% for crop production, and 3% for livestock. Local food production is carried out by local communities and farmers' organizations within targeted programmes that also secure their livelihood and sustain the environment. The question that remains to be answered is whether there would be a shift in agricultural aid towards supporting local food production or not. This is crucial because policy conditionality attached to aid could simply change its appearance from aid-tying to a more tailored liberalization tool such as 'aid for trade', which again results in limited support for local agriculture.

Lately, humanitarian assistance has been increasing, especially the component of food aid which can actually forestall agricultural development. Cheap, subsidized, or free US grains undercut prices of the locally produced food, driving local farmers out of business and into the cities. In Somalia in 1992, food aid poured in and reduced local prices by 75%. The USA provided funds but only if food was bought from US farms. As a result, many farmers in Somalia abandoned their farms and joined the queues for imported food aid. Food aid distorts local food markets, drives farmers off the land, and creates long-term dependency on imported food. Such factors need to be taken into account if aid to agriculture is to be effective, in the sense of producing the maximum positive impact on agriculture and the poor who depend on it.

An exceptional case regarding aid effectiveness is Ethiopia's Productive Safety Nets Programme (PSNP) which was financed by a number of partners to the tune of US$4.4 billion over a period of nine years (2005–14). The objective of this programme was to support a sustainable system that improved food security for at least 5 million people. Activities financed included environmental regeneration, which revived the water tables and vegetation cover, increased carrying capacity of livestock, small irrigation, farming, training and other activities that lead to a decrease in food insecurity. The programme was evaluated by the Independent Evaluation Group (IEG) of the World Bank (2011), which concluded that it was effective, pragmatic, and flexible. An earlier review in 2008 also indicated that households who have access to both a productive social safety nets programme and an agricultural support package are more likely to be food secure (Gilligan et al. 2008).

11.5.6 Scaling Up

The above discussed measures are important, but more is needed to achieve increased agricultural production and to make aid effective. This is why a systems approach is necessary (ASI 2012). A project could be established in one location, but if it addresses different aspects such as infrastructure, provision of credit, and participation in local and national markets, its overall effect would be much enhanced if it is scaled up.

One option is to focus on regionalization as suggested by Eicher (2003) and pursued by the Comprehensive Africa Agriculture Development Programme (CAADP). The African Union and NEPAD initiated this in

2003. Its aim is to achieve economic growth by promoting multiagricultural activities (Tibbett 2011). Members of CAADP have committed to increase allocation to agriculture and rural developments to at least 10% of their budget and raise agricultural activities by at least 6% (Tibbett 2011; Brown 2009). CAADP focuses on a regional strategy of regional integration and cooperation to benefit economically from common resources, infrastructure, and other social, cultural, environmental, and political similarities. CAADP works in four areas, namely land and water management, market access, food supply, and hunger and agricultural research, focusing predominantly on policy issues. Here, it takes advantage of the membership of African countries to achieve policy and institutional changes in order to promote agricultural activities (Tibbett 2011). Regionalization is particularly helpful when addressing environmental issues, as environmental degradation knows no political or geographical boundaries.

IFAD, for example, which is a relatively small organization, has a limited outreach and its effects are somewhat limited (NEPAD 2003). Thus its focus is not predominantly on achieving economic growth at a national level but rather on increasing people's incomes and improving their livelihoods by promoting agricultural productivity. Therefore, to make its programmes effective on a national scale, IFAD reached the conclusion that it was not enough to support agricultural productivity in a local area. Rather, it is necessary to upgrade local initiatives, and enhance quality standards of marketing and promotional services, transport and communication infrastructure, especially in rural areas, and improve relevant technologies in order to facilitate national and international trade. Current aid programmes are based on partnerships between different organizations that operate on different levels; that is, partnerships between NEPAD and UN organizations to provide assistance to international public funding sources, or technical support to enhance regional organizations' capacities to promote intra-regional trade in farm products, or national programmes to expand agricultural products.

Economic policy environment is of critical importance to the success of investment in agriculture. Aid from donors has shifted from direct lending to agriculture to policy-based lending. Apart from financial support, accompanying economic stabilization policies and land reform also feature highly in aid supported policy advice. It is found that donor support to land policy issues has contributed to a better understanding of property rights regimes and their importance for agricultural development, and by implication contribution to broad-based economic growth, for example the Rural Land Management Project in Côte d'Ivoire (World Bank 2007: 62). Projects with components addressing multiconstraints such as institutional capacity and credit provision are also good candidates for scaling up. This is because of shared issues of a lack of institutional capacity and access to credit in the majority of African countries. Thus projects which provide training—in establishing early warning systems for drought, improving monitoring and evaluation capacities, or developing information systems to assist in better planning, as well as providing credit to small-scale farmers—should be scaled up if their contribution is to be maximized.

11.5.7 Transferability

Similar to scaling up, it is not enough to initiate a programme in one country or one region. In the past, aid and development projects have often been applied to different social, political, environmental, cultural, and economic situations, thereby neglecting their different circumstances (i.e. one size fits all) (UK Food Group 2008). As a result, the World Bank is now promoting the idea of 'agriculture for development agendas', tailored to individual contexts, which is establishing or modifying development policies to reflect both national priorities and satisfy regional needs, which could also be replicated. Similar examples are the Consultative Group on International Agricultural Research (CGIAR), the programme financed by IFAD and the Global Environment Facility (GEF).

One of the success stories that should be a candidate for transferability is the experience of CGIAR, an amalgam of 15 international agricultural research centres (only four of which are African). CGIAR—a donor-funded group—is dubbed a success story by the Independent Evaluation Group of the World Bank in its 2007 report titled 'Sub-Saharan African Agriculture'. It is thought that CGIAR has contributed immensely to the development of improved varieties of many crops in Africa over the past 20 years (World Bank 2007: 43). Examples of individual projects in the area of production and multiplication of the seeds of major crops that are mentioned in World Bank 2007 include the Togo National Agricultural Service Project (1998) and Ethiopia's National Fertilizer Sector Project (1995). To extend the experience, donors at the Tokyo CGIAR meeting in 1985 decided to create the Special Program for African Agricultural Research (SPAAR).

One of the main constraints to agricultural productivity in Africa has been found to be low soil fertility. It is thought that only 6% of land on the continent has high agricultural potential. An evaluation of the World Bank's interventions in the African agricultural sector between 1990 and 2006 found that only 60% of projects were rated as satisfactory in outcome (World Bank 2007). It was concluded that this is attributable to political instability and weak institutional capacity, in addition to other constraints such as soil infertility and lack of access to credit. This points to the importance of research that develops crop varieties that are suitable for poor soil or develops fertilizers that are affordable for poor African farmers.

Given the importance of constraints to productivity, such as access to credit, projects that include credit or institutional capacity components should also be transferable to other countries, since the issues they address are common to most African countries. Examples from the World Bank's interventions in the agricultural sector include Rwanda's Agriculture and Rural Markets Development Project (1995) and Mali's Agricultural Competitiveness and Diversification Project (2006) (World Bank 2007).

IFAD, unlike the World Bank, supports partnerships at community level to translate local efforts into global environmental benefits (IFAD 2012b). It operates with international partnerships to transfer positive outcomes and frameworks to similar situations where applicable. IFAD aims to work with international standards and guidelines from international conventions, such as the Rio Convention, for institutional and policy changes, as well as the removal of barriers to trade (IFAD 2012a, b). Furthermore, IFAD says it follows a tailor-made approach; that is, flexible approaches to respond more effectively to the needs of individual countries. It claims that this is possible because of its flexible lending and non-lending instruments (IFAD 2012b, c). At a macro-level, IFAD intends to expand policy engagement and strengthen partnerships with national and international organizations, as well as public and private donors (IFAD 2012a).

Another prominent institution is GEF. This independent financial organization supports developing countries in carrying out programmes to achieve environmental protection. Since its establishment, GEF has provided more than US$6.2 billion in grants and US$20 billion in cofinancing projects to over 1800 projects (IFAD 2012b). Similar to the World Bank, IFAD and CAADP, GEF promotes tailored programmes, such as the National Adaptation Programmes of Action (NAPA), which are financed by global institutions but administered by national and local focal points to address individual countries' needs.

One example is that of Sudan, which has signed 16 multilateral environmental agreements (MEAs), the majority of which are based on aid provision for development and LDCs. The best funded MEAs are the climate change (United Nations Framework Convention on Climate Change, UNFCCC) and biodiversity (Convention on Biological Diversity, CBD) conventions, which are also funded by GEF (UNEP 2007). According to GEF (2012) Sudan, since joining GEF, has received a total of US$19.14 million for environmental programmes and projects. Through GEF-5 (July 2010–June 2014) Sudan will receive an indicative allocation of US$8.88 million to execute projects related to climate change, biodiversity, and land degradation.

A good example of scaling up and transferability is the GEF-funded NAPA, though the allocated funds are very limited. The NAPA Priority Intervention is to build resilience in the agriculture and water sectors to mitigate the adverse impacts of climate change in Sudan. It started in January 2010 with GEF funding of US$3 million, US$500,000 from UNDP Khartoum-Sudan Office, and an equivalent of US$3.3 million in local currency from the Government of Sudan. The project is an integral part of the Least Developed Countries Fund (LDCF) supported NAPA follow-up projects. The project aimed at achieving greater resilience of the most vulnerable communities through its various activities.

Among the main achievements of the NAPA follow-up project in Sudan are the following (Elhag et al. 2012):

- In situ water harvesting through terraces: earth bunds and deep ploughing in Gedarifand and South Darfur led to substantial increase in crop productivity. Yields increased from 50 to 150%, benefiting 730 households in Gedarif and 420 households in South Darfur states.
- In River Nile and North Kordofan states: water-efficient irrigation of crops and shelterbelts were provided using both conventional and solar pumps (supporting a switch to solar-powered water pumps) to irrigate 317 hectares planted with different crops, with a result of 20 to 60% increase in productivity.
- Early maturing and drought-resistant varieties were developed and used in South Darfur and Gedarif states to ensure higher crop productivity.
- New cash crops were introduced in all states: increase in household incomes in the four states; net profits ranged between US$500 per

household per season for tomato growers in the River Nile and US$1207 per season for cucumber planters in South Darfur.
- Micro-fencing in four villages in North Kordofan for sand dunes fixation delivered impressive results for yields and increased overall land productivity. Seedlings were planted inside the fences and tended (supplementary irrigation delivered during summer benefited 56 farmers); women have had an essential role in these adaptation measures, especially in establishing nurseries and tree planting. Women in North Kordofan were active in committees and sand dunes fixation activities. More than 800 women benefited from the project through crop cultivation, butane gas provision, animal husbandry, restocking, and breed improvement.

These achievements encouraged the Canadian International Development Agency (CIDA) to fund the Sudan NAPA projects, both to scale up the existing activities in the same location and/or transfer them to other places. Total CIDA funding is US$2.8 million, and the government allocated an equal amount to this both in kind and in local currency.

11.6 Recommendations

From the above discussion, the following recommendations can be made:

- The observed trend of donor renewed interest in agriculture should continue and be translated into increased volume of aid to agriculture, especially to those aspects directly related to agricultural productivity.
- Donor countries should increase aid directed to investments in the prime movers of development, such as human capital, technologies, and institutional innovation to increase farm production and hence agricultural growth. With more financial resources, NEPAD should focus on increasing African and donor investment in genetic and agronomic research on the major food staples to reduce food prices, which is an important aspect of poverty reduction strategies (Eicher 2003).
- Donors should provide adequate finance to infrastructure to help reduce the costs of transportation of food, locally and nationally. This requires a stronger partnership of different organizations to coordinate the areas of commitment. Since aid programmes are

increasingly focusing on a systems approach, it is clear that one initiative cannot address all aspects of a system. Therefore, greater and more efficient coordination between aid and development organizations and donor countries is essential to address different aspects of agricultural productivity.

- The donor community should provide access to markets. This requires that developed and developing countries work together to give the latter greater and fairer access to global markets and support their self-sufficiency efforts by reducing reliance on imported food (USAID 2012). They should also reduce subsidies which have been harming exports of low-income countries over the years. This situation has led the World Bank, among others, to lobby for a genuinely supportive Doha Round and for the elimination of OECD agricultural subsidies in international forums, but admittedly with limited success (World Bank 2007: xxvii). This also involves increasing developing countries' roles in decision-making positions in international politics (UK Food Group 2008; UNDP 2013).
- Less developed countries should change perception of agriculture among the youth. Agricultural activities are considered to be a low-status livelihood with low incomes, especially compared with life in urban areas. UNDP (2012a) recommends that countries need to make agriculture more attractive to young people, economically, socially, and culturally. This should also include the participation of local and national non-governmental organizations (NGOs), as they tend to have greater access to local communities and greater knowledge of local and national circumstances.
- Aid strategies should focus on raising people's incomes and livelihoods by strengthening local production systems, local markets, and fair trade. This also requires the protection of markets by introducing safety nets, strengthening social services, and empowering women and small-scale farmers, building people's capabilities.
- The recipient countries should earmark more resources to increase their capacities to absorb and manage aid-financed projects.
- Foreign aid should increasingly reflect the interest of recipient countries and thus should be translated into less tied aid. But in general foreign aid to agriculture should be increased substantially to reflect the importance of agriculture and agriculture productivity in poverty alleviation in low-income counties.

11.7 Conclusions

The recent history of foreign aid starts with the Marshall Plan (US aid to Europe) in 1948. Since then developed countries have committed themselves to give small percentages of their national incomes to support development effort in developing countries. However, aid also reflects the interests of donor countries, including military, political, or commercial interests. In fact, many critics have expressed their concern over the costs incurred by developing countries as a result of agricultural policies of developed countries. Figures for such costs go up to US$50 billion a year. Others estimate that costs are five times the aid granted. The idea is that some aid money, instead of helping poor countries, has effectively been used to fund the trade deficits of less developed countries that have resulted from trade liberalization supported by aid money.

Total aid flows increased from around US$58 billion in 2000 to US$133 billion in 2011. But aid to agriculture remained low, around 5%, although some major providers such the World Bank's IDA gave a larger share, 9%. Yet aid to agriculture declined between 1981 and 2001 (by up to 60%) before picking up, especially since the world food crisis of 2007–8. The main reasons behind the decline were both external and internal. They included, among others, such factors as donor shift to social sector funding and the perception that agricultural investment is risky and its returns are low. This perception is corroborated by the weak institutional and human capacity of recipient countries to design, administer, and implement projects. That said, agriculture remains the backbone of the economies of many African countries, and its importance in alleviating poverty is beyond question. Research results show that for the poorest 10%, a 1% growth in GDP increases income by more than 2.5%.

Recent debates on aid, agriculture, and poverty alleviation have focused on the concept of sustainable agriculture. This involves including different dimensions in project planning and implementation to enable it to be sustainable over the long run. It also includes looking beyond the dichotomy of producers and consumers by including aspects of policies, environment, and opinion as well as the interests of different stakeholders in the preparation and implementation of projects. Numerous projects, programmes, and action plans have been initiated over the past years, many of which have been unsuccessful because they were not sustainable after the termination of the programmes and the drying up of funding. Sustainable agriculture also includes meeting people's social needs and maintaining environmental protection, which requires the coordination of several actors/donors on multiple levels.

Aid effectiveness has recently became a top agenda item in donors' priorities because of concerns about its overall impact on poor countries, and also because of budget pressures in donor countries as well as queries raised by their taxpayers. However, principles developed to gauge aid effectiveness have focused on maximizing efficiency, transparency and the like, which are less relevant to recipient countries. Factors that have adversely affected aid effectiveness have been food aid and the shift to policy-based lending; and both need to be given serious attention by donors to measure aid effectiveness more meaningfully.

Despite scepticism about aid effectiveness and the negative spillover effects on the economies of recipient countries, there exist successful experiences in aid-supported projects that could be candidates for both scaling up and/or transferring across countries. Prominent amongst these are donor-supported agricultural research institutions such as CGIAR; or projects with elements that address major constraints to African agriculture, for instance those addressing access to credit and institutional capacity-building (Ethiopia, Togo, Rwanda, Mali) or climate change-related projects such as GEF's National Adaptation Programmes of Action (Sudan).

Acknowledgements We are grateful to Samah El-Bakri who provided valuable research assistance, and contributed greatly to the outcome of this paper through her insightful comments.

Notes

1. President Truman's inaugural speech, 20 January 1948.
2. Address by Mark Malloch Brown, UNDP Administrator, Makerere University, Kampala, Uganda, 12 November 2002.
3. UN Office of the Special Advisor on Africa (2010).

References

Amekawa, Y. 2010. *Towards Sustainable Agriculture in the Developing World. Theoretical Perspectives and Empirical Insights.* Graduate Theses and Dissertations. Paper 11294. Ames: Iowa State University.

ASI Agricultural Sustainability Institute at UC Davis. 2012. *What Is Sustainable Agriculture?*. Davis: University of California Davis. http://www.sarep.ucdavis.edu/sarep/about/def. Accessed 27 October 2012.

Assal, M.A.M. 2008. *Is It the Fault of NGOs Alone? Aid and Dependency in Eastern Sudan*. Sudan Working Paper 5. Bergen: Chr. Michelsen Institute. http://www.cmi.no/publications/file/3067-is-it-the-fault-of-ngos-alone.pdf. Accessed 27 October 2012.

Bovard, J. 1986. *The Continuing Failure of Foreign Aid*. CATO Institute Policy Analysis No 65. Washington, DC: Cato Institute.

Brown, M. 2009. *Rapid Assessment of Aid Flows for Agricultural Development in Sub-Saharan Africa*. FAO Investment Centre Division Discussion Paper 2009/09. Rome: Food and Agriculture Organization of the UN.

Cohen, J.M. 1987. *Integrated Rural Development: The Ethiopian Experiences and the Debate*. Uppsala: The Scandinavian Institute of African Studies.

Coppard, D. 2010. *Agricultural Development Assistance. A Summary Review of Trends and the Challenges of Monitoring Progress*. San Pedro: One Development Group. http://homepageone.s3.amazonaws.com/cmspercent2F20100518044145-Agricultural_Development_Assistance.pdf. Accessed 04 November 2012.

Eicher, C.K. 2003. *Flashback: Fifty Years of Donor Aid to African Agriculture*. Paper Presented at International Policy Conference on 'Successes in African Agriculture: Building for the Future', 1–3 December 2003. Pretoria: South Africa. http://www.ifpri.org/events/conferences/2003/120103/papers/paper16.pdf. Accessed 27 October 2012.

Elhag F., M. Elhassan, and A. Khatir. 2012. *NAPA Best Practices in Sudan*. Documentation Study 2012/09. Khartoum: Higher Council for Environment and Natural Resources.

Elliot, K. 2011. *US Super Committee in Subsidy Saving for Equity, Efficiency and Development*. Aid Effectiveness, Global Development Tags, 14 November. Washington, DC: Center for Global Development.

Elliot, K., and E. Collins. 2012. *Assessing Quality of Aid for Agriculture*. Policy Paper 010, 2012/08. Washington, DC: Center for Global Development.

Elliot, K. 2012. *Can We Assess Agricultural Aid Quality for Agriculture*. Aid Effectiveness Tags, QuODA. Washington, DC: Center for Global Development.

GEF. 2012. *Sudan and GEF*. Country Fact Sheet Paper 2012/10. Washington, DC: Global Environment Facility.

IFAD. 2012a. *Enabling Poor Rural People to Improve Their Food Security and Nutrition, Raise Their Incomes and Strengthen Their Resilience*. Rome: International Fund for Agricultural Development. http://www.ifad.org/sf/index.htm. Accessed 10 November 2012.

———. 2012b. *The Global Environment Facility (GEF)*. Rome: International Fund for Agricultural Development. http://www.ifad.org/operations/gef/index.htm. Accessed 10 November 2012.

———. 2012c. *What We Do. In.* Rome: International Fund for Agricultural Development. http://www.ifad.org/operations/index.htm. Accessed 10 November 2012.

———. 2012d. *Who We Are. In.* Rome: International Fund for Agricultural Development. http://www.ifad.org/governance/index.htm. Accessed 10 November 2012.

Gilligan, D.O., J. Hoddinott, and A.S. Taffesse. 2008. *Impact of Ethiopia's Productive Social Safety Nets Programme and Its Linkages.* IFPRI Discussion Paper 00839. Washington, DC: International Food Policy Research Institute.

Islam, N. 2011. *Foreign Aid to Agriculture: Review of Facts and Analysis.* IFPRI Discussion Paper 01053. Washington, DC: International Food Policy Research Institute.

Kassie, M., and P. Zikhali. 2009. *Brief on Sustainable Agriculture.* Paper Presented at Expert Group Meeting on 'Sustainable Land Management and Agricultural Practices in Africa. Bridging the Gap between Research and Farmers', 16–17 April 2009. Gothenburg: Sweden.

Khan, W. 2011. *Feeding the Developing World with Sustainable Agricultural Methods. Possibilities, Constraints, and Proposals.* Paper Presented at 'Conference of the Association for Heterodox Economics', 6–9 July 2011. Nottingham: United Kingdom. http://www.hetecon.net/documents/ConferencePapers/2011Non-Refereed/Khan_AHE2011052P.pdf. Accessed 31 October 2012.

Kuyvenhoven, A. 2008. Africa, Agriculture, Aid. *NJAS–Wageningen Journal of Life Sciences* 55 (2): 93–112.

NEPAD. 2003. *Comprehensive Africa Agriculture Development Programme.* New Partnership for Africa's Development Publication 2003/07. Midrand: NEPAD. http://www.nepad.org/system/files/caadp.pdf. Accessed December 2012.

Ngambeki, D.S. 2003. *Towards a Green Revolution in Africa: Harnessing Science and Technology for Sustainable Modernisation of Agriculture and Rural Transformation.* Sustainable Development Division Paper 2003/06. Addis Ababa: UN Economic Commision for Africa. http://repository.uneca.org/bitstream/handle/10855/3810/bib-29687_I.pdf?sequence=1. Accessed 01 November 2012.

Oxfam. 2002. *Europe's Double Standards: How the EU Should Reform Its Trade Policies with the Developing World.* Oxfam Briefing Papers 2002/01. Brussels: Oxfam International.

Sen, A. 1985. *Commodities and Capabilities.* Oxford: Oxford University Press.

———. 1989. Development as Capability Expansion. *Journal of Development Planning* 19: 41–58.

Spagnoli, F. 2010. Statistics on International Development Aid. http://filipspagnoli.wordpress.com/stats-on-human-rights/statistics-on-poverty/statistics-on-international-development-aid/. Accessed February 2013.

Tibbett, S. 2011. *CAADP. A Toolkit for Civil Society Organization, Engagement and Advocacy*. Johannesburg: ActionAid International. http://www.actionaid.org/sites/files/actionaid/caadp_toolkit_to_print.pdf. Accessed 17 November 2012.

UK Food Group. 2008. *More Aid for African Agriculture. Policy implications for Small-scale Farmers*. London: UK Food Group.

UN Office of the Special Advisor on Africa. 2010. *Aid to Africa 2010*. Policy Brief 1, October. New York: UN Office of the Special Advisor on Africa.

United Nations Department of Economic and Social Affairs. 2012. *Food Security and Nutrition and Sustainable Agriculture*. New York: UN Division for Sustainable Development. http://sustainabledevelopment.un.org/index.php?menu=258. Accessed 03 November 2012.

United Nations Development Programme. 2012a. *Towards a Food Secure Future*. Africa Human Development Report. New York: United Nations Development Programme.

———. 2012b. *Towards a Food Secure Future*. Africa Human Development Report: Report Summary. New York: United Nations Development Programme.

———. 2013. *The Rise of the South: Human Progress in a Diverse World*. Human Development Report. New York: United Nations Development Programme.

United Nations Environment Programme. 2007. *Sudan's Post-Conflict Environmental Assessment*. Nairobi: Earthprint.

USAID. 2012. *Investing in Sustainable Agriculture*. Washington, DC: USAID. http://www.usaid.gov/what-we-do/agriculture-and-food-security/investing-sustainable-agriculture. Accessed 27 October 2012.

World Bank. 2003. *Reaching the Rural Poor—A Renewed Strategy for Rural Development*. Washington, DC: World Bank.

———. 2007. *World Bank Assistance to Agriculture in Sub-Saharan Africa*. An IEG Review. Washington, DC: World Bank.

———. 2008. *Agriculture for Development*. World Development Report. Washington, DC: World Bank.

———. 2010. *Growth and Productivity in Agriculture and Agribusiness*. Washington, DC: World Bank.

———. 2011. *Increased Productivity and Food Security, Enhanced Resilience and Reduced Carbon Emissions for Sustainable Development, Opportunities and Challenges for a Converging Agenda: Country Examples*. Washington, DC: World Bank.

CHAPTER 12

The Global Partnership on Foreign Aid for Sustainable Development

Yongfu Huang and Muhammad G. Quibria

12.1 Introduction

Expanding populations and economies in the current world are placing considerable pressure on the environment while the widespread ageing of the populace and rapid technological change are placing great stress on social equity and cohesion. Environmental sustainability and social inclusion have become major political priorities, especially for developing countries. This research focuses on the global partnership in areas such as foreign aid for sustainable development, which is one important vision articulated in the Millennium Declaration.[1]

Since high growth performance does not necessarily bring about high levels of development, sustainable development has been increasingly

This chapter was originally published as "The Global Partnership for Sustainable Development" in the *Natural Resources Forum*, Blackwell Publishing, 2015, Volume 39, Issue 3–4, pp. 157–74.

Y. Huang (✉)
ICC, National Development and Reform Commission of PRC, Beijing, China

M.G. Quibria
Department of Economics, Morgan State University, Baltimore, USA

Y. Huang, U. Pascual (eds.), *Aid Effectiveness for Environmental Sustainability*, https://doi.org/10.1007/978-981-10-5379-5_12

regarded as the primary objective in many countries. As an alternative approach to the traditional growth path that concentrates only on economic advancement, sustainable development pursues a balance between economic development, social equity and environmental protection, not as conflicting goals but as pillars which complement each other.[2]

At a time when the world is faced with environmental degradation and rising inequity and poverty, developing countries are much more vulnerable to adverse shocks than the developed nations. This is due to various reasons such as low adaptation capacity, weak regulatory systems and disproportionate dependency on natural resources. They need financial assistance from developed countries to support their efforts towards a sustainable future. In this respect, foreign aid by developed countries has played an important role in the global arena to boost prosperity in developing countries.

The history of foreign aid dates back to the days immediately after the Second World War, when aid was used to address the impacts of war in Europe as well as other reconstruction efforts. Since the 1950s, the objective of foreign aid has been to promote economic growth and combat poverty and inequality in developing countries. In recent decades, as environmental degradation and income inequality have reached alarming proportions, the purpose of aid has broadened to include multiple goals—such as the United Nations (UN) Sustainable Development Goals (SDGs) that focus on poverty, environment, literacy, health, women's rights and so on. A number of foreign aid projects and programmes have been designed and established to integrate environmental sustainability and social inclusion into agendas of development cooperation.[3]

The total volume of official development assistance (ODA) in the recent post-economic crisis period reached a peak in 2010, but declined in 2011 for the first time since 1997. This widened the gap between actual disbursements and the amounts committed in accordance with the UN target of 0.7% of donor country gross national income (GNI). The recent fiscal austerity and economic challenges have increased pressure on the traditional donors to produce tangible results for their governments and taxpayers, as there is widespread scepticism about aid effectiveness.

The literature on aid effectiveness is voluminous; and it focuses primarily on the impact of foreign aid in advancing economic growth. Foreign aid has always had its proponents and opponents.[4] Proponents argue that aid is the driver of continued economic growth in developing countries, and that it generally leads to technological advances and the accumulation of human capital that can sustain economic growth (Hansen and Tarp 2001; Stiglitz 2003; Sachs 2006). Its opponents, however, disparage most aid as unproductive and even counter-productive—a sheer waste of money. For example, aid has been criticized as undermining democracy and freedom, retarding economic development and contributing to larger bureaucracies and less efficient governments (Friedman 1995; Rajan and Subramanian 2008; Easterly 2009). However, systematic research into aid effectiveness for sustainable development is hugely lacking.

This chapter aims to explore the role of a renewed global partnership in promoting sustainable development, with a specific focus on foreign aid, which is highly significant and policy relevant for developing countries.[5] The underlying argument is that with foreign aid, these countries are in an improved financial situation to take care of their natural resources, protect the environment and develop more equitable societies. To achieve sustainable development, the Millennium Declaration calls for international cooperation to go beyond aid to encompass trade, investment, governance and so on. This implies that fundamental changes in the global partnership are required to address current and emerging challenges in such areas as climate change mitigation and adaptation, employment and migration.

This research starts with an illustrative growth model, based on the traditional Solow Model, where foreign aid contributes to sustainable development protecting the environment. It then moves on to empirical analysis based on annual data for 70 aid recipient countries covering the period 1985–2010. Factor-based instrumental variable estimator (factor-IV) and factor-based generalized method of moments estimator (factor-GMM) according to Bai and Ng (2010) and Kapetanios and Marcellino (2010) are used to estimate a dynamic panel data model with endogenous regressors. It investigates three potential channels through which foreign aid could stimulate sustainable development—namely economic growth, natural resource exploitation and energy intensity.

Our research provides evidence that foreign aid has boosted sustainable development of aid recipient countries. This is measured with three different indicators: genuine savings, the ecological footprint/bio-capacity ratio and sustainability-adjusted Human Development Index (HDI). It further suggests that foreign aid has a significant effect on sustainable development through such channels as growth, natural resources and technology (in connection with energy intensity). Although other sources of international financing for development such as non-Development Assistance Committee (DAC) aid and private philanthropy continue to grow, foreign aid remains the main source of funding for development cooperation. We believe this research has important implications for an enhanced global partnership in areas such as foreign aid in the 2030 development agenda for achieving a sustainable future.

The structure of the chapter is as follows. The theoretical model is given in Sect. 12.2. Section 12.3 outlines the empirical framework, followed by the data and stylized facts in Sect. 12.4. The evidence is discussed in Sects. 12.5, and 12.6 concludes the chapter.

12.2 A Theoretical Model

This section provides an illustrative barebones model to motivate the empirical investigation that follows. As this is an illustrative model, we limit our discussion of sustainability to a simple notion of environmental sustainability. This choice was dictated by the fact that a model of sustainability, in its broadest sense, would be analytically virtually intractable.

In the following, we develop a simple Solow model, which follows Brock and Taylor (2010). However, we believe that the present formulation, which deviates from Brock and Taylor in a number of important ways, is analytically somewhat simpler and intuitively more straightforward; yet it yields all the major results, including the environmental Kuznets curve (EKC).

The production function is assumed to be Cobb-Douglas and is given by:

$$X = AK^{a}L^{(1-a)} \tag{12.1}$$

where X, K, L, A represent output, capital, labour and total factor productivity, respectively. Equation (12.1) can be expressed in intensive form:

$$x = Ak^{a} \tag{12.2}$$

where $x = X/L$, is gross output per worker; and $k = K/L$, capital per worker. Finally, as is well known, $0 < a < 1$, which implies that there are diminishing returns to output per worker.

The (net) output is defined by:

$$y = x(1-\lambda) \tag{12.3}$$

Where Y= net output and $y = Y/L$, net output per worker; and λ= a fixed proportion of the domestic (gross) output devoted to emission control.

It is assumed that both domestic savings rate and the foreign savings rate are proportionately related to net output; and national savings rate is sum of domestic and foreign savings. Thus, positing, domestic savings = $s_1 y$ and foreign savings rate = $s_2 y$, national savings rate is given by:

$$s = (s_1 + s_2)y \tag{12.4}$$

The capital accumulation equation is given by:

$$dk / dt = sAk^a (1 - \lambda) - (\delta + n)k \tag{12.5}$$

where dk/dt = change in capital per worker. The first term on the right-hand side, $sAk^a(1 - \lambda)$, represents gross investment; the second term, $\delta + n$, is the sum of the depreciation rate and the labour force growth rate. In other words, we have assumed that:

$$\hat{L} \equiv (1 / L)(dL / dt) = n.$$

With respect to pollution, we have assumed the following emission function:

$$e = \theta x / Az, \; 0 < \theta < 1 \tag{12.6}$$

where, as in the rest of the chapter, we have expressed emission, e, in per worker units.

The following provides some explanatory observations with respect to the emission Eq. (12.5):

1. First, it is assumed that emission varies proportionately with gross output x, the scale of activity. The proportion is given by ϕ. This is a standard assumption in the literature, used, among others, by Brock and Taylor (2010).

2. Second, we assume that abatement of emissions varies inversely with technology. As Reis (2001) suggests, a higher value of A indicates cleaner technology. We have further assumed that technological progress takes place exogenously at a rate π. In other words, $\hat{A} = \pi$.
3. Finally, it is assumed that emissions decrease as more resources are targeted toward abatement. We have assumed that a fixed proportion of gross output, λx, is devoted to abatement; λ is a policy parameter, subject to change. The abatement function is given by:

$$z = (\lambda x)^{\mu} \text{ with } 0 < \mu < 1 \tag{12.7}$$

Equation (12.7) states that resource expenditures for pollution control have a positive but diminishing impact on abatement. This assumption is both plausible and consistent with the existing literature.

The balanced growth path can be derived as follows.

Equation (12.4) would imply:

$$\hat{k} \equiv \dot{k} / k = sAk^{a-1}(1-\lambda) - (\delta + n) = 0 \tag{12.8}$$

Thus, the steady-state solution k^* is given by:

$$k^* = \{sA(1-\lambda)/(\delta + n)\}^{1/(1-a)}$$

This expression shows that the higher the proportion of output devoted to abatement (say, in improving technology), the lower the steady-state k^*. As k^* decreases, y^*, the steady-state per capita income decreases. This, however, does not affect the steady-state growth rate. On the other hand, an increase in the savings rates—both domestic and foreign—increase k^*, the steady capital–labour ratio, and long-run net output, y^*. However, note that foreign aid does not affect the long-run growth rate, though it affects the long-run income level.

Next, we explore the implications of inflow of foreign assistance on the environment. First, we investigate how the Solow steady-state relates to the environmental Kuznets curve (EKC). However, to do so, let us consider Eq. (12.6). Substituting $z=(\lambda x)^{\mu}$ from Eq. (12.7) into Eq. (12.6), denoting $\theta^{*}\equiv\theta a^{(1-\mu)}$ and simplifying, we can derive:

$$e=\theta^{*}k^{(1-\mu)a}\,/\,A^{\mu}\lambda^{\mu} \tag{12.9}$$

The equation can be rewritten in the proportionate rate of change form, noting that $\hat{\theta}^{*}=0$.

$$\hat{e}=(1-\mu)a\hat{k}-\mu\hat{A}-\mu\hat{\lambda} \tag{12.10}$$

This can be rewritten by:

$$\hat{e}=(1-\mu)a\hat{k}-\mu\pi-\mu\hat{\lambda} \tag{12.11}$$

As is evident from Eq. (12.11):

- Growth in emissions is negatively related to technological progress as well as increases in the rate of expenditures in abatement.
- Other things remaining the same, the emission curve, Eq. (12.11), exactly mirrors the Solow fundamental equation of growth Eq. (12.8); Eq. (12.11) produces the EKC, which is a linear transformation of Eq. (12.8), as it is obvious from inspection of both the equations.
- When $\hat{k}=\hat{\lambda}=0$, $\hat{e}=-\mu\pi<0$. This implies that the EKC reaches its downward slope before the model reaches the Solow steady-state solution if there is technological progress, assuming other parameters remain the same. However, the maximum point of the EKC will approach faster if expenditures for abatement increase or if there is an improvement in technology that reduces the emission parameter related to output, θ.
- When $\hat{e}=(1-\mu)a\hat{k}-\mu\pi=0$, that is, when growth in emissions stops, it can be seen: $\hat{k}=\mu\pi/(1-\mu)a>0$. In other words, the emissions growth rate reaches zero—that is, the highest point EKC curve—at a capital per worker or income level that lies below the corresponding Solow steady-state levels.

- Finally, it can be easily demonstrated that $d\hat{e}/ds_2 = (1-\mu)a(d\hat{k}/dk)(dk/ds_2) < 0$. This means that the inflow of foreign aid (indicated by s_2)—through its impact on economic growth and technology—helps reduce the growth of emission.

The above relationship between the Solow steady-state solution and the EKC can be seen from the following geometric exposition. From Eq. (12.8), we can define the steady solution as follows:

$$k^* = \{k : \hat{k} = 0\} \tag{12.12}$$

Assuming $\hat{\theta}^* = \hat{\lambda} = 0$ and substituting $\hat{k}$ from Eq. (12.8) into Eq. (12.11), we can define k^{**}, the capital per worker where the EKC reaches its maximum, as follows:

$$k^{**} = \{k : \hat{e} = (1-\mu)a\hat{k} - \mu\pi = 0\} \tag{12.13}$$

1. Now the above discussion can be summarized as follows: When $\hat{e} = 0$, then $\hat{k} > 0$.
2. When $\hat{k} = 0$, then $\hat{e} = -\mu\pi < 0$.

Together, (1) and (2) imply that $k^* < k^{**}$. In this process shortening the EKC, foreign aid can play an important part.

As we have demonstrated above, the above illustrative model is consistent with important stylized facts of environment–growth interrelationship. It also shows that appropriate domestic policies, such as higher expenditures on pollution abatement or technological innovations in green technology can help usher in a greener phase of the EKC faster than a stance of benign neglect. This illustration also highlights the importance of foreign aid, or similar form of foreign finance, in helping environmental sustainability by promoting development and new technology, such as low carbon technology.

12.3 An Empirical Framework

This section formulates the empirical model and outlines estimation methods developed by Bai and Ng (2010) and Kapetanios and Marcellino (2010) for the linear dynamic panel data model with fixed effects and endogenous regressors when the cross-sectional dimension (N) and time series dimension (T) are large.

Section 12.2 provided an illustrative model to explore the links between foreign aid and a concept of sustainable development that focuses on environmental sustainability. This section examines a broader concept of sustainable development which encompasses economic development, social development and environmental protection.

The sustainable development process, denoted by SD_{it}, can be modelled as a function of its lag ($SD_{i,t-1}$), foreign aid indicator (AID_{it}), a number of control variables of 'beyond aid' ($BEYONDAID_{it}$) and transmission channels ($CHANNEL_{it}$), for country i ($i = 1, 2, \ldots, 70$) at time period t ($t = 1, 2, \ldots, 26$), as follows:

$$SD_{it} = \gamma_i + \alpha_1 SD_{i,t-1} + \alpha_2 AID_{it} + \alpha_3 CHANNEL_{it} + \alpha_4 AID \times CHANNEL_{it} + \alpha_5 BEYONDAID_{it} + \varepsilon_{it}\; i = 1, 2, \ldots, 70;\; t = 2, \ldots, 26$$

γ_i is fixed effects and ε_{it} is the error term. Independent variables (AID_{it}, $CHANNEL_{it}$, $AID_{it} \times CHANNEL_{it}$, $BEYONDAID_{it}$) are assumed to be endogenous with respect to the error ε_{it}, owing to possible measurement error and/or simultaneity. Data and measures for the dependent variable and independent variables are discussed in the next section.

To estimate the above dynamic panel data model with fixed effects, we demean the data for each country to control for fixed effects at the first place. The above model can be simplified as follows:

$$y_{it} = \alpha y_{i,t-1} + x_{it}'\beta + \theta_{it} \quad i = 1, 2, \ldots, 70;\; t = 2, \ldots, 26 \tag{12.14}$$

Where y_{it} is the demeaned SD_{it}, x_{it} is a vector of demeaned endogenous regressors (AID_{it}, $CHANNEL_{it}$, $AID_{it} \times CHANNEL_{it}$, $BEYONDAID_{it}$) and θ_{it} is the demeaned error ε_{it}. More specifically, $y_{it} = SD_{it} - T^{-1} \sum_{T}^{t=1} SD_{it}$. $y_{i,t-1}$, x_{it} and θ_{it} are defined in the same manner. We assume that $E(x_{it}\theta_{it}) \neq 0$ for all i and t.

For a dynamic panel data model with endogenous regressors, it is common practice to use the past values of observed variables as instruments to estimate the parameters of the model following, for example, the well-known GMM approaches by Arellano and Bond (1991) and Blundell and Bond (1998). However, these approaches are typically effective for the case of fixed T panel data setting.

To estimate a dynamic panel data model with endogenous regressors when N and T are both large, Bai and Ng (2010) and Kapetanios and Marcellino (2010) propose using estimated factors as instruments for endogenous regressors. Using a standard instrument and strong factor asymptotics in a data-rich environment where many instruments are weakly exogenous for endogenous regressors, Bai and Ng (2010) propose the factor-based instrumental variable estimator, denoted by factor-IV. Bai and Ng (2010) note that 'if the variables in the system are driven by common sources of variations, then the ideal instruments for endogenous variables in the system are their common components'. They suggest using the estimated common factors as instrumental variables for endogenous regressors. More specifically, they assume that regressors are driven by a small number of unobservable common factors as follows:

$$x_{it} = \Lambda_i^{'} F_t + u_{it} \tag{12.15}$$

Where Λ_i is a $r \times K$ matrix of factor loadings with fixed and bounded components (r is the number of common factors and K is the number of regressors). F_t is a $r \times 1$ matrix of common factors, which are assumed to be uncorrelated with θ_{it}. u_{it} is the error term, which is assumed to be correlated with θ_{it}.

These researchers (2010) show that the common component, $\Lambda_i^{'} F_t$, is the ideal instrument for x_{it}, and is more effective than F_t in terms of convergence rate and the mean squared errors of the estimator. However, $\Lambda_i^{'} F_t$ is not observable, and needs to be estimated.

Bai and Ng (2004, 2010) suggest using a principal component analysis on the observed data on endogenous regressors to estimate Λ_i and F_t by solving the following minimization problem[6]:

$$\begin{aligned} V(k) &= (NT)^{-1} \sum_{N}^{i=1} \sum_{T}^{j=1} \left(x_{it} - \Lambda_i^{'} F_t \right)^2 \\ s.t.\ & \frac{\Lambda_i^{'} \Lambda_i}{N} = I_r ; \frac{F_t^{'} F_t}{T} = I_r \end{aligned} \tag{12.16}$$

Let $X_i = (x_{i1}, x_{i2}, \ldots, x_{iT})'$ be the $T \times K$ matrix of endogenous regressors for the ith cross-sectional unit, so we have the following $T \times (NK)$ matrix for all cross-sectional units:

$$X = (X_1,,, X_2,,, \ldots,,, X_N)$$

The principal component estimate of factor matrix, denoted by $\tilde{F}_t$, can be expressed as $\sqrt{T}$ times the eigenvectors corresponding to the r largest eigenvalues of the $T \times T$ matrix XX'. Given F_t, the estimated factor loading matrix, denoted by $\tilde{\Lambda}_i$, can be computed by $\frac{X' F_t}{T}$. The estimated common factors, $\widetilde{\Lambda_i' F_t}$, are the ideal instruments for x_{it}. A remaining issue now is how to determine the number of common factors, r.

For the approximate factor model such as $x_{it} = \Lambda_i' F_t + u_{it}$, Bai and Ng (2002) develop a method to estimate the number of factors using information criteria, which could be the only rigorous method available so far.[7] They suggest (ibid.) using a principal component analysis on the observed data to calculate the number of factors by minimizing[8]:

$$IC(r) = \ln\left[(NT)^{-1} \sum_{N}^{i=1} \sum_{T}^{j=1} \left(x_{it} - \Lambda_i^{r'} F_t^r\right)^2\right] + rg(N,T) \qquad (12.17)$$

With respect to $r \in \{0, 1, \ldots, r_{\max}\}$ for some fixed $r_{\max}$. The above criterion function captures a trade-off between a measure of fit captured by the first term and a penalty function, $g(N, T)$, that depends on the size of panel. When the number of factors increases, the fit must improve, but the penalty goes up. Among the many criterion functions proposed by Bai and Ng (2002), $IC_{p2}(r)$ is used since it has the largest penalty on the fitted factor number, where $g(N,T) = \left(\frac{N+T}{NT}\right) \ln\left(\frac{NT}{N+T}\right)$. The estimation of $\tilde{F}_t^r$ and $\tilde{\Lambda}_i^r$ are the same as above. The integer that minimizes a criterion function is the estimated number of factors.

With $\widetilde{\Lambda_i' F_t}$ as instruments, the following pooled two-step least squares estimator has been proposed:

$$\hat{\beta}_{PFIV} = \left(\sum_{N}^{i=1} \sum_{T}^{t=1} \tilde{\Lambda}_i' \tilde{F}_t x_{it}'\right)^{-1} \sum_{N}^{i=1} \sum_{T}^{t=1} \tilde{\Lambda}_i' \tilde{F}_t y_{it} \qquad (12.18)$$

Bai and Ng (2010) show that when T and N are of comparable magnitudes, $\hat{\beta}_{PFIV}$ is $\sqrt{T}$ consistent and asymptotically normal. They suggest that the factors contracted from the endogenous regressors are not only valid but also more strongly correlated with endogenous regressors than each individually observed instrument. Accordingly, factor-IV estimation is more efficient than standard IV or GMM estimation that uses a large number of observed variables as instruments. The factor-IV estimator is consistent even if the number of instruments exceeds the sample size. It is also consistent even when the observed variables are invalid instruments, as long as the unobserved factors driving the economy are valid instruments.

Based on the assumption that there are many or weak instruments having a weak factor structure, Kapetanios and Marcellino (2010) propose the factor-based GMM approach, denoted by factor-GMM, to estimate this type of model. They argue that the new penalty function, $g(N, T) = \ln[\min(N, T)]^{-1}$, ensures consistency of the estimated number of factors even in the case of weak-factor structure, while the criteria by Bai and Ng (2002) tend to underestimate the number of factors. They also argue that variable preselection based on their correlations with endogenous variables, in comparison to using a large number of variables from a large dataset such as in Stock and Watson (2005), can effectively alleviate the problems of weak instrument and weak-factor structure when constructing instruments. However, Bai and Ng (2010) find that valid instruments can be constructed from endogenous regressors, which are a small number of variables selected from a large dataset. Kapetanios and Marcellino (2010) also show that factor-GMM estimation is more efficient than the standard GMM estimation that applies the observed variables as instruments.

12.4 Data and Stylized Facts

12.4.1 *Measures and Data on Sustainable Development*

A number of indicators have been proposed in the literature for measuring sustainable development, such as the sustainability-adjusted human development index (SHDI), adjusted net savings, total wealth, and ecological footprint. The following summarizes the sustainability indicators used in this study[9]:

Sustainability-Adjusted Human Development Index (SHDI)
The HDI is the most widely used overall measure for human progress covering four aspects: life expectancy, literacy, education and standard of living (UNDP 2011). However, 'the HDI does not take into account sustainability variables in a broader sense', as Pineda (2012) notes. He subsequently suggests imposing a loss function to a country's human development achievements, given its unfair use of the environment, according to planetary boundaries. The following representations have been proposed:

$$SHDI^i = \left(1 - G^i\right)^* HDI^i$$

$$G^i = \frac{1}{p}\sum_{p}^{j=1} G_j^i$$

$$G_j^i = \min\left\{1, \frac{POP_i}{POP - POP_i} \frac{\left[S_j^i - \overline{S}_j\right]_+}{\overline{S}_j}\right\}$$

Where G_j^i and S_j^i are the loss function and the level of environmental use for environmental indicator j ($j=1, 2, \ldots, p$) and country i, respectively. POP_i is the population of country i and POP is the population in the world. $\overline{S}_j$ is the global planetary boundary for environmental indicator j. The operator $[\,]_+$ is defined as $[x]_+ = \max[x, 0]$. $\frac{POP_i}{POP - POP_i}$ is the global responsibility term which implies that the larger a country's population, the greater its responsibility for the use of the environment. $\frac{\left[S_j^i - \overline{S}_j\right]_+}{\overline{S}_j}$ is the fair share of the environment term, which captures the situation when a country's environmental utilization exceeds its fair share.

Following Pineda (2012), this analysis considers three environmental indicators to compute SHDI: carbon dioxide emissions per capita, natural resource depletion and the share of permanent cropland.[10] For the lower threshold of the planetary boundary for these environmental indicators, this research uses one standard deviation above the mean. Data on population, total carbon dioxide emissions per capita, forest depletion (per cent of GNI), mineral depletion (per cent of GNI), energy depletion (per cent of GNI) and permanent cropland (per cent of land area) are from the World Development Indicators (WDI) Database (World Bank 2012). Annual data on HDI for the period 1980–2010 are taken from the United Nations Development Programme (UNDP) (2011).[11]

Genuine Savings or Adjusted Net Saving (GSAV)
This sustainability indicator, developed by the World Bank under its work programme on the wealth of nations, is based on stock accounting (total wealth) and flow accounting (genuine or adjusted new savings). Genuine or adjusted new savings measure changes in total wealth over time, taking into account natural resources depletion, pollution damage and investment in human capital. More specifically, the series of adjusted net savings provided by the WDI Database (World Bank 2012) are equal to net national savings (gross savings less the value of depreciation of produced assets) plus the value of investment in human capital (education expenditure) and minus the value of resource depletion (energy depletion, mineral depletion, net forest depletion) and environmental degradation (carbon dioxide). It measures the extent to which countries use the income generated from produced and nature capital to invest in education to increase their total wealth over time. It is the true savings rate of an economy in terms of generating and maintaining total wealth, including produced capital, human capital and natural capital.[12] Data on adjusted net savings, excluding particulate emission damage (per cent of GNI) are taken from the WDI Database (World Bank 2012).

Ratio of Ecological Footprint to Bio Capacity (EFBIO)
This indicator, proposed by Moran et al. (2008), measures environmental sustainability: the higher the indicator, the lower the level of sustainable development. Labelling it the 'earth-equivalent ratio', Moran et al. (2008) calculate the ratio of ecological footprint per capita to globally available bio-capacity per capita. They argue that this ratio measures 'the minimum number of earth-equivalent planets that would be required to support the current human population if a given country's level of consumption were universal'. An increasing earth-equivalent ratio would imply that man is consuming more of the earth's natural resources and adding to the acceleration of environmental degradation, while a decreasing ratio would indicate that we are approaching sustainability. Since a ratio above 1 indicates that ecological goods and services are consumed faster than the rate of

biosphere regeneration, Moran et al. (2008) argue that if sustainability is to be achieved, the minimum requirement is an earth-equivalent ratio no greater than 1. In other words, development and resource use can be sustainable only if the demand on the biosphere stays within the regenerative capacity of the planet over time.

The framework of ecological footprint and bio-capacity, first proposed by Rees (1992), remains a leading biological accounting tool in comparing man's demands on the present-day ecosystems with the planet's gross ecological capacity to sustain human life.[13] More specifically, ecological footprint addresses the aggregate demand of an economy on ecosystems by measuring how much land and water areas are needed to support the consumption of a given population and to assimilate the corresponding wastes. It is a consumption-based indicator, equal to the sum of the ecological footprint of production and imports of ecologically embedded goods minus the exports of ecologically embedded goods. Bio-capacity describes the supply side of an economy in providing a flux of biological resources and services useful to humanity by calculating the total area of ecologically productive land. The unit of the two measurements is the global hectare per capita.

The Global Footprint Network has gathered facts annually since 2007 from Food and Agriculture Organization of the United Nations (FAO) and UNDP to develop a database of ecological footprint and bio-capacity data for 241 countries for the period 1961 to 2008. Our analysis utilizes the data from the Global Footprint Network (2012).[14]

12.4.2 Measures and Data on Independent Variables

The key independent variable is foreign aid, denoted by *AID*. Foreign aid is the international transfers of capital, goods or services from a country or international organization for the benefit of a recipient country or its population. It can be humanitarian or development aid, official or private or non-governmental aid, and bilateral or multilateral. Development aid was defined by the Organisation for Economic Co-operation and Development's (OECD's) DAC in 1969 as the 'flows of official financing administered with the promotion of the economic development and welfare of developing countries as the main objective and which are concessional in character with a grant element of at least 25 per cent'.[15] Development aid usually consists of ODA, official assistance and private voluntary assistance. ODA, accounting for the bulk of total development aid, refers to grants or loans to countries and territories on the DAC list of recipients (developing countries) and to multilateral agencies that meet certain conditions.

This research considers two indicators: the ratio of net aid transfers to gross domestic product (GDP), denoted by *NAT*, and the ratio of ODA received to GDP, denoted by *ODA*. Data on net aid transfers and net ODA received are taken from Roodman (2006).[16] Data for total GDP by purchasing power parity (PPP) (constant 2005 international dollars) are from the WDI Database (World Bank 2012).

The following control variables for the 'beyond aid' scenario are used in this analysis:

- gross domestic savings (*GDS*);
- gross national income per capita (*GNIPC*);
- trade openness (*TRADE*);
- financial depth measured by the ratio of M2 to GDP (*M2*)
- institutional quality measured by polity indicator (*POLITY*);
- urbanization (*URBAN*); and
- population growth rate (*POPGR*).

Data for gross domestic savings (per cent of GDP), GNI per capita (constant 2000 US$), trade (per cent of GDP), M2 (money and quasi money, per cent of GDP), urban population (per cent of total population) and population growth rate (annual per cent) are from the WDI Database (World Bank 2012). Data for the polity indicator, polity2, are taken from the PolityIV Database (Marshall and Jaggers 2012). The polity indicator is often used to measure institutional quality based on freedom of suffrage, operational constraints, balance of executives, and respect for other basic political rights and civil liberties.

Three potential channels are investigated, namely economic growth, nature resource exploitation and energy intensity.[17] Economic growth, denoted by *GR*, is the GDP per capita growth (annual per cent); nature resource exploitations, denoted by *NRENT*, is measured by total natural resources rents (per cent of GDP), which is the sum of rents from oil, natural gas, coal (hard and soft), minerals and forests that are generated from the exploitations of those natural resources. Data for annual GDP per capita growth rate and total natural resources rents (per cent of GDP) are taken from the WDI Database (World Bank 2012). Energy intensity, denoted by *EINTEN*, is measured by the final energy intensity of GDP at purchasing power parities. Data for energy intensity are from the Enerdata's Global Energy Market Data (2012).

The sample includes 70 aid recipient countries in 1985–2010, as listed in Table 12.6. We exclude the countries that have fewer than 15 annual observations for the dependent variables, ODA-to-GDP ratio, GNI per capita, polity indicator or natural resource rents.[18] Variable definitions and sources are given in Table 12.5.

12.4.3 *Stylized Facts*

Before presenting the panel data evidence, we review a few stylized facts on the sustainable development process of the past decades. Figure 12.1, which plots the kernel density evolution path of sustainability-adjusted HDI and genuine savings,[19] shows that in 1985 the sustainability-adjusted HDI ranged between 0.2 to 0.8, with the most concentrated value less than 0.6. Since then, this sustainability indicator has been increasing, reaching its most concentrated value at about 0.7 in 2010. Although the most concentrated value of genuine savings was almost the same in 1985 and 2010, its distribution in 1985 was dispersed mainly between −20 to 20% while in 2010 it was about −5 to about 40%. This figure shows in general a trend towards increased sustainability over the past decades. Here we attempt to investigate whether foreign aid has played any significant role in this process.

12.5 Evidence

This section presents econometric evidence on the sustainability effects of foreign aid for 70 aid recipient countries over the period 1985–2010. It then examines the potential channels through which foreign aid can stimulate sustainable development. Both the factor-IV and factor-GMM estimates are based on Bai and Ng (2010) and Kapetanios and Marcellino (2010). The number of common factors is determined as one, using the new penalty function, $g(N, T) = \ln[\min(N, T)]^{-1}$, from Kapetanios and Marcellino (2010). For the factor-GMM estimates, the lagged values of estimated factors from t-1 to t-10 are used as instruments. The panel-robust standard errors based on Arellano (1987) are reported in brackets to adjust for serial correlation.

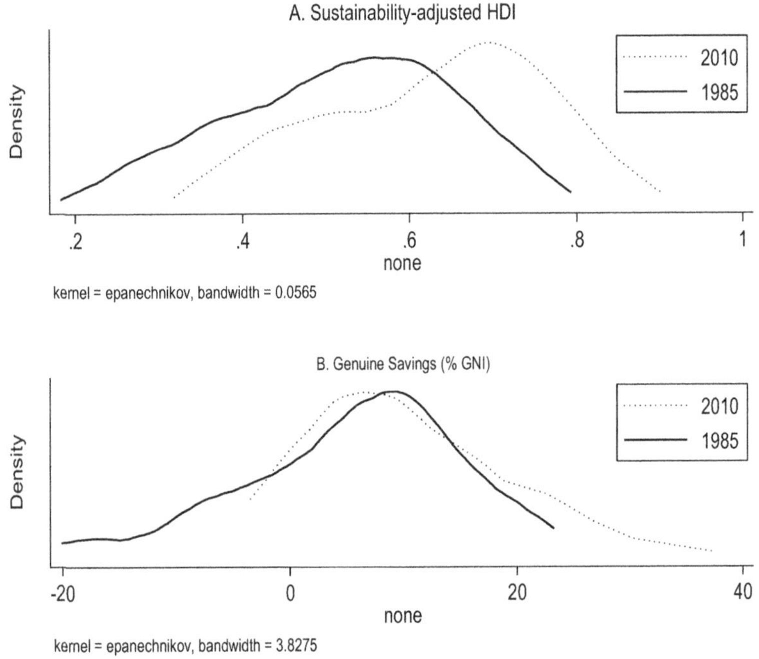

Fig. 12.1 Kernel density evolution of sustainable indicators. Note: This figure shows the kernel density plots of the distribution of two sustainability indicators in 1985 and 2010. Variables and data are described in the text. Source: Authors' own calculation

12.5.1 Baseline Models

Using the three different sustainability indicators, we examine here whether foreign aid has increased sustainability in recipient countries. Table 12.1 gives the results when *NAT* (net aid transfers, per cent of GDP), is used to measure foreign aid. In this table, factor-GMM estimates are preferable for genuine savings and the footprint/bio-capacity ratio, as endogeneity tests clearly reject the null and Hansen J tests cannot reject the null for factor-GMM estimates. This suggests that regressors in this context are endogenous and instruments constructed from these regressors are valid. For the sustainability-adjusted HDI indicator, factor-IV estimates are preferable, as indicated by the p-value of endogeneity test. *NAT* is found to exert a positive impact on sustainable development, as

Table 12.1 The effects of foreign aid (measured by NAT) on various sustainability indicators, 1985–2010

Dependent variable	*Sustainability-HDI*		*Genuine savings*		*Footprint/bio-capacity ratio*	
Method	*Factor-IV*	*Factor-GMM*	*Factor-IV*	*Factor-GMM*	*Factor-IV*	*Factor-GMM*
Lag dependent var.	0.966***	0.984***	−0.576	0.004***	0.758***	0.493***
	[0.010]	[0.007]	[0.738]	[0.001]	[0.140]	[0.041]
Net aid transfers (% GDP)	0.020*	0.017***	13.870*	7.472***	−0.006	−0.112***
	[0.012]	[0.003]	[7.781]	[0.247]	[0.038]	[0.017]
'Beyond aid'						
Domestic gross savings	0.000***	0.000***	−0.008	0.001	0.000	0.000***
	[0.000]	[0.000]	[0.007]	[0.000]	[0.000]	[0.000]
GNI per capita	0.000	0.000	−0.011	−0.007***	0.000	0.000***
	[0.000]	[0.000]	[0.009]	[0.002]	[0.000]	[0.000]
Trade openness	0.007***	0.001	1.425	0.576***	−0.003***	−0.000
	[0.001]	[0.001]	[1.018]	[0.073]	[0.001]	[0.001]
Financial depth	−0.005	−0.005***	0.477	0.360***	0.006*	0.009***
	[0.004]	[0.001]	[0.419]	[0.087]	[0.003]	[0.001]
Governance	0.015	0.039***	−2.117	0.475	0.005	−0.019
	[0.011]	[0.006]	[2.487]	[0.446]	[0.007]	[0.012]
Urbanization	0.043**	0.031***	2.363	0.915***	−0.004	−0.017***
	[0.021]	[0.010]	[2.059]	[0.329]	[0.009]	[0.006]
Population growth	0.088	−0.086*	27.016	13.780***	−0.011	0.097***
	[0.106]	[0.051]	[21.717]	[1.443]	[0.046]	[0.030]
Constant	0.439***	0.422***	3.966**	0.542*	0.040*	0.020***
	[0.029]	[0.017]	[1.940]	[0.288]	[0.022]	[0.006]
R-squared	0.969	0.967	0.941	0.982	0.779	0.612
Hansen J	0.00	23.16	0.00	20.12	0.00	16.53
Endogeneity (*P*-value)	0.01	0.12	0.55	0.00	0.20	0.07

(*continued*)

Table 12.1 (continued)

Dependent variable	*Sustainability-HDI*		*Genuine savings*		*Footprint/bio-capacity ratio*	
Method	*Factor-IV*	*Factor-GMM*	*Factor-IV*	*Factor-GMM*	*Factor-IV*	*Factor-GMM*
Number of countries	70	70	70	70	70	70
Observations	1750	1610	1750	1610	1725	1587

Note: This table reports factor IV/GMM estimates based on Bai and Ng (2010) and Kapetanios and Marcellino (2010). See text for definitions of variables and data sources. Hansen J test examines the null that the instruments are valid. Endogeneity test examines the null that the specified endogenous regressors can actually be treated as exogenous. Panel-robust standard errors based on Arellano (1987) are reported in brackets

*Significant at 10%; **significant at 5%; ***significant at 1%

Source: Authors' own calculations

measured by both sustainability-adjusted HDI and genuine savings. As we noted earlier, 'the higher this indicator, the lower the level of sustainable development' if sustainable development is measured by the footprint/bio-capacity ratio, and we expect therefore that *NAT* has a negative impact. Factor-GMM estimates clearly support this assumption for *EFBIO*.

Table 12.7 presents evidence for the robustness of the results, when ODA (per cent of GDP) is used to measure foreign aid. The pattern in Table 12.7 is fairly similar to that of Table 12.1.

12.5.2 Full Models

Whereas the previous section reported the results for the baseline models, here we examine the full models (with interaction terms). Tables 12.2, 12.3 and 12.4 present our investigation of the existence of three potential channels through which foreign aid may stimulate sustainable development: economic growth, natural resource exploitations and energy efficiency; these and their interaction terms with *NAT* have been added to the baseline models in which three different sustainability indicators are still used.

Divergent patterns emerge for the sustainability indicators when economic growth and its interaction term with *NAT* are included (Table 12.2). The factor-GMM estimates suggest that for sustainability-adjusted HDI,

Table 12.2 Transmission channel: economic growth

Dependent variable	*Sustainability-HDI*		*Genuine savings*		*Footprint/bio-capacity ratio*	
Method	*Factor-IV*	*Factor-GMM*	*Factor-IV*	*Factor-GMM*	*Factor-IV*	*Factor-GMM*
Lag dependent var.	0.974***	0.974***	1.001***	0.960***	0.527**	0.677***
	[0.011]	[0.012]	[0.016]	[0.018]	[0.225]	[0.080]
Net aid transfers (NAT)	0.016	0.011	−0.007	0.362**	−0.121	−0.049**
	[0.012]	[0.007]	[0.124]	[0.146]	[0.116]	[0.025]
Economic growth (GR)	0.073***	0.076***	0.069	0.168**	−0.000	0.006
	[0.011]	[0.010]	[0.047]	[0.088]	[0.012]	[0.010]
NAT × GR	−0.001	−0.001**	0.008	−0.035*	0.012	−0.002
	[0.001]	[0.001]	[0.005]	[0.018]	[0.024]	[0.002]
'Beyond aid'						
Domestic gross savings	−0.000	0.000	−0.000	0.001	0.000	0.000
	[0.000]	[0.000]	[0.000]	[0.000]	[0.000]	[0.000]
GNI per capita	0.000	0.000	−0.000	0.000**	0.000*	0.000***
	[0.000]	[0.000]	[0.000]	[0.000]	[0.000]	[0.000]
Trade openness	0.002	0.001	0.019***	0.012	0.002	−0.002
	[0.002]	[0.002]	[0.006]	[0.010]	[0.005]	[0.002]
Financial depth	−0.003	−0.002	−0.010	0.025**	0.008	0.007
	[0.003]	[0.003]	[0.006]	[0.011]	[0.006]	[0.004]
Governance	0.012	0.019	0.019	0.041	−0.019	−0.007
	[0.010]	[0.012]	[0.033]	[0.062]	[0.021]	[0.009]
Urbanization	0.030*	0.035**	−0.020	0.082*	−0.025	−0.010
	0.015]	[0.016]	[0.031]	[0.043]	[0.026]	[0.007]
Population growth	0.131*	0.128	0.595**	1.505***	0.021	0.035
	[0.073]	[0.096]	[0.248]	[0.366]	[0.115]	[0.067]
Constant	0.456***	0.443***	0.077	0.055	0.024	0.028***
	[0.025]	[0.027]	[0.053]	[0.074]	[0.017]	[0.008]
R-squared	0.982	0.981	0.993	0.990	−0.062	0.709
Hansen J	0.00	31.61	0.00	38.78	0.00	26.55
Endogeneity (*P*-value)	0.16	0.64	0.36	0.97	0.31	0.73
Number of countries	70	70	70	70	70	70

(*continued*)

Table 12.2 (continued)

Dependent variable	*Sustainability-HDI*		*Genuine savings*		*Footprint/bio-capacity ratio*	
Method	*Factor-IV*	*Factor-GMM*	*Factor-IV*	*Factor-GMM*	*Factor-IV*	*Factor-GMM*
Observations	1718	1578	1718	1578	1713	1575

Note: This table reports the evidence for the channel of economic growth, GDP per capita growth (annual %), using NAT (% GDP) to measure foreign aid. See Table 12.1 for more notes

Source: Authors' own calculations

Table 12.3 Transmission channel: natural resource exploitations

Dependent variable	*Sustainability-HDI*		*Genuine savings*		*Footprint/bio-capacity ratio*	
Method	*Factor-IV*	*Factor-GMM*	*Factor-IV*	*Factor-GMM*	*Factor-IV*	*Factor-GMM*
Lag dependent var.	0.938***	0.981***	0.884***	0.888***	0.760***	0.405**
	[0.025]	[0.011]	[0.027]	[0.032]	[0.147]	[0.169]
Net aid transfers (NAT)	0.040	0.019*	0.828***	0.512***	−0.004	−0.146***
	[0.071]	[0.010]	[0.201]	[0.130]	[0.040]	[0.053]
Natural resources rents (NRENT)	0.011	0.014	0.080	−0.155	0.004	0.003
	[0.010]	[0.009]	[0.049]	[0.101]	[0.006]	[0.006]
NAT × NRENT	−0.034**	−0.001	−0.056	−0.303*	0.001	0.022**
	[0.016]	[0.009]	[0.064]	[0.159]	[0.002]	[0.010]
'Beyond aid'						
Domestic gross savings	0.000***	0.000**	−0.000	−0.001**	0.000	0.000
	[0.000]	[0.000]	[0.001]	[0.001]	[0.000]	[0.000]
GNI per capita	0.000	0.000	−0.000	0.000*	0.000	0.000***
	[0.000]	[0.000]	[0.000]	[0.000]	[0.000]	[0.000]
Trade openness	0.009*	−0.001	−0.015	0.067*	−0.003***	−0.001
	[0.005]	[0.002]	[0.023]	[0.036]	[0.001]	[0.003]

(*continued*)

Table 12.3 (continued)

Dependent variable	*Sustainability-HDI*		*Genuine savings*		*Footprint/bio-capacity ratio*	
Method	*Factor-IV*	*Factor-GMM*	*Factor-IV*	*Factor-GMM*	*Factor-IV*	*Factor-GMM*
Financial depth	−0.005	−0.007*	0.005	0.004	0.006	0.012*
	[0.005]	[0.004]	[0.013]	[0.012]	[0.004]	[0.006]
Governance	0.016	0.048**	0.171**	−0.148	0.006	−0.028
	[0.014]	[0.019]	[0.083]	[0.191]	[0.008]	[0.020]
Urbanization	0.059*	0.031	0.087	0.103	−0.004	−0.021
	[0.035]	[0.019]	[0.056]	[0.068]	[0.008]	[0.013]
Population growth	0.056	−0.111	0.274	1.463	−0.012	0.103
	[0.154]	[0.114]	[0.676]	[1.071]	[0.048]	[0.138]
Constant	0.448***	0.430***	0.350	0.209	0.041*	0.035***
	[0.033]	[0.030]	[0.291]	[0.191]	[0.023]	[0.012]
R-squared	0.948	0.967	0.991	0.987	0.779	0.584
Hansen J	0.00	25.68	0.00	27.82	0.00	20.56
Endogeneity (*P*-value)	0.17	0.34	0.04	0.37	0.65	0.94
Number of countries	70	70	70	70	70	70
Observations	1731	1590	1731	1590	1725	1586

Note: This table reports the evidence for the channel of natural resources rents (% GDP) using NAT (% GDP) to measure foreign aid. See Table 12.1 for more notes

Source: Authors' own calculations

NAT is no longer significant; however, both economic growth and its interaction term with *NAT* were significant in entering the model. This implies that the level of economic growth is a crucial factor in achieving sustainable development. It also implies that the effect of foreign aid on sustainable development is likely to be transmitted through economic growth as spurred growth; however, once the growth rate reaches a certain level, foreign aid could be detrimental to sustainable development process in aid recipient countries.

As suggested by factor-GMM estimates, both *NAT* and economic growth are significant in the model for genuine savings, as is their interaction term. This indicates that *NAT* could have both direct and indirect effects on genuine savings, and the indirect effect of *NAT* is likely to be

Table 12.4 Transmission channel: energy intensity

Dependent variable	*Sustainability-HDI*		*Genuine saving*		*Footprint/ bio-capacity ratio*	
Method	*Factor-IV*	*Factor-GMM*	*Factor-IV*	*Factor-GMM*	*Factor-IV*	*Factor-GMM*
Lag dependent var.	0.960***	0.939***	0.996***	0.597***	0.659***	0.640***
	[0.015]	[0.026]	[0.034]	[0.049]	[0.091]	[0.084]
Net aid transfers (NAT)	−0.002	−0.079	0.051	−0.054	−0.039**	−0.042**
	[0.017]	[0.048]	[0.155]	[0.160]	[0.019]	[0.020]
Energy intensity (EINTEN)	−0.027***	−0.042***	−0.015	−0.207***	0.004	0.004
	[0.006]	[0.009]	[0.023]	[0.055]	[0.003]	[0.003]
NAT × EINTEN	0.000	0.006	0.004	−0.06***	0.001	−0.002
	[0.002]	[0.006]	[0.015]	[0.019]	[0.001]	[0.002]
'Beyond aid'						
Domestic gross savings	−0.000	0.000	−0.001**	0.001	−0.000	−0.000
	[0.000]	[0.000]	[0.000]	[0.001]	[0.000]	[0.000]
GNI per capita	0.000	0.000	−0.000	−0.000	0.000***	0.000***
	[0.000]	[0.000]	[0.000]	[0.000]	[0.000]	[0.000]
Trade openness	0.003	0.005*	0.017***	0.033*	−0.003	−0.003*
	[0.002]	[0.003]	[0.006]	[0.017]	[0.002]	[0.002]
Financial depth	−0.006	−0.006**	−0.013**	−0.016	0.007*	0.007*
	[0.004]	[0.003]	[0.005]	[0.013]	[0.004]	[0.004]
Governance	0.016	0.013	0.025	−0.054	0.003	0.002
	[0.010]	[0.012]	[0.032]	[0.074]	[0.006]	[0.008]
Urbanization	0.038*	0.047**	−0.012	−0.036	−0.007	−0.007

(*continued*)

Table 12.4 (continued)

Dependent variable	*Sustainability-HDI*		*Genuine saving*		*Footprint/ bio-capacity ratio*	
Method	*Factor-IV*	*Factor-GMM*	*Factor-IV*	*Factor-GMM*	*Factor-IV*	*Factor-GMM*
	[0.019]	[0.021]	[0.032]	[0.058]	[0.007]	[0.007]
Population growth	0.045	0.084	0.435	0.312	0.020	0.036
	[0.101]	[0.094]	[0.288]	[0.556]	[0.057]	[0.062]
Constant	0.449***	0.409***	0.083	0.075	0.023***	0.028***
	[0.026]	[0.033]	[0.057]	[0.086]	[0.006]	[0.009]
R-squared	0.975	0.975	0.984	0.419	0.735	0.695
Hansen J	0.00	24.12	0.00	20.97	0.00	32.80
Endogeneity (*P*-value)	0.23	0.42	0.02	0.04	0.18	0.98
Number of countries	70	70	70	70	70	70
Observations	1694	1552	1694	1552	1692	1552

Note: This table reports the evidence for the channel of energy intensity (measured by final energy intensity of GDP at purchasing power parities). See Table 12.1 for more notes

Source: Authors' own calculations

transmitted through economic growth. The indirect effect is likely to be positive when growth rates are low; however, as growth rates rise, the indirect effect of foreign aid on sustainable development fades and finally becomes negative. This is suggested by the signs of the three coefficients.

With respect to the footprint/bio-capacity ratio, the factor-GMM estimates show no evidence for either economic growth or its *NAT* interaction term, although *NAT* remains significant in the model. Foreign aid could be conducive to a reduction of the ecological footprint; but economic growth seems to have no direct or indirect role in preventing it.

In addition, GNI per capita is noted to be negatively associated with sustainable development while trade openness, financial depth, urbanization and population growth are found to be significantly positively linked to sustainable development.

Table 12.3 attempts to determine whether foreign aid has worked through the natural resource exploitation channel, which is measured by natural resource rents (per cent of GDP). For the marker sustainability-adjusted HDI, the factor-IV estimates show evidence of a significantly negative effect for the interaction term between *NAT* but no evidence is found for *NAT* and natural resource rents. Based on a comparison of column 1 in Table 12.1, we expect that foreign aid could exert an indirect effect on sustainable development via natural resource exploitations.

For the indicators genuine savings and footprint/bio capacity ratio, the factor-GMM estimates suggest both *NAT* and its interaction term with natural resource rents enter the models as significant. Natural resource rents are insignificant in four models. This implies that foreign aid could have a significant direct impact on sustainable development; moreover, it stimulates the process of sustainable development via discouraging the exploitation of natural resources. Furthermore, we observe that domestic gross savings has a negative impact on sustainable development while governance plays a positive role in this process.

In Table 12.4 we investigate whether energy intensity also constitutes a transmission channel through which foreign aid promotes sustainable development. For the sustainability-adjusted HDI and genuine savings, energy intensity enters the models as significant while the significance of *NAT* fades. The effects of foreign aid on sustainable development are likely to be picked up by energy intensity. When the indicator genuine savings is utilized, the factor-GMM estimates further reveal that energy intensity is indeed a transmission path through which the effect of foreign aid is channelled. More specifically, foreign aid could be used to incentivize technological development that could lead to reduced energy intensity and increased sustainable development. The higher the level of energy intensity is, the slower the sustainable development process will be. As energy efficiency improves because of aid-financed technological developments, the effect of foreign aid on sustainable development diminishes.

Both factor-IV and factor-GMM estimates show no evidence on the part of the footprint/bio-capacity ratio for either energy intensity or its *NAT* interaction term, while *NAT* remains negative in the model. Energy efficiency seems to have no role in limiting our ecological footprint.

Columns 3 and 4 further confirm the significantly negative effects of domestic gross savings and financial depth, and a positive impact of trade openness for sustainability. In sum, this analysis shows that the effect of foreign aid on sustainable development has materialized in 70 aid recipient countries over the past two to three decades. It also finds evidence of an effect for GNI per capita, governance, trade openness, financial depth, urbanization and population growth, which are all closely linked to sustainable development. Three transmission paths were identified through which the effect of foreign aid is channelled to sustainable development—economic growth, natural resource exploitations and/or energy intensity.

12.5.3 Policy Discussion

There have been considerable doubts about the effectiveness of foreign aid for sustainable development. For example, UNEP (2007) and Purvis (2003) argue that some foreign aid programmes in the developing countries can give rise to unsustainable development, because of their pace—by worsening pollution and accelerating the exploitation of the aid recipient's natural resources.[20] Over the past decades, and especially after adoption of the Millennium Development Goals (MDGs) in 2000, the international community has taken steps to improve aid effectiveness. Major efforts include the 2005 Paris Declaration, 2008 Accra Agenda for Action and 2011 Busan Partnership for Effective Development Cooperation, all of which are important platforms for discussions for meeting the development goals and increasing aid effectiveness.[21]

However, progress with the Paris Declaration principles and targets has been disappointing.[22]

Our results here are consistent with some recent research that suggests that international assistance programmes promote not only economic but also sustainable development in countries that lack sufficient financial resources to embark on a sustainable development path on their own (e.g. Arvin et al. 2009).[23] The emergence of some developing countries as growth poles and important sources of non-aid development finance has in recent years eroded the relative importance of foreign aid as a source of development finance. Nevertheless, foreign aid, in the era of global sustainability, will continue to play an important role in assisting developing countries implement policies and programmes that facilitate the attainment of sustainable development goals. But more efforts are still warranted in this regard.[24]

Although foreign aid remains important for development finance, achieving positive sustainable development outcomes needs more than just finance. Policies are also essential in such areas as trade openness, financial depth, urbanization and governance to create an environment that is conducive to sustainable development. In the 2030 development agenda ‘beyond aid’, the renewed global partnership should identify effective mechanisms for mobilizing ODA and other development assistance, build more equitable multilateral trading and financial systems, maximize the potential benefits of greater labour mobility and work towards an inclusive and equitable system of global governance—with an increasing voice for and representation by the developing countries.

Our chapter makes a contribution to the existing aid-growth literature, where fierce debates exist regarding the role of foreign aid in economic development. Even though there is a large body of literature that suggests the positive impact of foreign aid on economic development, some recent papers have expressed scepticism with about aid’s effectiveness—for example, Rajan and Subramanian (2008) and Doucouliagos and Paldam (2008). Mekasha and Tarp (2013) apply different meta-analysis techniques on a database of 68 studies on the aid-growth link employed by Doucouliagos and Paldam (2008). They find evidence of the positive and significant weighted average effect of aid on growth and no evidence for the existence of publication bias.[25] As pointed out by Temple (2010), the lack of evidence does not necessarily imply the absence of evidence. With this in mind, empirical results should be interpreted with caution. The finding of this analysis with respect to the growth channel adds to the current literature on aid effectiveness.

The finding with respect to the natural resource channel for aid effectiveness points to the central issue of sustainable development. Natural resources, land and ecological systems provide the goods and services vital for the economy, society and all living creatures. Sustainable development is a pattern of resource use that aims to meet human needs while preserving the environment so that these needs can be met not only by the present generation, but also by future generations. More specifically, it means that human activity must be limited to the carrying (finite) capacity of ecosystems. Man must not exploit nature beyond the point of natural replenishment of resources. If nature's resources (natural capital) are used up faster than they can be replenished, the result is 'unsustainability' of the environment. It is important to take measures to conserve natural resources and prevent their overexploitation. These measures need to include finance and investment, and in this regard foreign aid can play a crucial role. In 2010 the International Monetary Fund (IMF) launched the Topical Trust Fund on Managing Natural Resource Wealth, a multidonor trust fund, for 15–20 low-income and lower middle-income countries, endowed with oil, gas and minerals, to finance technical assistance on manage their natural resource wealth.

Our findings with regard to the energy intensity channel are consistent with Kretschmer, Hübler and Nunnenkamp (2010) who report that aid has been effective in reducing the energy intensity of GDP in 80 recipient countries over 1973–2005. Thus substantial ODA, other forms of development assistance and domestic investment should be earmarked for renewable energy and efficiency projects in developing countries; this could help avoid the otherwise resultant higher greenhouse gas emissions.

12.6 Conclusion

The analysis in this chapter, which is based on careful theoretical and econometric work, finds that an enhanced global partnership with deeper international collective action could effectively contribute to a sustainable development process. It further suggests that foreign aid has significant effects on sustainable development through channels relating to growth, natural resources and a technology with respect to energy intensity. We first present a theoretical framework that illustrates the role of foreign aid promoting environmental sustainability; then we move on to a dynamic panel data study based on annual data for 70 countries from 1985 to 2010 with three indicators to measure sustainable development. Special attention was given to possible channels through which this effect could be transmitted. To address the issue of endogeneity in a large *T* dynamic panel data model, this research applied factor-IV and factor-GMM methods according to Bai and Ng (2010) and Kapetanios and Marcellino (2010).

This research has shed some light on the interaction between development finance, economic growth, natural resource conservation and energy technological progress in the process of global sustainability. The positive effect of aid on sustainability is shown to work through natural resource conservation and/or energy intensity where foreign aid is used to encourage natural resource conservation and technological innovation of energy systems. Growth provides a concrete foundation for environmental protection and social development, and equips governments financially and technologically to fight climate change and stimulate social inclusiveness and development. This research has significant implications for both research and practice.

All this suggests that an enhanced global partnership in areas such as foreign aid, trade, investment, migration (from rural to urban areas) and governance could play a crucial role in the process of global sustainability. Although efforts at the national level are crucial for sustainable development, action at the global level is also needed to provide support for diverging national needs and circumstances. The global partnership for sustainable development should be strengthened to encompass explicit commitments by all countries to the various goals and internationally coordinated measures that strive to create an enabling environment for development, to address the causes of climate change and income inequality, to facilitate sustainable management of the global commons and to achieve economic and financial stability.

Appendix

Table 12.5 Variables

Variable	*Description*	*Source*
SHDI	Sustainability-adjusted human development indicator, based on Pineda (2012)	Own calculations based on annual data on HDI from UNDP
GSAV	The adjusted net savings excludes particulate emission damage (% of GNI)	Calculated on the basis of World Bank (2012)
EFBIO	The ratio of ecological footprint per capita to globally available bio-capacity per capita (also known as earth-equivalents ratio)	Annual data from Global Footprint Network (2012)
NAT	Net aid transfer (NAT) (% of GDP)	Calculated based on aid data from Roodman (2006, revised in 2013) and GDP data from the World Bank (2012)
ODA	Net ODA received (% of GDP)	Calculated based on aid data from Roodman (2006, revised in 2013) and GDP data from the World Bank (2012)
GR	GDP per capita growth (annual %)	World Bank (2012)
NRENT	Total natural resources rents (% of GDP), the sum of rents from oils, natural gas, coal (hard and soft), minerals and forests	World Bank (2012)
EINTEN	Final energy intensity of GDP at purchasing power parities	Enerdata (2012)
GDS	Gross domestic savings (% of GDP)	World Bank (2012)
GNIPC	Gross national income per capita	World Bank (2012)
TRADE	Trade openness (% GDP)	World Bank (2012)
M2	Money and quasi money (M2) (% of GDP)	World Bank (2012)
POLITY	Polity indicator "polity2"	Marshall and Jaggers (2012)
URBAN	Urban population (% of total)	World Bank (2012)
POPGR	Population growth (annual %)	World Bank (2012)

Source: Compiled by authors

Table 12.6 Names and country codes for the 70 aid recipient countries considered in this study

Country name	*Country code*
Albania	ALB
Argentina	ARG
Armenia	ARM
Benin	BEN
Bangladesh	BGD
Bulgaria	BGR
Bolivia	BOL
Brazil	BRA
Botswana	BWA
Chile	CHL
China	CHN
Côte d'Ivoire	CIV
Cameroon	CMR
Congo, Rep.	COG
Colombia	COL
Costa Rica	CRI
Cyprus	CYP
Dominican Republic	DOM
Ecuador	ECU
Egypt, Arab Rep.	EGY
Gambia, The	GMB
Guatemala	GTM
Honduras	HND
Hungary	HUN
Indonesia	IDN
India	IND
Israel	ISR
Jordan	JOR
Kazakhstan	KAZ
Kenya	KEN
Kyrgyz Republic	KGZ
Korea, Rep.	KOR
Sri Lanka	LKA
Morocco	MAR
Moldova	MDA
Mexico	MEX
Mali	MLI
Mozambique	MOZ
Mauritius	MUS
Malaysia	MYS
Namibia	NAM

(*continued*)

Table 12.6 (continued)

Country name	*Country code*
Nicaragua	NIC
Pakistan	PAK
Panama	PAN
Peru	PER
Philippines	PHL
Poland	POL
Paraguay	PRY
Romania	ROU
Russian Federation	RUS
Rwanda	RWA
Sudan	SDN
Senegal	SEN
El Salvador	SLV
Slovak Republic	SVK
Swaziland	SWZ
Syrian Arab Republic	SYR
Togo	TGO
Thailand	THA
Trinidad and Tobago	TTO
Tunisia	TUN
Turkey	TUR
Tanzania	TZA
Uganda	UGA
Ukraine	UKR
Uruguay	URY
Venezuela, RB	VEN
Vietnam	VNM
South Africa	ZAF
Zambia	ZMB

Source: Compiled by authors

Table 12.7 The effects of foreign aid (measured by ODA) on various sustainability indicators, 1985–2010

Dependent variable	*Sustainability-HDI*		*Genuine savings*		*Footprint/bio-capacity ratio*	
Method	*Factor-IV*	*Factor-GMM*	*Factor-IV*	*Factor-GMM*	*Factor-IV*	*Factor-GMM*
Lag dependent var.	0.966***	0.984***	−0.590	0.006***	0.762***	0.498***
	[0.010]	[0.008]	[0.807]	[0.002]	[0.135]	[0.048]
ODA (% GDP)	0.021*	0.019***	14.331*	7.633***	−0.003	−0.110***
	[0.012]	[0.004]	[8.696]	[0.281]	[0.035]	[0.019]
'Beyond aid'						
Domestic gross savings	0.000***	0.000***	−0.007	0.001**	0.000	0.000***
	[0.000]	[0.000]	[0.007]	[0.000]	[0.000]	[0.000]
GNI per capita	0.000	0.000	−0.011	−0.006**	0.000	0.000***
	[0.000]	[0.000]	[0.009]	[0.002]	[0.000]	[0.000]
Trade openness	0.007***	0.001	1.446	0.578***	−0.003***	−0.000
	[0.001]	[0.001]	[1.099]	[0.083]	[0.001]	[0.001]
Financial depth	−0.005	−0.006***	0.588	0.416***	0.006*	0.009***
	[0.004]	[0.001]	[0.502]	[0.092]	[0.003]	[0.001]
Governance	0.015	0.035***	−2.122	0.606	0.005	−0.021*
	[0.011]	[0.006]	[2.605]	[0.447]	[0.007]	[0.011]
Urbanization	0.043**	0.032***	2.186	0.672*	−0.003	−0.016**
	[0.021]	[0.011]	[2.142]	[0.384]	[0.008]	[0.006]
Population growth	0.089	−0.090*	27.843	14.601***	−0.010	0.109***
	[0.107]	[0.051]	[23.730]	[1.647]	[0.046]	[0.033]
Constant	0.439***	0.423***	3.961*	0.517*	0.040 *	0.021***
	[0.029]	[0.017]	[2.163]	[0.294]	[0.022]	[0.006]
R-squared	0.969	0.967	0.935	0.979	0.778	0.599
Hansen J (*P*-value)	0.00	23.51	0.00	20.42	0.00	17.77
Endogeneity (*P*-value)	0.01	0.21	0.56	0.02	0.18	0.10
Number of countries	70	70	70	70	70	70
Observations	1750	1610	1750	1610	1725	1587

Note: This table makes use of ODA (% of GDP). See Table 12.1 for more notes

Source: Authors' own calculations

Table 12.8 The transmission channel of economic growth (using ODA)

Dependent variable	*Sustainability-HDI*		*Genuine savings*		*Footprint/bio-capacity ratio*	
Method	*Factor-IV*	*Factor-GMM*	*Factor-IV*	*Factor-GMM*	*Factor-IV*	*Factor-GMM*
Lag dependent var.	0.974***	0.975***	0.995***	0.963***	0.538**	0.680***
	[0.011]	[0.012]	[0.015]	[0.017]	[0.222]	[0.080]
ODA	0.014	0.012*	0.040	0.346**	−0.114	−0.047*
	[0.012]	[0.007]	[0.120]	[0.139]	[0.114]	[0.024]
Economic growth (GR)	0.072***	0.076***	0.076	0.161*	0.002	0.007
	[0.011]	[0.011]	[0.046]	[0.087]	[0.013]	[0.010]
ODA × GR	−0.001	−0.001*	0.009*	−0.034*	0.013	−0.002
[0.001]	[0.001]	[0.005]	[0.018]	[0.024]	[0.002]	
'Beyond aid'						
Domestic gross savings	−0.000	0.000	−0.000	0.001	0.000	0.000
	[0.000]	[0.000]	[0.000]	[0.000]	[0.000]	[0.000]
GNI per capita	0.000	0.000	−0.000	−0.000**	0.000*	0.000***
	[0.000]	[0.000]	[0.000]	[0.000]	[0.000]	[0.000]
Trade openness	0.003	0.001	0.019***	0.013	0.002	−0.002
	[0.002]	[0.002]	[0.006]	[0.010]	[0.005]	[0.002]
Financial depth	−0.003	−0.002	−0.009	0.025**	0.008	0.007
	[0.003]	[0.003]	[0.006]	[0.011]	[0.006]	[0.004]
Governance	0.012	0.018	0.020	0.044	−0.020	−0.008
	[0.010]	[0.013]	[0.031]	[0.064]	[0.022]	[0.009]
Urbanization	0.030*	0.035**	−0.014	0.079*	−0.023	−0.009
	[0.015]	[0.016]	[0.031]	[0.043]	[0.025]	[0.007]
Population growth	0.130*	0.129	0.618**	1.534***	0.023	0.038
	[0.072]	[0.096]	[0.245]	[0.387]	[0.117]	[0.067]
Constant	0.455***	0.443***	0.082	0.048	0.025	0.030***
	[0.025]	[0.027]	[0.051]	[0.075]	[0.017]	[0.008]
R-squared	0.982	0.981	0.993	0.990	−0.105	0.706
Hansen J (*P*-value)	0.00	31.74	0.00	39.87	0.00	28.96
Endogeneity (*P*-value)	0.31	0.57	0.41	0.93	0.30	0.59
Number of countries	70	70	70	70	70	70
Observations	1718	1578	1718	1578	1713	1575

Note: This table uses ODA (% of GDP) to measure foreign aid. See Tables 12.1 and 12.3 for more notes

Source: Authors' own calculations

Table 12.9 The transmission channel of natural resources exploitation (using ODA)

Dependent variable	*Sustainability-HDI*		*Genuine savings*		*Footprint/bio-capacity ratio*	
Method	*Factor-IV*	*Factor-GMM*	*Factor-IV*	*Factor-GMM*	*Factor-IV*	*Factor-GMM*
Lag dependent var.	0.907***	0.981***	0.898***	0.896***	0.764***	0.416**
	[0.177]	[0.011]	[0.027]	[0.024]	[0.141]	[0.169]
ODA	0.025	0.020*	0.747***	0.578***	−0.003	−0.139***
	[0.128]	[0.011]	[0.178]	[0.130]	[0.037]	[0.052]
Natural resources rents (NRENT)	0.009	0.013*	0.060	−0.072	0.004	0.002
	[0.017]	[0.007]	[0.048]	[0.064]	[0.005]	[0.006]
ODA × NRENT	−0.099	−0.002	−0.047	−0.222*	0.000	0.017**
	[0.268]	[0.011]	[0.063]	[0.116]	[0.002]	[0.008]
'Beyond aid'						
Domestic gross savings	−0.001	−0.000***	−0.000	−0.001	0.000	0.000
	[0.001]	[0.000]	[0.001]	[0.000]	[0.000]	[0.000]
GNI per capita	0.000	0.000	−0.000	−0.001*	0.000	0.000***
	[0.000]	[0.000]	[0.000]	[0.000]	[0.000]	[0.000]
Trade openness	0.010	−0.000	−0.027	0.070*	−0.003***	−0.001
	[0.015]	[0.002]	[0.020]	[0.035]	[0.001]	[0.002]
Financial depth	−0.008	−0.008*	0.003	0.013	0.006*	0.011*
	[0.006]	[0.004]	[0.012]	[0.013]	[0.003]	[0.006]
Governance	0.003	0.043**	0.185**	−0.147	0.006	−0.030
	[0.045]	[0.019]	[0.076]	[0.182]	[0.008]	[0.020]
Urbanization	0.075	0.032*	0.068	0.131*	−0.004	−0.020
	[0.123]	[0.019]	[0.055]	[0.072]	[0.008]	[0.013]
Population growth	−0.091	−0.112	0.092	1.896*	−0.013	0.088
	[0.340]	[0.099]	[0.668]	[1.133]	[0.048]	[0.130]
Constant	0.460***	0.430***	0.354	0.265	0.041*	0.035***
	[0.058]	[0.029]	[0.295]	[0.202]	[0.022]	[0.012]
R-squared	0.829	0.967	0.992	0.990	0.778	0.577
Hansen J (*P*-value)	0.00	23.86	0.00	26.22	0.00	20.24
Endogeneity (*P*-value)	0.22	0.25	0.05	0.98	0.77	0.74
Number of countries	70	70	70	70	70	70
Observations	1732	1591	1732	1591	1725	1586

Note: This table uses ODA (% of GDP) to measure foreign aid. See Tables 12.1 and 12.4 for more notes

Source: Authors' own calculations

Table 12.10 The transmission channel of energy intensity (using ODA)

Dependent variable	*Sustainability-HDI*		*Genuine savings*		*Footprint/bio-capacity ratio*	
Method	*Factor-IV*	*Factor-GMM*	*Factor-IV*	Factor-GMM	*Factor-IV*	*Factor-GMM*
Lag dependent var.	0.960***	0.949***	0.991***	0.599***	0.670***	0.644***
	[0.016]	[0.023]	[0.031]	[0.049]	[0.090]	[0.084]
ODA	−0.008	−0.058	0.124	−0.125	−0.026	−0.035
	[0.022]	[0.041]	[0.150]	[0.197]	[0.017]	[0.021]
Energy intensity (EINTEN)	−0.024***	−0.035***	−0.017	−0.178***	0.003	0.004
	[0.007]	[0.007]	[0.021]	[0.048]	[0.003]	[0.003]
ODA × EINTEN	0.000	0.002	0.008	−0.042***	−0.000	−0.003*
	[0.002]	[0.003]	[0.014]	[0.014]	[0.001]	[0.001]
'Beyond aid'						
Domestic gross savings	−0.000*	0.000	−0.001**	0.001	0.000	−0.000
	[0.000]	[0.000]	[0.000]	[0.001]	[0.000]	[0.000]
GNI per capita	0.000	0.000	−0.000	−0.000	0.000***	0.000***
	[0.000]	[0.000]	[0.000]	[0.000]	[0.000]	[0.000]
Trade openness	0.004*	0.004*	0.018**	0.034**	−0.003*	−0.003*
	[0.002]	[0.002]	[0.007]	[0.017]	[0.002]	[0.002]
Financial depth	−0.005	−0.005*	−0.013**	−0.010	0.007*	0.007*
	[0.004]	[0.003]	[0.005]	[0.012]	[0.004]	[0.004]
Governance	0.015	0.018	0.024	−0.046	0.004	0.004
	[0.010]	[0.012]	[0.032]	[0.074]	[0.005]	[0.007]
Urbanization	0.039*	0.045**	−0.005	−0.039	−0.005	−0.006
	[0.020]	[0.020]	[0.030]	[0.060]	[0.007]	[0.007]
Population growth	0.064	0.077	0.425	0.510	0.012	0.034
	[0.106]	[0.079]	[0.283]	[0.530]	[0.056]	[0.058]
Constant	0.447***	0.423***	0.081	0.076	0.025***	0.030***
	[0.027]	[0.030]	[0.059]	[0.080]	[0.007]	[0.009]
R-squared	0.974	0.977	0.984	0.430	0.740	0.701
Hansen J (*P*-value)	0.00	25.55	0.00	21.24	0.00	34.18
Endogeneity (*P*-value)	0.12	0.44	0.01	0.04	0.12	0.55
Number of countries	70	70	70	70	70	70
Observations	1694	1553	1694	1553	1692	1553

Note: This table uses ODA (% of GDP) to measure foreign aid. See Tables 12.1 and 12.4 for more notes

Source: Compiled by the authors

Notes

1. The Millennium Declaration calls for global policies and measures to ensure that 'globalization becomes a positive force for all the world's people'. It focuses on development, poverty eradication, environmental protection, peace, security, disarmament, human right, good governance and protecting the vulnerable. Its broad vision is encapsulated as inclusive and sustainable development.
2. Formally introduced by the World Commission on Environment and Development (WCED) (1987) or Brundtland Report, the most widely accepted definition of sustainable development is 'development which meets the needs of the present without compromising the ability of future generations to meet their own needs'.
3. For example, in 1992 the World Conference on Development and Environment initiated the Global Environment Facility (GEF), a mechanism to facilitate aid for environment. The Organisation for Economic Co-operation and Development's (OECD's) Green Growth Strategy, announced in 2009, is aimed at helping developing countries achieve economic growth, job creation, environmental protection and the development of more equitable societies than those exist today.
4. See Temple (2010) for a review.
5. See UN System Task Team on the Post-2015 UN Development Agenda, UN/DESA, UNDP and others, 'A renewed global partnership for development', March 2013.
6. Typically, when $T < N$, the normalization of $\frac{F_t' F_t}{T} = I_r$ is used. In case of $T > N$, the normalization that $\frac{\Lambda_i' \Lambda_i}{N} = I_r$ is used.
7. Recent research estimates the number of factors for static factor models.

   For example, Kapetanios (2010) develops a method for this purpose based on the behaviour of eigenvalues of a large sample covariance matrix. Onatski (2010) proposes the edge distribution estimator using differenced eigenvalues while Ahn and Horenstein (2013) propose a new approach by maximizing the ratio of two adjacent eigenvalues.
8. Bai and Ng (2002) provide six criterion functions, $IC_{p1}(r)$, $IC_{p2}(r)$, $IC_{p3}(r)$, $PC_{p1}(r)$, $PC_{p2}(r)$ and $PC_{p3}(r)$. In general, $IC(r)$ are easier to use since they do not involve the estimation of a penalty function.
9. Data on sustainable development used in this paper are downloadable from www.yongfu-huang.net/research.html.
10. One of the three indicators used by Pineda (2012) is fresh water withdrawals. However, data on this indicator are largely missing in the WDI; therefore this analysis uses instead natural resource depletion, the sum of forest

depletion (per cent of GNI), mineral depletion (per cent of GNI) and energy depletion.

11. A multiplication factor of 100 is applied on the calculated SHDI.
12. Although this indicator has advantages over GDP as a measure of economic progress, it has been criticized (Stiglitz et al. 2009). For example, it has been said to have narrow economic view of human capital and wealth and to ignore social capital such as trust, respect, altruism, culture and institutions. It focuses on producers that produce and export using their natural resources, rather than consumers that consume natural resources. As such, most developing countries that depend on natural resources exploitation are unsustainable whereas developed nations are fairly sustainable. Moreover, its assumptions on consumption growth, discount rates and asset lifetimes are viewed as unrealistic. Nevertheless, it has been widely used in the literature.
13. Although the indicators of ecological footprint and bio-capacity are appealing, they are not exempt from reproach. The conversion methods that transform energy, food, timber consumption per capita into land units have been heavily criticized as having limited scope with some important consumption and emission aspects not included. They are also assessed to have limited relevance to policy- and decision-makers.
14. A multiplication factor of 100 is applied on the ratio of ecological footprint of consumption per capita to bio-capacity per capita.
15. OECD (n.d.) 'Glossy of Statistical Terms'. Available at: //stats.oecd.org/glossary/detail.asp?ID=6043
16. Net ODA is a capital flow concept while NAT is a net transfer concept. Net ODA data are from DAC-OECD. *NAT* is the net ODA minus old non-aid loan cancellations and interest payments received from developing countries on outstanding concessional loans. A multiplication factor of 1 million is applied on data for net ODA and NAT.
17. The channel of economic growth is motivated by the extensive literature on aid and growth. The selection of nature resource exploitation and energy intensity is due to the fact that they are closely related to environmental sustainability. It is also motivated by a number of papers such as Huang (2012) which presents evidence that natural resource depletion, the key component of sustainability measured by genuine savings, is often affected by some economic variables, output volatility or shocks.
18. Missing data are predicted by using linear approximations based on real GDP per capita (constant prices: chain series) from Heston, Summers and Aten (2012).
19. Since the Global Footprint Network (2012) has annual data on the ecological footprint of consumption per capita and bio-capacity per capita only until 2008, we are unable to report the kernel density distribution in 2010 for *EFBIO*.

20. The United Nations Environment Programme (UNEP) (2007) evaluation of the environmental impacts of 661 humanitarian, recovery and development aid programmes in Sudan in 2006 reports that the vast majority of foreign aid programmes had no positive impact on the environment while three projects actually had adverse environmental effects.
21. The 2005 Paris Declaration on Aid Effectiveness adopted five principles to strengthen aid effectiveness and 13 targets to measure their implementation which were to be achieved by 2010. The principles and targets set out in Paris were reinforced in Accra in 2008. The Fourth High-level Forum on Aid Effectiveness in Busan in 2011 shifted from focusing purely on aid effectiveness to a more holistic approach that looks at the contribution that cooperation can make to overall development effectiveness, marking a turning point in the international consideration of development cooperation.
22. OECD's final report (2012) on the implementation of the Paris principles shows that only Target 4 in terms of coordinated technical cooperation was met at the global level.
23. Arvin, Dabir-Alai and Lew (2006) observe that for the full sample, aid overall had a positive impact on environmental protection. More specifically, they note that in the full sample and in the subsample of upper income nations, a bidirectional causality link existed between foreign aid and pollution. In the lower-income country subsample, only one directional causal relationship existed; environmental pollution decreased with the increase of foreign assistance.
24. For example, environmental aid still attracts only a small percentage of the total international assistance funding. According to a report by UK's National Audit Office (2011), the majority of the overseas aid provided by the UK in 2009–10 was earmarked to such traditional causes as economic development and humanitarian aid. Only very limited funding was given primarily for environmental conservation and climate change mitigation.
25. Mekasha and Tarp (2013) argue further that the homogeneity assumption of fixed effect model is unrealistic in the aid-growth literature; accordingly a random effects model is to be preferred. They find that, with and without non-linear term, the weighted average effect of aid is positive and significant in the random effects model. Furthermore, publication bias is not a problem once heterogeneity is controlled for.

References

Ahn, S.C., and A.R. Horenstein. 2013. Eigenvalue Ratio Test for the Number of Factors. *Econometrica* 81 (3): 1203–1227.

Arellano, M. 1987. Computing Robust Standard Errors for Within-Groups Estimators. *Oxford Bulletin of Economics and Statistics* 49 (4): 431–434.

Arellano, M., and S. Bond. 1991. Some Tests of Specification for Panel Data: Monte Carlo Evidence and an Application to Employment Equations. *Review of Economic Studies* 58 (2): 277–297.

Arvin, M., P. Dabir-Alai, and B. Lew. 2006. Does Foreign Aid Affect the Environment in Developing Economies? *Journal of Economic Development* 31 (1): 63–87.

Arvin, B.M., Z. Kayani, and M.A. Scigliano. 2009. Environmental Aid and Economic Development in the Third World. *International Journal of Applied Econometrics and Quantitative Studies* 9 (1).

Bai, J., and S. Ng. 2002. Determining the Number of Factors in Approximate Factor Models. *Econometrica* 70 (1): 191–221.

———. 2004. A Panic Attack on Unit Roots and Cointegration. *Econometrica* 72 (4): 1127–1177.

———. 2010. Instrumental Variable Estimation in a Data Rich Environment. *Econometric Theory* 26 (06): 1577–1606.

Blundell, R., and S. Bond. 1998. Initial Conditions and Moment Restrictions in Dynamic Panel Data Models. *Journal of Econometrics* 87 (1): 115–143.

Brock, W., and M. Taylor. 2010. The Green Solow Model. *Journal of Economic Growth* 15 (2): 127–153.

Doucouliagos, H., and M. Paldam. 2008. Aid Effectiveness on Growth: A Meta Study. *European Journal of Political Economy* 24 (1): 1–24.

Easterly, W. 2009. Can the West Save Africa? *Journal of Economic Literature* 47 (2): 373–447.

Enerdata. 2012. *Global Energy Intelligence's Global Energy Market Data* (Updated on 1 March 2012).

Friedman, M. 1995. *Foreign Economic Aid: Means and Objectives.* 1st ed. Washington, DC: Hoover Institution Press.

Global Footprint Network. 2012. *National Footprint Accounts, 2011 Edition.* www.footprintnetwork.org

Hansen, H., and F. Tarp. 2001. Aid and Growth Regressions. *Journal of Development Economics* 64 (2): 547–570.

Heston, A., R. Summers, and B. Aten. 2012. *Penn World Tables, Version 7.1.* Philadelphia, PA: Center for International Comparisons of Production, Income and Prices at the University of Pennsylvania.

Huang, Y. 2012. Is Economic Volatility Detrimental to Global Sustainability? *World Bank Economic Review* 26 (1): 128–146.

Kapetanios, G. 2010. A Testing Procedure for Determining the Number of Factors in Approximate Factor Models with Large Datasets. *Journal of Business & Economic Statistics* 28 (3): 397–409.

Kapetanios, G., and M. Marcellino. 2010. Factor-GMM Estimation with Large Sets of Possibly Weak Instruments. *Computational Statistics & Data Analysis* 54 (11): 2655–2675.

Kretschmer, B., M. Hübler, and P. Nunnenkamp. 2010. *Does Foreign Aid Reduce Energy and Carbon Intensities in Developing Countries?* Kiel Working Paper 1598.

Marshall, M.G., and K. Jaggers. 2012. *The Polity IV Project: Political Regime Characteristics and Transitions, 1800–2010.* Vienna, VA: Center for Systematic Peace.

Mekasha, T.J., and F. Tarp. 2013. Aid and Growth: What Meta-analysis Reveals. *Journal of Development Studies* 49 (4): 564–583.

Moran, D.D., M. Wackernagel, J.A. Kitzes, S.H. Goldfinger, and A. Boutaud. 2008. Measuring Sustainable Development: Nation by Nation. *Ecological Economics* 64 (3): 470–474.

OECD. 2012. *Aid Effectiveness 2011: Progress in Implementing the Paris Declaration.* Pairs: OECD Publishing.

Onatski, A. 2010. Determining the Number of Factors from Empirical Distribution of Eigenvalues. *Review of Economics and Statistics* 92 (4): 1004–1016.

Pineda, J. 2012. *Sustainability and Human Development: A Proposal for a Sustainability Adjusted HDI (SHDI).* MPRA Paper 42636.

Purvis, N. 2003. *Greening US Foreign Aid through the Millennium Challenge Account.* Brookings Policy Brief Series, 119.

Rajan, R.G., and A. Subramanian. 2008. Aid and Growth: What Does the Cross-Country Evidence Really Show? *Review of Economics and Statistics* 90 (4): 643–665.

Rees, W.E. 1992. Ecological Footprints and Appropriated Carrying Capacity: What Urban Economics Leaves Out? *Environment and Urbanization* 4 (2): 121–130.

Reis, A.B. 2001. Endogenous Growth and the Possibility of Eliminating Pollution. *Journal of Environmental Economics and Management* 42 (3): 360–373.

Roodman, D. 2006. *An Index of Donor Performance.* CGD Working Paper 67 (Revised February 2013).

Sachs, J. 2006. *The End of Poverty: Economic Possibilities for Our Time.* New York: Penguin Books.

Stiglitz, J.E. 2003. *Globalization and Its Discontents.* 1st ed. New York: W. W. Norton & Company.

Stiglitz, J., A. Sen, and J.P. Fitoussi. 2009. *Report of Commission on the Measurement of Economic Performance and Social Progress.* Paris: OECD.

Stock, J.H., and M.W. Watson 2005. *Implications of Dynamic Factor Models for VAR Analysis.* NBER Working Papers 11467.

Temple, J.R.W. 2010. Chapter 67: Aid and Conditionality. In *Handbook of Development Economics*, ed. D. Rodrik and M. Rosenzweig, vol. 5, 4415–4523. Amsterdam: Elsevier.

UNDP. 2011. *Human Development Report, 2011: Sustainability and Equity: A Better Future for All.* New York: UNDP.

UNEP. 2007. *Sudan, Post-conflict Environmental Assessment: Synthesis Report.* UNEP/Earthprint.

WCED (World Commission on Environment and Development). 1987. *Our Common Future.* New York: Oxford University Press.

World Bank. 2012. *World Development Indicators.* Washington, DC: World Bank.

CHAPTER 13

Conclusion

Yongfu Huang and Unai Pascual

The human battle against environmental degradation and climate change requires enormous efforts and determination. Research presented in this volume has suggested that foreign aid has played an essential role in catalyzing global efforts to address environmental concerns. Foreign aid is clearly vital to many developing countries, especially for the poorest countries, in tackling emerging additional challenges such as climate change and food price increases. It can be used to finance global social, economic and environmental goals, which are often not financed by the private sector. It can also be used to leverage private finance in areas that promote social goals, such as climate mitigation/adaptation.

Resources to address climate change need to be scaled up considerably over the next few decades both in developed and developing countries. Although in recent years the proportion of foreign aid in total financing flows to developing countries is diminishing, foreign aid will not be obsolete; instead it is expected to remain an important source of development finance for many countries in the 2030 development agenda. However, aid will need to be used in a more focused and catalytic manner. It is increasingly important for multilateral development banks to fully

Y. Huang (✉)
ICC, National Development and Reform Commission of PRC, Beijing, China

U. Pascual (✉)
Basque Centre for Climate Change (BC3), Leioa, Spain

Y. Huang, U. Pascual (eds.), *Aid Effectiveness for Environmental Sustainability*, https://doi.org/10.1007/978-981-10-5379-5_13

utilize their catalytic role and to leverage their potential to mobilize additional financing from diverse sources. Furthermore, the current range of available funds and potential of climate capital flows imply that a combination of sources is needed to address the tough financing task for climate actions, which includes strengthening the existing funding sources, exploring alternative finance and encouraging private capital flows.

While foreign aid is clearly important as a source of development finance, making development progress is not just about resources. It is also about policies and public goods that create a favorable environment for development. Appropriate governance and institutional arrangements at the national level need to be in place for efficient, effective and sustainable financing of mitigation/adaption measures, and an enabling environment at the national level can ensure efficient implementation of funds and risk reduction, using international resources and national funds as well as national development and financial institutions (IPCC 2014).

Foreign aid represents a form of global collective action for financing global development goals. Although efforts at the national level are crucial for sustainable development, action at the global level is also needed to provide support for diverging national needs and circumstances. An enhanced global partnership in areas such as foreign aid, trade, investment, migration (from rural to urban areas) and governance could play a crucial role in the process of global sustainability (Huang and Quibria 2017). Huang and Quibria argue that "the global partnership for sustainable development should be strengthened to encompass explicit commitments by all countries to the various goals and internationally coordinated measures that strive to create an enabling environment for development, to address the causes of climate change and income inequality, to facilitate sustainable management of the global commons and to achieve economic and financial stability".

As one of the most essential facets of global efforts to tackle the climate change crisis, foreign aid represents a wise investment in terms of building a cleaner, greener and more sustainable future for us all, and it will continue to play an essential role in spurring green growth in developing countries. In accordance with Article 9 of the "Paris Agreement under the United Nations Framework Convention on Climate Change" signed in November 2015, developed countries intend to continue their existing collective mobilization goal through 2025. The Agreement is very weak unfortunately in the sense that rich countries are not required to assume compulsory aid donations compatible with their stage of development, although the Agreement claims to stick to the principle of "common but differentiated responsibilities". Given the critical situation of environmental degradation and the limited resources available to address it, improving aid effectiveness can play an important role in addition to attracting new sources of finance. Further research into aid effectiveness for sustainable development in some specific sectors, or in general at national, regional and global level, will be needed.

References

Huang, Y., and M. Quibria. 2017. *The Global Partnership on Foreign Aid for Sustainable Development*. Chapter 12 of this book.

IPCC. 2014. *Climate Change 2014: Mitigation of Climate Change*. IPCC Working Group III Contribution to AR5. http://mitigation2014.org/

Index[1]

NUMBERS & SYMBOLS
3H Basin, *see* Huang-Huai-Hai River Basin
1992 Rio Earth Summit, 262
2005 Paris Declaration and Accra Forum, 223
2008 UN Resolution 62/98, 261
2010 UNFCCC 16th Conference of Parties, 283

A
Abu Dhabi, 139
ACCCRN, *see* Asian Cities Climate Change Resilience Network
Access foreign aid in sustainable energy and climate mitigation, 113–115
Accra Agenda for Action (2008), 2, 188, 383, 423
Accra High Level Forum on Aid Effectiveness, 169
Adaptation Coalition Framework, 28
Adaptation Fund, 34, 44n23, 333, 334
Adaptive capacity, 78n3, 280, 284, 344
Adelman, C.C., 224, 230, 244
Adequate capacity, 56, 57
Advisory Group on Energy and Climate Change (AGECC), 82
Affordability, 98, 99, 109
  of modern energy services, 99
Africa, 20, 369
  humanitarian aid, 250n2
Africa Adaptation Programme, 38, 44n22
African Development Bank, 163, 174n43
African universities, 322, 323
AGECC, *see* Advisory Group on Energy and Climate Change
Agence Française de Développment (AFD), 91, 118n6, 192

[1] Note: Page numbers followed by 'n' refer to notes.

Y. Huang, U. Pascual (eds.), *Aid Effectiveness for Environmental Sustainability*, https://doi.org/10.1007/978-981-10-5379-5

Agricultural Competitiveness and Diversification Project (Mali), 387
Agricultural insurance, subsidized, 10, 348, 352
Agriculture, 20, 38, 41
  agricultural production, 161, 283, 339, 349, 351, 373, 377–379, 384–386, 389, 390
  agrobiodiversity, 10
  cash crops, 309, 388
  financing and aid to, 332–337
  food security, 283, 284, 329, 330, 334, 343, 346, 347, 349, 350, 358–361, 377, 378, 381, 382, 384
  irrigation, 341–345, 347, 351, 356–358, 377, 378, 388, 389
  subsistence agriculture, 282, 309, 323
  sustainable, 329–362, 369–392
  swidden, 284
Agriculture and Rural Markets Development Project (Rwanda), 387
Ahn, S.C., 434n7
Aid
  bilateral, 5, 9, 11, 98, 217, 218, 226, 331, 333, 334, 371
  development, 19, 411
  multilateral, 5, 9, 17, 82, 217, 218, 226, 331, 333, 334, 371
  multiple goals with multiple measures, 245
  non-governmental, 8, 146, 147, 217, 262
  official, 3, 5, 201, 202, 217
  policy and implementation level focus, 245
  private, 5, 217
  promoting information transparency, 246
  sustainability on projects and, 245
  systematic and integral approach, 246
  as tool for filling gaps, 246
  See also Foreign aid
Aid effectiveness, 2, 3, 15n1, 26–28, 34, 39, 100, 136, 138, 148–150, 187–194, 198, 203, 204, 217, 222, 223, 245, 332, 371, 377, 382–384, 392, 398, 399, 423, 425, 436n21, 443
Aid paradox, 11, 40, 41
Aid pessimists, 25
Aid policies, 240, 310, 325
Alternate Wetting and Drying (AWD), 342, 343
Amazon Fund, 272
Anderson, E., 221
Angola, 153
Animal husbandry, 309, 389
Annex finance, 288
Arellano, M., 406, 413
Arndt, C., 97, 220
Arvin, B.M., 194, 225
Arvin, M., 436n23
Asafu-Adjaye, J., 226
Asia countries or territories receive foreign aid, 218
Asian Cities Climate Change Resilience Network (ACCCRN), 41, 42
Asian Development Bank (ADB), 87, 89, 106, 112, 155, 174n43
Assal, M.A.M., 375
Aten, B., 435n18
Australia, 154, 274
  market-based REDD+ mechanism, 274
Australian Agency for International Development, 234
Ausubel, J.H., 317, 323
AWD, *see* Alternate Wetting and Drying

B

Bai, J., 399, 404, 406–408, 413, 426, 434n8
Balderas Torres, A.B., 280
Bali, 4, 90, 93
Bali Action Plan, 25
Ban Ki-Moon, 149, 158
Banco Estado, BE (Chile), 118n6
Banco Nacional de Desenvolvimento Econômico e Social, BNDES (Brazil), 118n6
Bancoldex S. A. (Colombia), 118n6
Banerjee, A., 221
Bangladesh, 31, 62, 197, 313, 362n1
Barcelona Development Agenda, 98
Barren land, 232, 233
Base of Pyramid (BoP) Model, 96, 119n7
  in renewable energy program, 97
Bauer, P., 135
Bayer, P., 44n19
Beichuan City, China, 13, 232, 235, 238, 242, 243
Beichuan Library, 238
Bhavnani, R.R., 136
Bilateral aid, *see* Aid
Bilateral assistance, 17, 23, 58, 112
Bilateral funds, climate change, 17, 23
Bill and Melinda Gates Foundation, 19, 172n20, 346, 347
Bio-capacity, 399, 410, 411, 414–423, 430–433, 435n13, 435n15, 435n19
Biodiversity, 1, 4, 7–10, 13, 28, 63, 118n3, 215, 224, 227, 233, 259, 260, 262, 263, 265, 267, 272, 278, 281, 288, 307, 311, 315, 316, 320, 325, 388
Biodiversity conservation, 5–10, 257, 262–264, 266, 269, 271, 275, 280, 281, 288, 290
Biomass depletion, 258
Birner, S., 156
Black Sea Trade and Development Bank, BSTDB (Greece), 118n6
Bloomberg New Energy Finance, 91
Blundell, R., 406
Bond, S., 406
Boone, P., 136, 197, 221
Bosnia, 76
Brazil, 12, 13, 28, 144, 152, 153, 219, 354, 360
  Curitiba, sustainable urban development in, 238
Bretton Woods institutions, 135, 381
Brock, W., 400, 401
Brown, M.M., 370
Brundtland Commission, 186
Brundtland Report, *see* World Commission on Environment and Development
Buchner, B., 43n8
Bulgaria, 58
Burkina Faso, 347
Burnell, P., 222
Burnside, C., 136, 187, 197
Bus Rapid Transit, 37, 191
Busan Partnership for Effective Development Cooperation (2011), 2, 423, 436n21
Bush, George W., 40
Business models, 96, 97, 105, 109, 110, 115, 224
  effectiveness of foreign aid in sustainable energy and climate mitigation, 96–97
Buss, T.F., 222

C

CAADP, *see* Comprehensive Africa Agriculture Development Programme

CADU, *see* Chilalo Agricultural Development Union
Caisse de Dépôt et de Gestion, CDG (Morocco), 118n6
Cambodia, 62
Cambodia Climate Change Alliance (CCCA), 62
Canadian International Development Agency (CIDA), 389
Cancun Agreement, 220
Capacity
  definition of, 55, 78n3
  levels of, 55, 56, 78n5
  See also specific entries
Capacity building, 5, 6, 11, 51–78, 96, 154, 157, 158, 221, 222, 229, 316, 348
  accountability, 272
  to address climate change, foreign aid for, 17–42
  awareness, 28, 32, 37, 272, 357
  country-driven and demand-driven approaches, 60–61
  data and resources sharing, 63
  decision making, improving, 63–64
  education, 61, 62, 310, 317, 321
  local, 358–360
  results monitoring, 62
  scaled up aid, 68–74
  training, 21, 26, 61, 285, 349
Capacity Building in Development of Policy Framework for Promotion of Low Carbon Emission Societies in Central Asia, 73
Capacity development, *see* Capacity building
Capacity Development for Policymakers to Address Climate Change, 64
Capital investment, 216
CAR, *see* Central African Republic, 373
Carbon dioxide emissions, 152, 191, 195–202, 204, 207–211, 226, 315, 337, 341, 348
  reduction of, 13, 151, 187, 351, 353–356
Carbon finance, 37, 96, 167, 284, 355
Carbon Fund, 91, 272, 288
Carbon Partnership Facility (CPF), 91, 274
Carbon sequestration, 10, 259, 260, 264, 269, 280, 312, 315, 325, 339, 341, 348, 353–356, 361
Caribbean, 28, 38, 76, 152
Carnahan, K., 280
Cash crop systems, 309, 388
Cayan, D., 44n24
CBD, *see* Convention on Biological Diversity
CCCA, *see* Cambodia Climate Change Alliance
CDKN, *see* Climate and Development Knowledge Network
CDM, *see* Clean development mechanism
CDS, *see* City development strategies
Central African Republic (CAR), 373
Central American Bank for Economic Integration, BCIE/CABEI (Honduras), 118n6
Central Asia, 42, 73, 151, 190
CERs, *see* Certified emissions reduction
Certified emission credits, 6, 84
Certified emissions reduction (CER), 34
CGIAR, *see* Consultative Group on International Agricultural Research
Chilalo Agricultural Development Union (CADU), 375
China, 12, 13, 27, 67, 141, 143–145, 154, 155, 165, 191, 219,

343–345, 349–351, 354, 356, 358–360
in developing sustainable-city, 227
in urbanization process, 231
China Development Bank, CDB (China), 118n6
CIDA, *see* Canadian International Development Agency
CIFAL Curitiba project, 240
CIFs, *see* Climate Investment Funds
Cities Alliance, 225
urban dwelling improvement and policies, 225
City development strategies (CDS), 224, 232
City planning, 226, 244, 248
specific fields, 226
Clausen, V., 43n11
Clean Development Mechanism (CDM), 6, 27, 28, 32, 34–37, 39–41, 83, 92, 277
Clean energy markets, 87
Clean Technology Fund (CTF), 83, 91, 190, 193
Clemens, M.A., 43n10, 43n11, 136
Climate and Development Knowledge Network (CDKN), 61, 62
Climate change, 6, 7, 11, 78n3, 189, 191–194, 197, 219
adaptation, vii, 18, 22, 23, 26, 27, 31, 32, 37–39, 41, 42, 52, 57, 61, 64, 67, 68, 77, 132, 191, 220, 259, 264, 275, 280, 283, 330–334, 336, 337, 340, 341, 343–345, 348, 353, 356, 358, 361, 362, 399
bilateral funds, 17, 23
capacity building to address, foreign aid for, 42
clean development mechanism, 27, 28, 32, 34, 35, 37, 39–41, 53, 76, 277
climate smart agriculture, 9–10
financing, 174n44
foreign aid and urbanization, 226
mitigation, vii, 18, 22, 23, 25–27, 29, 31, 32, 37, 38, 41, 52, 57, 61, 64, 67, 68, 74, 77, 81–119, 131, 132, 138, 150, 190, 194, 220, 236, 249, 259, 263, 290, 325, 331–334, 336, 337, 339, 342, 345, 347, 353, 354, 357, 359–362, 388, 399, 441
multilateral funds, 17, 23
nationally appropriate mitigation actions, 25
professionals, training, 73–74
Climate Change Initiative, Norway, 271
Climate change mitigation-related aid, defined, 42n7
Climate finance, 15n3, 21, 25, 27, 29–37, 203, 334–338
bilateral funds, 333, 334
compared with aid and climate-change related aid, 35
defined, 33–34
microfinance, 238
multilateral funds, 333, 334
and UNFCCC, 92–93
Climate Investment Funds (CIFs), 3, 23, 83, 91, 131, 160, 193
Climate mitigation, *see* Climate change
Climate Policy Initiative, 31, 37
Climate smart agriculture, 5–11, 14
ClimateWise, 32
Clinton Global Initiative, 132
Cobb-Douglas production function, 400
Cohen report, 375
Collier, P., 43n12, 150, 222
Commission for Africa (2005), 381
Commission for Sustainable Development, 377

Common but differentiated
responsibilities, 443
Community cooperatives, 93
Community forestry, 311
Community institutions, 98
Community ownership, of micro-
energy projects, 104
Comprehensive Africa Agriculture
Development Programme
(CAADP), 384, 385, 387
Concessional credits, 218
Conference of Parties (COP) of
UNFCCC, 92–93
Congo Basin Forest Fund (CBFF),
272
Conrad, B., 345
Consultative Group on International
Agricultural Research (CGIAR),
347, 360, 386, 392
Convention on Biological Diversity
(CBD), 262, 265, 285, 388
Cooking stove, 14, 99, 317–319
COP-13, 4, 90, 93
COP-15, at Copenhagen, 4, 93
COP 21, 131
Copenhagen, 4, 5, 31, 93
Copenhagen Accord, 17, 18, 33, 41,
270
Corruption, 27, 40, 101, 197, 226,
265, 310, 317, 319, 320,
323–325, 376
Costa Rica, 269, 274, 313
market-based REDD+ mechanism,
274
Côte d'Ivoire, 375, 386
Countries or territories receive foreign
aid, 218
Middle East, Africa, Latin America,
Asia, Afghanistan, Pakistan,
Indonesia, Palestine, 218
Country programmable aid (CPA), 4,
21, 42n5
Creditor reporting system (CRS), 84,
118n3, 134, 137, 331
Creed, A, 288
Crop rotation, 356, 378
Cross-border pollution, 226
CRS, *see* Creditor reporting system
CTF, *see* Clean Technology Fund
Cuba, 153, 313
Curitiba, in Brazil, 13, 232, 238
sustainable urban development,
238–243
Czech Republic, 57, 58

D

Dabir-Alai, P., 225, 436n23
DAC, *see* Development Assistance
Committee
Dalgaard, C.-J., 43n11
Davis, C., 279
DBS, *see* Direct budgetary support
De Gryzde, S., 280
Decentralization, 376
Deforestation, 8, 9, 28, 63, 258–260,
262, 264, 265, 267, 268, 272,
273, 281, 294, 295n3, 307, 309,
311, 313, 314, 319, 324, 325
Deforestation and forest degradation
(REDD+), reducing emissions,
312–314
Degradation and global forest lost,
257–259
Demand side management (DSM),
156, 157
Democratic Republic of Congo
(DRC), 141, 143, 250n2, 281,
373
humanitarian aid, 250n2
Denmark, viii, 313
Descriptive statistics, 205
Developing nations, 10, 52–54,
57, 58, 61, 63, 66, 67, 72, 76,
125, 172n15, 250n1, 334,
361
Developing world, 9, 13, 52–54, 96,
97, 147, 215, 216, 226, 309,

330, 332, 339, 340, 345, 348, 352, 355, 359–361
Development
expenditure, 319, 373
green growth, 54, 220, 232, 442
income, 227, 346, 355, 357, 371, 373, 377, 382, 388
investment, 128–130, 132, 141, 143, 150–154, 163, 167, 168, 225, 232, 312, 325, 330, 331, 333, 334, 336–338, 341, 343, 345–353, 357–362, 370, 372, 373, 375–377, 385, 389, 391
livelihoods, 324, 325, 344–346, 355, 373, 383, 385, 390
poverty, 127, 133, 134, 136, 149, 150, 159, 169, 216, 324, 346, 347, 360, 371, 373, 374, 389–391
Development aid, 19, 411
Development assistance, foreign aid, 216
Development Assistance Committee (DAC), 19–23, 42n1, 42n3, 42n5, 84, 217
Development Bank of Southern Africa, DBSA (South Africa), 118n6
Development cooperation, 2, 3, 15n1, 126, 132, 140, 146, 148–150, 155, 163, 220, 398, 399, 436n21
Direct budgetary support (DBS), 3, 83
vs. traditional support, 159–160
Disease control, 217
Dollar, D., 136, 187, 197, 222
Domestic capital markets, 106
Donor agencies, 65, 220, 231
good practice guidance notes, 220
Donor aid, 263, 375
Donor community, 68, 73, 74, 82, 158, 221, 222, 249, 390
Donor countries, 11, 27, 42n5, 66, 71, 98, 100, 101, 141, 218, 310, 317, 325, 369, 371, 389–392, 398
Doucouliagos, H., 43n10, 43n11, 424
DRC, *see* Democratic Republic of Congo
Drought Tolerant Maize for Africa (DTMA) project, 345, 346
DTMA, *see* Drought Tolerant Maize for Africa project
Durschinger, L., 280
Dynamic panel data model, 14, 399, 404–406, 426

E
Earth-equivalent ratio, 410, 411
Easterly, W., 97, 98, 135, 189, 221
Eberstadt, N., 224, 230, 244
Ecocities, development plans, 248
Ecological Footprint to Bio Capacity (EFBIO) ratio, 410, 411, 416
Eco-neighbourhoods, 233
Economic growth, 3, 26, 27, 40, 78n2, 97, 152, 157, 193, 216, 220, 222, 225, 229, 239, 247, 373–375, 385, 398, 399, 404, 412, 416–419, 422, 431
Ecosystem services, 8, 9, 13, 257, 259, 275, 278, 283, 307, 309, 314, 315, 317, 319, 321, 325
Millennium Ecosystem Assessment, 315
payment for ecosystem services, 261, 263, 268, 269
soil protection, 258
Eco-valley corridors, 233
Education, capacity building, 321
EFBIO, *see* Ecological Footprint to Bio Capacity ratio
Eicher, C.K., 384

EKC, *see* Environmental Kuznets curve
Eliasch Review, 273
Emergency aid, 216
Enerdata, 413
Energy, 20, 27
 crisis, 130
 efficiency, 20, 133, 139, 149, 151, 153, 154, 156–158, 167
 electricity, 6, 20, 83, 94, 95, 99–101, 103–105, 107, 110, 111, 119n9, 119n10, 127, 129, 131, 139, 152, 155–157, 162, 191, 192, 197, 198, 201, 204, 235, 282, 308, 310, 317–319, 324, 351, 376
 fossil fuels, 20, 37, 128, 157, 331, 349
 intensity, 399, 412, 413, 420–423, 425, 426, 433
 International Energy Agency, 20, 127, 129, 139, 151
 renewable, 20, 81, 128, 129, 133, 139, 149, 153–155, 157, 158, 165, 167, 189, 190, 192–194, 197, 201–204, 235, 238, 312
 sector, 20
 security, 151
 supply, 189–190
 supply *vs.* service, 161–163
 sustainable energy, 125–170
Energy Policy, 162
Energy Sector Management Assistance Programme (ESMAP), 138, 151, 160, 161, 174n47, 178n94, 234
Energy+ Initiative, 131, 160, 172n15
Environmental degradation, 2, 54, 186, 223, 385, 398, 410, 441, 443
Environmental forestry, 311
Environmentalists, 219
Environmental Kuznets curve (EKC), 400, 402–404
Environmental sustainability, 3, 10, 11, 219, 397, 398, 400, 404, 405, 410, 426, 435n17
Eritrea, 373
ERR, *see* Expected rate of return
ESMAP, *see* Energy Sector Management Assistance Programme
Ethanol programme (Brazil), 152, 153
Ethiopia, 193, 250n2, 352, 362n1, 373, 375, 384, 386, 392
 humanitarian aid, 250n2
EU, *see* European Union
Europe, 2, 104, 135, 151, 190, 216, 322, 369, 391, 398
European Investment Bank (EIB), 91, 186
European Union (EU), 42n1, 61–63, 66–68, 72, 74, 76, 131, 160, 163, 164, 218, 274, 277, 347, 370, 376, 380
 REDD+, 274
Evaluation of effectiveness, foreign aid in sustainable energy and climate mitigation, 97–115
Expected rate of return (ERR), 107

F

Factor-GMM, 399, 408, 413, 414, 416, 419, 422, 423
Factor-IV, 399, 406, 408, 413, 414, 422, 423
FAO, *see* Food and Agriculture Organization
Farm forestry, 311
Fast Start Finance programme, Japan, 131
FCPF, *see* Forest Carbon Partnership Facility
FDI, *see* Foreign direct investment
Feed-in laws, 104

Feed-in-tariff (FIT), 94, 103, 104, 109, 110, 156, 177n84
Feffer, J., 236
Finance, 190, 194, 203
  international financial institutions, 82, 112, 174n43, 374
  multilateral development banks, 21, 441
Financial resources, 15n3, 33, 52, 63, 68, 92, 93, 165, 218, 220, 245, 276, 389, 424
  foreign aid form, 218
Finland, ix, 63
FIP, *see* Forest Investment Programme
FIT, *see* Feed-in-tariff
Food aid, 2, 19, 362, 373, 384, 392
Food and Agriculture Organization (FAO), 78n9, 258, 314, 330, 342, 358, 359, 411
Food security, 9, 14, 42n3, 78n1, 90, 131, 161, 165, 283, 284, 329, 330, 334, 343, 346, 347, 349, 350, 358–361, 377, 378, 381, 382, 384
Foreign aid
  aid paradox, 11, 40, 41
  for capacity building to address climate change, 42
  challenges for, 127–134
  Climate Investment Funds, 23, 131, 160, 193, 334
  effectiveness, 187–189
  evolution of, 135–139
  foreign assistance, 17–19, 36, 37, 39–41, 370
  history of, 135–139, 398
  humanitarian assistance, 19, 135, 147, 373, 384
  private voluntary assistance, 19
  reforms, 26–30
  sustainable energy and climate mitigation, 83–116
  See also Aid
Foreign direct investment (FDI), 3, 12, 83, 84, 90, 132, 141, 143, 153, 158, 195, 373
Forest, *see specific entries*
Forest Carbon Partnership Facility (FCPF), 272, 279, 285, 287
Forest certification schemes, 262, 263
Forest conservation, 8–10, 13, 14, 259, 260, 282, 285, 289, 290
  and management, 260, 263–265, 273, 276
Forest degradation (REDD+) and deforestation, reducing emissions, 312
Forest governance and institutions, 276, 279, 280
Forest Investment Programme (FIP), 83, 272
Forestry, 28
  deforestation, 28
  forest conservation, 8, 9, 259, 260, 263
  forest degradation, 28
  industrial forestry, 282, 311
  reducing emissions from deforestation and forest degradation REDD+, 273–278, 280–282
  social forestry, 311
  sustainable forest management, 8, 13, 257–295
  university role in, 321
Forestry sector, 3, 4, 8, 13, 307, 309–311, 315, 324
  foreign aid, 257–295
Forest transition, 236, 309, 312–314, 316, 319, 321, 324, 325
Forum on Forests and the International Tropical Timber Organization, 262
Fossil fuels, 6, 20, 37, 84, 99, 113, 157, 190, 331

Fragile states, 149, 218
France, 42n3, 72, 155, 313
Friedman, M., 135
Fundação Cultural de Curitiba, 239

G

G8 countries, 3, 20, 130, 381
G20 countries, 130
Gambia, 313, 373
Ganapati, S., 43n18
Gardner, A., 222
Gates Foundation, The, 19, 162
GDP, *see* Gross domestic product
GEA, *see* Global Energy Assessment
Gender, 54, 108, 114, 131, 149, 161, 382
General Motors, 250n4
Generalized moments method (GMM), 187, 195, 196, 198–200, 399, 406, 408, 413–423, 426, 430–433
Genuine saving or adjusted net saving (GSAV), 410
German International Climate Initiative, 34
Germany, 28, 68, 139
  Kreditanstalt fur Wiederaufbau (KfW), 91, 118n6
Ghana, 66, 101, 153, 229
  Millennium Villages project, 229
GHG, *see* Greenhouse gas
Giant Panda Nature Reserve restoration, 237
Gillies, J., 230
Global Climate Change Alliance of the European Union, 131
Global Energy Assessment (GEA), 99, 127–129, 171–177n5, 178n101
Global Energy Fund (GEF), 37, 38, 83, 90–92, 109, 153–157, 176n74, 176n75, 176n76, 191, 193, 219, 234–236, 246, 262, 263, 272, 333, 334, 386–388, 392, 434n3
Global Energy Market Data, 413
Global Environment Facility (GEF), 3, 15n2, 37, 44n23, 83, 91, 153–157, 191, 262, 333, 344, 386–388, 434n3
  National Adaptation Programmes of Action, 392
Global Footprint Network, 411, 435n19
Global forest lost and degradation, 257–259
Global Fund to Fight AIDS, Tuberculosis and Malaria, 20
Global greenhouse gas (GHG) emissions, 186, 258
Global Tracking Framework, 128, 129, 165
Globalization, 324, 375, 376, 379, 380, 387, 390
Globe Sustainable City Award, 232
GMM, *see* Generalized moment method
Good governance, 40, 100, 169, 221, 222, 224, 248, 279, 309, 375, 376, 379, 380, 434n1
Good institutions, 27
Good practices, scaling up and transferring, 193–194
Governance, 35, 198, 201
  bureaucracy, 197, 226, 248
  command-and-control, 268
  corruption, 101, 265, 310, 317, 323, 376
  economic incentives, 108
  forest, 265
  improved, 379–381
  institutions, 88, 279
  land tenure, 265, 312

norms, 258
policy, 289, 320
rules, 258
weak, 372
See also Good governance
Government effectiveness index, 202
Government-to-government cooperation, 247
Government-to-government transfers, 217
Graham, C., 222
Grants, 2, 19, 37, 38, 83, 92–94, 103, 107, 118n1, 147, 163, 164, 167, 218, 233, 234, 262, 307, 360, 387, 411, 412
Green cities, 7, 12, 13, 186–196, 201–204, 215–250
Green cities and urbanization, 215–250
Green Climate Fund (GCF), 220, 226, 272, 278
Green Fund, 72
Green growth, 52, 54, 61, 78n2, 90, 92, 220, 232, 442
Green Growth Strategy, 15n2, 220, 434n3
Green technology diffusion, 72–73
Green transportation, 227, 238
Greenhouse gas (GHG), 5–8, 42–43n7, 54, 56, 57, 68, 78n7, 79n10, 91, 92, 99, 116, 128, 134, 178n93, 260, 266, 273, 331, 337, 342, 343, 349–351, 353, 361
emissions, 185, 186, 234, 331, 337, 338, 342, 343, 349–351, 353, 361
Greening cities, 13, 204
Gross domestic product (GDP), 44n23, 197, 198, 201, 202, 308, 373, 379, 391, 412–416, 418, 419, 421, 422, 425, 435n12, 435n18
GSAV, *see* Genuine saving or adjusted net saving
Guangdong Province, 236
Guinea, 373
Guterres, A., 149

H

Habitat fragmentation, 258
Hadjimichael, M.T., 43n11
Hansen, H., 43n11, 197
HARITA, *see* Horn of Africa Risk Transfer for Adaptation
Harmonization, 188, 190, 221–224, 230, 242, 244, 246, 285, 290, 383
Harris, N., 227, 230
Harvard University, 321
Hatzipanayotou, P., 226
Hayward, F.M., 322
He, R., 221
Health, 14, 19, 20, 25, 26, 29, 38, 54, 90, 98, 105, 160, 192, 223, 225, 235, 374, 377–379, 381
AIDS, 20
Tuberculosis, 20
Malaria, 20
healthcare, 54, 218, 233, 282, 308, 325
Health issues, 131, 220, 382
Helsinki package, 42n4, 133, 148
Herzegovina, 76
Heston, A., 435n18
Hitachi, property developers, 233
Hoffman, K., 222, 227
Home photovoltaic (PV) systems, 93
Hong Kong Red Cross, 237
Horenstein, A.R., 434n7

Horn of Africa Risk Transfer for Adaptation (HARITA), 352–353
Howard G. Buffett Foundation, 346
Huang, J., 10, 14
Huang, Y., viii, 6, 10, 14, 42, 435n17, 442
Huang-Huai-Hai River Basin (3H Basin), 344, 345, 356–358
Hübler, M., 425
Human Development Index, 399
Human Development Report, 381
Human rights, 221, 222, 248, 249
Humanitarian aid, 19
  Africa, Sudan, Ethiopia, Democratic Republic of Congo, 250n2
  foreign aid, 216
Humanitarian assistance, 19, 135, 373, 384

I

IASSTD, *see* International Assessment of Agricultural Knowledge, Science and Technology for Development
IDFC, *see* International Development Finance Club
IE Singapore, 234
IFAD, *see* International Fund for Agricultural Development
IFIs, *see* International financial institutions
IMF, *see* International Monetary Fund
Impact Evaluation Group, 221
INDCs, *see* Intended Nationally Determined Contributions
India, 12, 144, 219, 359, 360, 362n1
Indonesia, 68, 197, 362n1
  market-based REDD+ mechanism, 274
  REDD+ planning, 281
Indonesia Exim Bank (Indonesia), 118n6
Industrial agriculture, 378
Industrial Development Bank of Turkey, TSKB (Turkey), 118n6
Industrial forestry, 282, 311
Industrialized countries, 119n9
Inequality, 216
Innovative Insurance Products for Climate Change Adaptation, 66, 75
Intended Nationally Determined Contributions (INDCs), 172n16
Inter-American Development Bank (IDB), 153, 174n43, 240
Intergovernmental Panel on Climate Change (IPCC), 32, 39, 73, 330, 343
Internal macroeconomic imbalances, 217
International aid, 72, 97, 129, 220, 227, 244–246, 249
  See also Foreign aid
International Assessment of Agricultural Knowledge, Science and Technology for Development (IASSTD), 382
International Chamber of Commerce, 139
International Climate Fund, of UK, 131
International Conference on Financing for Development, in Monterret, 308
International Development Finance Club (IDFC), 92
International Energy Agency (IEA), 20
International Energy Initiative, *see* Energy+ initiative
International Finance Corporation, 238

International financial institution policymakers, 217
International financial institutions (IFIs), 83
International Fund for Agricultural Development (IFAD), 347, 382, 385–387
International Institute for Tropical Agriculture, 346
International Maize and Wheat Improvement Center, 346
International Monetary Fund (IMF), 381, 425
International negotiation capacity, boosting, 71–72
International Renewable Energy Agency(IRENA), 139
International Rice Research Institute (IRRI), 342, 350, 359
International Tropical Timber Organization (ITTO), 258
IPCC, *see* Intergovernmental Panel on Climate Change
IRENA, *see* International Renewable Energy Agency
IRRI, *see* International Rice Research Institute
Irrigation, 341–345, 347, 351, 356–358, 377, 378, 384, 388, 389
Isaac, C., 229
Isham, J., 43n16
Islam, N., 371
Italy, 67, 73
ITTO, *see* International Tropical Timber Organization

J
Jamaica, 157
Japan, 38, 91, 131
Japan International Cooperation Agency, JICA (Japan), 91, 118n6
Johnston, T.A., 68
Jones, S., 220
Joss, M, 228

K
Kablan, 187
Kapetanios, G., 399, 404, 406, 408, 413, 426, 434n7
Kauppi, P.E., 312
Kenya, 150, 362n1
Keohane, R.O., 43n17
kernel density evolution, of sustainable indicators, 414
KfW Bankengruppe (Germany), 118n6
Kisumu, Kenya, 224
Knack, S., 150, 188, 189
Korea Finance Corporation, KoFC (South Korea), 118n6
Kostandini, G., 347
Krasner, S.D., 43n10
Kretschmer, B., 100, 187, 195, 201, 212n1, 425
Kyoto Protocol, 6, 76, 79n10, 83, 84, 92, 101, 133, 333
  Adaptation Fund, 34, 44n23
Kyoto Statement, 32

L
Lahiri, S., 226
Lake Victoria City, CDS, 224
Lancaster, C., 150
Land tenure, 265, 312
Laos, 354
Large scale economic trade off and REDD+, 282
La Rovere, R., 346

Latin America, 28, 72, 152, 218, 270, 370
LDCF, *see* Least Developed Countries Fund
Leakage, 295n3
Least Developed Countries Fund (LDCF), 333, 334
Lensink, R., 43n11
Lew, B., 225, 436n23
Li, Q., 345
Liberia, 373
Limits to Growth, 219
Liu, L., 43n18
Low-carbon energy, 108
Low-density residence, 239

M

MACC, *see* Mainstreaming Adaptation to Climate Change
Madagascar, 354, 373
Mahama, A., 113
Mainstreaming Adaptation to Climate Change (MACC), 38
Mainstreaming Climate Change Adaptation in Irrigated Agriculture, 345
Malawi, 373
Malaysia, 144
Malaysia Energy Center, 190
Mali, 162, 354, 387, 392
Marcellino, M., 399, 404, 406, 408, 413, 426
Market- and private sector-led agricultural growth, 378–379
Market incentives, 95
  effectiveness of foreign aid in sustainable energy and climate mitigation, 94
Market penetration of sustainable energy technologies, 114
Marshall Plan, 135, 369, 391
Martinot, E., 109, 156
McDermott, C.L., 261
MDBs, *see* Multi-lateral development banks
MDGs, *see* Millennium Development Goals
MEAs, *see* Multilateral environmental agreements
Measurable, reportable and verifiable (MRV), 25
Mekasha, T.J., 424, 436n25
Methane emissions
  mitigation of, 342, 343
  reduction of, 357–358
Mexico, 151, 157
Michael, M.S., 226
Micro-credits, 93, 110, 112
Micro-entrepreneurs, 93
Micro-fencing, 389
Micro-finance, 93, 94, 157
Middle East, 152, 218
Military aid, 19
Millennium Challenge Corporation, 40
Millennium Declaration, 397, 399, 434n1
Millennium Development Goals (MDGs), 2, 81, 130, 131, 136, 149, 151, 423
Millennium Ecosystem Assessment (MEA), 315
Millennium Villages project, in Ghana, 229
Mitigation, *see specific entries*
Monterrey consensus, 100
Moran, D.D., 410
Morgenthau, H., 135
Morocco, 360
Mozambique, 153, 373
MRV, *see* Measurable, reportable and verifiable
Multi-function platforms, 161

Multi-lateral agencies, 91, 92, 118n1, 204
Multilateral aid, *see* Aid
Multi-lateral assistance, 58
Multi-lateral development banks (MDBs), 82, 83
See also Multi-lateral agencies
Multilateral environmental agreements (MEAs), 388
Multilateral financial institutions, 174n43
Multilateral funds, climate change, 17, 23
Mwega, F.M., 150

N

Nacional Financiera, NAFIN (Mexico), 118–119n6
NAFORMA, *see* National Forest Resources Monitoring and Assessment programme
Nakhooda, S., 44n20, 288
Namibia, 360
NAPA, *see* National Adaptation Programmes of Action
National Adaptation Programmes of Action (NAPA), 387–389
Priority Intervention, 388
National Audit Office (UK), 436n24
National Bank of Egypt, 192
National Climate Change Committee programme, 62
National Fertilizer Sector Project (Ethiopia), 386
National forest managers, 280
National Forest Resources Monitoring and Assessment programme (NAFORMA), 320
Nationally appropriate mitigation actions (NAMAs), 25
Natural disasters, 220
Natural gas, 191
Natural habitats, 215
Natural resource exploitation, 399, 418, 419, 422
Natural resources
biodiversity, 28, 118n3, 260, 263, 266, 307, 315, 316, 388
forests, 37
soil, 309, 337, 339, 347–349, 353–356, 358, 361, 378, 386, 387
water, 341–345, 347, 356–358, 361, 376, 377, 384, 385, 388
NECTAR project, 71, 72
NEPAD, *see* New Partnership for Africa's Development
Nepal, 63
The Netherlands, 61, 154
Neuhoff, K., 43n9
New green revolution for Africa, 378
New Partnership for Africa's Development (NEPAD), 379, 380, 384, 385, 389
Ng, S., 399, 404, 406–408, 413, 426, 434n8
NGO, *see* Non-governmental organization
Nickell, S., 195
Nieusma, 95
Nigeria, 362n1
Non-governmental aid, *see* Aid
Non-governmental organization (NGO), 3, 4, 12, 58, 61, 83, 92, 93, 95, 101, 104, 125, 126, 139, 146, 147, 171n2, 217, 218, 225, 236, 262, 276, 287, 352, 390
North Africa, 67
North America, 190
North Kordofan, 388, 389
Norway, 72, 131, 159, 172n15
Climate Change Initiative, 271

Nunnenkamp, P., 425
Nygaard, I., 162

O
OA, *see* Official assistance
Ocean acidification, 78n1
ODA, *see* Official development assistance
OECD, *see* Organisation for Economic Co-operation and Development
Official aid, *see* Aid
Official assistance (OA), 19
Official development assistance (ODA), 3–7, 12, 19–23, 30, 33, 34, 37, 39, 44n23, 83, 126, 140, 142, 143, 187, 190, 194, 196, 203, 204, 218, 219, 259, 308, 371–373, 376, 380, 383, 398, 411–413, 416, 424, 425, 430–433, 435n16
  effectiveness of foreign aid in sustainable energy and climate mitigation, 90–91
  net, 196, 198, 202
O'Hanlon, M., 222
Olandes Sewage Treatment Plant, 192
Onatski, A., 434n7
Organisation for Economic Co-operation and Development (OECD), 26, 42n1, 42n3, 43n7, 60, 84, 126, 147, 198, 217, 308, 333, 370, 380, 390, 411, 434n3
  Development Assistance Committee (OECD-DAC), 2, 19–23, 42n1, 42n3, 42n5, 125, 126, 132, 133, 136–138, 140–142, 144, 148, 168, 169, 175n54
Our Common Future report, 219
Overseas aid, 436n24
  See also Foreign aid
Oxfam Hong Kong, 236

P
Package approach, renewable technology, 95
Pakistan, 31
PAKLIM project, 68
Paldam, A., 43n10
Paldam, M., 43n11, 424
Paris Agreement, 131, 171n11, 172n16
Paris Declaration on Aid Effectiveness (2005), 2, 15n1, 26, 27, 29, 30, 34, 40, 43n13, 96, 101, 144, 148, 149, 169, 175n54, 188, 423, 424, 436n21
Paris High Level Forum on Aid Effectiveness, 26
Parks, B., 54
Pascual, U., viii
Pastreich, E., 236
Payment for ecosystem services (PES), 261, 263, 268
PDA, *see* Private development assistance
Peak farmland, 323
Performance-based financing *vs.* traditional financial assistance, 160–161
Persson, R., 311, 321
PES, *see* Payment for ecosystem services
Philippines, 157, 192, 342, 343, 362n1
Philipsi, property developers, 233
Pilot Programme for Climate Resilience (PPCR), 83
Pineda, J., 409, 434n10
Poland, 58, 157
PolityIV Database, 412
Pollution and waste treatment, 192
Population growth, 321, 383, 422, 423
Post-2015 development agenda, 400, 424

Poverty, 83, 218, 220, 227, 312, 324
abatement, 315
alleviation, 217, 218, 229, 271, 280
reduction, 27, 38, 97, 193, 224
Power purchase agreement (PPA), 94
PPA, *see* Power purchase agreement
PPCR, *see* Pilot Programme for Climate Resilience
PPPs, *see* Private–public partnerships
Practical Action, 147, 175n58
Pray, C., 359
Prince Claus Fund, 238
Private aid, *see* Aid
Private and non-governmental organizations, aid by, 146–148
Private development assistance (PDA), 147
Private–public partnerships (PPPs), 83
Private voluntary assistance, 19
Productive Safety Nets Programme (PSNP), 384
Programme aid, 19
Project aid, 19
Prosol Industrial project, 73
PSNP, *see* Productive Safety Nets Programme
Public housing, 233
Public policy, 32, 103, 189, 218, 352, 379
and urban designs, 190–191
Public–private partnerships (PPP), 105
Purvis, N., 423

Q

Qiang minority, 238
Qiu, B., 227
Quibria, M.G., 6, 14, 442

R

Radelet, S., 43n10, 136
Rahman A., 150, 189
Rajan, R.G., 43n10, 424
Recipient country, 216
REDD+, 4, 8, 9, 13, 14, 273
diverse stakeholders, 279
financing, 288, 290, 291
forest carbon policies, 266, 267
in forestry, 270, 286
framework analyses, 274
and large scale economic trade off, 282
national level land use planning, 269
policies, 259, 260
public financing, 276–278, 280
Readiness Fund, 272
scalability and decision making, 286
scaling up, 285
and SFM objectives, 267
REEP, *see* Renewable Energy and Energy Efficiency Partnership
Rees, W.E., 411
Regional development banks, 83
Regional Gateway for Technology Transfer and Climate Change Action, 72
Regionalization, 384–385
Reis, A.B., 402
Renewable biomass, 312
Renewable energy, 20, 189, 190, 192–194, 197, 201–204
Renewable Energy and Energy Efficiency Partnership (REEP), 139
Renewable energy service companies (RESCOs), 111
Renewable energy technologies, 82
Renewable portfolio standard (RPS), 94
Replacement expenditure, 93
Republic of Korea, 144
RESCOs, *see* Renewable energy service companies
Results-based monitoring, 30
Riddell, A., 230

Riddell, R.C., 43n10
Riley, 95
Rio 20+ conference, 310
Rio Conventions (1992), 22, 84, 387
Rio markers, 84, 118n3
Rio Summit 1992, 82
Risk management, investment in, 352
River Nile, 388, 389
Roberts, J.T., 54
Robustness test, 202, 203, 207–211
Rockefeller Foundation, 359
Rogerson, A., 229
Romania, 58
Roodman, D., 412
RPS, *see* Renewable portfolio standard
Rua XV de Novembro, popular pedestrian walkway, 239
Rudel, T.K., 324
Rural household electrification, commercialization of, 96
Rural Land Management Project (Côte d'Ivoire), 386
Russian Federation, 144
Rwanda, 61, 373, 387, 392

S

Sachs, J., 136
Sachs, J.D., 43n12, 221
Sahel region, 311
Scalability
  definition, 228
  foreign aid in sustainable energy and climate mitigation, 111–112
  in foreign aid on urbanization and green cities, 228
Scaling, 30–39
Scaling-Up Renewable Energy Programme (SREP), 83, 91, 160
SCCF, *see* Special Climate Change Fund
Schürenberg-Frosch, H., 43n11
SDGs, *see* Sustainable Development Goals
SE4ALL, *see* Sustainable Energy for All
Second World War, 216
Sectoral breakdown, of climate change-related aid, 24
Senegal, 153, 347, 373, 374
Serbia, 67
Sewage, 13, 189, 192, 244
SFM, *see* Sustainable forest management
SGP, *see* Small Grants Programme
SHDI, *see* Sustainability-adjusted human development index
Shifting cultivation, 284
Siemens, property developers, 233
Sierra Leone, 360, 373
Sino-Singapore Tianjin Eco-City, 232, 233
Site-specific nutrient management, 350
Skutsch, M., 280
Slovakia, 58
Small Grants Programme (SGP), 92
Small Industries Development Bank of India, SIDBI (India), 119n6
Social forestry, 311
Social harmony, 233
Social inclusion, 397
Social services and empowerment, in Africa, 381
Social sustainability, 381
Soil degradation, 258
Soil protection, 258
Solar cell-battery system, 318
Solar cells, 317
Solar power, 192
Solow Model, 399
Somalia, 384
South Africa, 190, 362n1
South Asia, 192
South Korea, 73

South–south aid programmes, 217
South–South Cooperation (SSC) Initiative, 360
SPAAR, *see* Special Program for African Agricultural Research
Spain, 71, 72
Special Climate Change Fund (SCCF), 333, 334, 344, 345, 357
Special Program for African Agricultural Research (SPAAR), 386
SREP, *see* Scaling Up Renewable Energy Programme
SSC, *see* South–South Cooperation Initiative
ST Engineering, 233
Stern, N., 136
Stiglitz, J., 136
Stock, J.H., 408
Strategic Initiative to Address Climate Change, 64
Strategy on Agriculture and Rural Development, 380
Stren, R., 224
Subedi, M.S., 222
Subramanian, A., 43n10, 424
Sub-Saharan Africa, 224
Succession planning, 108
Sudan, 153, 250n2, 388, 392
Summers, R., 435n18
Sustainability
  agricultural, 377–378
  clean energy, 87
  ecological, 193
  economic, 193
  fiscal, 186
  sustainable development, 28, 52, 54, 56, 64, 72, 77, 87, 92, 113, 186, 216, 223, 397–433, 442, 443
  Sustainable Development Goals (SDGs), vii, 131, 149, 398
  transition, 236, 309, 312–314, 316, 319, 321, 324, 325
Sustainability-adjusted human development index (SHDI), 409–410
Sustainable city, 186, 215
Sustainable Development Goals (SDGs), 2, 3
Sustainable energy and climate mitigation, foreign aid, 102–106
Sustainable Energy for All (SE4All), 12, 82, 128, 131, 149, 158, 170, 171n12
Sustainable forest management (SFM), 8, 13, 257–295
Sustainable forestry, *see* Forestry; Sustainable forest management
Suzhou Industrial Park, 233
Sweden, viii, 156, 375
Swedish forestry, 310
Swedish International Development Cooperation Agency, 155
Swickard, N., 280
Swidden agriculture, 284
Swiss bank accounts, 97

T

Tanzania, 29, 320
Tarp, F., 43n11, 197, 220, 424, 436n25
Taylor, M., 400, 401
Technology, 192, 203
  low-carbon technology, 54, 61, 67, 73, 76, 77, 91, 107, 108, 350
  technical assistance, 19
  technology transfer, 27, 36, 53, 54, 56, 61, 67, 72, 73, 202, 333, 350, 359–361
Temple, J.R.W., 424, 434n4
Terrestrial biodiversity, 307
Thailand, 157, 359

Thornton, N., 27
Three Rivers Grassland Carbon Sequestration project, 354–355
Tianjin Eco-City project, 247
Tianjin, urbanization in, 232–235
Tied aid, 173n26
Timmons Roberts, J, 226, 227
Togo, 392
Togo National Agricultural Service Project, 386
Topical Trust Fund on Managing Natural Resource Wealth, 425
Tourism, 248, 316
Trade, 83, 141, 143, 144, 150, 168, 187, 370, 373–375, 379, 380, 385, 387, 390, 391
Transferability
  foreign aid in sustainable energy and climate mitigation, 110–111
  in foreign aid on urbanization and green cities, 230
Transport, 20, 23, 41, 186, 191, 319, 325, 351, 385, 389
Tree planting programme, 313
Tropical resource management, 258
Truman, H., 135, 369
Tsikata, T.M., 43n10
Tunisia, 73, 154, 354
Turkey, 144

U

Uganda, 60, 154, 373
UK, *see* United Kingdom
UK Department for International Development, 346
UK Food Group, 376, 380
UMP, *see* Urban Management Programme
UN, *see* United Nations
UN Energy, 138, 149
UN Framework Convention on Climate Change (UNFCCC), 56, 57, 61, 63, 64, 69, 72–74, 76, 77
  Green Climate Fund, 272
UN General Assembly, 134
UN Millennium Project, 220
UNCED, *see* United Nations Conference on Environment and Development
UNDP, *see* United Nations Development Programme
UNEP, *see* United Nations Environment programme
UNESCO, *see* United Nations Educational, Scientific and Cultural Organization
UNFCCC, *see* United Nations Framework Convention on Climate Change
UNFF, *see* United Nation Forum on Forests
United Kingdom (UK), 61, 159, 349
United Nation Forum on Forests (UNFF), 261
United Nations (UN), 42n6, 216, 314
  See also specific entries
United Nations Children's Fund, 217
United Nations Conference on Environment and Development (UNCED), 81, 82, 261
United Nations Conference on Human Environment, in Stockholm, 219
United Nations Conference on Sustainable Development (1992), 148
United Nations Development Programme (UNDP), 54–56, 77, 83, 131, 139, 148, 151, 153, 155, 162, 220, 377, 390, 410
  Africa Adaptation Programme, 38

United Nations Educational, Scientific and Cultural Organization (UNESCO), 153, 322
United Nations Environment Programme (UNEP), 74, 76, 190, 225, 262, 423
  Finance Initiative, 32
  Framework Convention on Climate Change, 262
United Nations Framework Convention on Climate Change (UNFCCC), 5, 6, 8, 26, 34, 128, 191, 333, 388
  18th Conference of Parties in Doha, 269
  and climate finance, 92
  multilateral adaptation programme, 44n23
United Nations General Assembly Resolution, 308
United Nations Institute for Training and Research, 240
United Nations University, ix
United Nations' Millennium Development, 217
United States (US), 41, 152, 156, 218, 274, 370, 373
University role, in forestry, 321–323
UNU World Institute for Development Economics Research (UNU-WIDER), viii, ix, 308, 314
UNU-WIDER, *see* UNU World Institute for Development Economics Research
Urban areas, 191
  ecocity development, 13, 228, 231–234, 236–238, 240–250, 250n3
  greening cities and, 7, 8, 185–204
  infrastructure, 37
  urbanization, 154, 165, 185, 186, 189, 193, 203
  urban planning, 37, 225, 240
Urban designs, public policy and, 190–191
Urban growth, 250n1
Urbanization
  definition of, 223
  and green cities, foreign aid in, 223
Urban Management Programme (UMP), 224
Urban productivity growth, 227
Urpelainen, J., 44n19
US Agency for International Development, 217
USAID, 346, 378

V
Van de Walle, N., 68
Victor, D.G., 44n21
Victor, M.A., 43n17
Vietnam, 63, 67, 155, 156
Village forestry, 311
Vincent, A., 228
Vnesheconombank, VEB (Russia), 119n6
Vulnerability, 56, 62
  hazard mitigation, 90, 228, 333, 344, 353, 361
  prevention of disasters, 218

W
Wang, Y., 9, 10, 14
Wassmann, R., 342
Waste treatment, 192
Water Programme for Environmental Sustainability (WPA), 67
Water supply, 191–192
Watson, M.W., 408

WB, *see* World Bank
WBG, *see* World Bank Group
WCED, *see* World Commission on Environment and Development
WDI, *see* World Development Indicators
Weather and flood forecasting system, 31
Wenchuan, China, and massive earthquake, 232
Wenchuan and Beichuan areas, after the earthquake, 235–238
Wenchuan Earthquake Recovery Project, 235
WEO, *see* World Energy Outlook
What Works Working Group, 26
White, H., 43n11
Whitfield, L., 228, 231
WHO, *see* World Health Organization
Whole Village Development Model, in renewable energy program, 96
Williamson, C., 189
Wind power, 192
Wolong area redevelopment, 237
Wolong National Nature Reserve, 250n5
Wood-fired stove, 317
World Bank, 28, 37, 61, 69, 74, 83, 88, 138, 148, 149, 151, 153, 155, 157, 160, 163, 169, 174n43, 185, 186, 192, 216, 217, 234, 295n1, 332, 344, 345, 347, 356, 374–376, 381, 386, 387, 390, 410
  Analytical and Advisory Activity, 357
  'Assessing Aid' study, 27
  "Building sustainability in an urbanizing world" report, 192
  Independent Evaluation Group (IEG), 384, 386
  International Development Association (IDA), 371
  Open Knowledge Repository, 138
  urban lending and grant programmes, 37
World Bank Group (WBG), 3, 83, 88
World Business Council on Sustainable Development, 139
World Commission on Environment and Development (WCED), 219, 434n2
World Conference on Development and Environment (1992), 15n2, 434n3
World development indicators (WDI), 198, 201, 410, 412
World Development Report (2008), 54, 370, 373, 380
World Energy Council, 139
World Energy Outlook (WEO), 127, 140
World Health Organization (WHO), 139, 236
World Resources Institute, 58, 331
World Summit on Sustainable Development (1992), 149, 308
Worldwide Federation for Nature, 190
WPA, *see* Water Programme for Environmental Sustainability

Y
Youth and United Nations Global Alliance Education Programme on Climate Change, 74

Z
Zambia, 373
Zimbabwe, 153
Zoellick, R., 374

The manufacturer's authorised representative in the EU is Springer Nature Customer Service Centre GmbH, Europaplatz 3, 69115 Heidelberg, Germany. If you have any concerns regarding our products, please contact ProductSafety@springernature.com

Printed and bound by CPI Group (UK) Ltd, Croydon, CR0 4YY

30/09/2026

02225444-0001